Industrial CFD and Fluid Modelling in Engineering, 2nd Edition

Industrial CFD and Fluid Modelling in Engineering, 2nd Edition

Guest Editors

Francesco De Vanna
Ernesto Benini

MDPI

Basel • Beijing • Wuhan • Barcelona • Belgrade • Novi Sad • Cluj • Manchester

Guest Editors

Francesco De Vanna
Department of Industrial Engineering
University of Padova
Padova
Italy

Ernesto Benini
Department of Industrial Engineering
University of Padova
Padova
Italy

Editorial Office
MDPI AG
Grosspeteranlage 5
4052 Basel, Switzerland

This is a reprint of the Special Issue, published open access by the journal *Fluids* (ISSN 2311-5521), freely accessible at: https://www.mdpi.com/journal/fluids/special_issues/OKUQIPY98O.

For citation purposes, cite each article independently as indicated on the article page online and as indicated below:

Lastname, A.A.; Lastname, B.B. Article Title. *Journal Name* **Year**, *Volume Number*, Page Range.

ISBN 978-3-7258-6738-7 (Hbk)
ISBN 978-3-7258-6739-4 (PDF)
https://doi.org/10.3390/books978-3-7258-6739-4

Contents

About the Editors

Francesco De Vanna

Francesco De Vanna is Assistant professor in machines fluid dynamics at the Dipartiment di of Industrial Engineering of the Università degli Studi di Padova, Italy. His research interests focus on computational fluid dynamics for industrial and engineering applications, with a particular emphasis on turbulence modeling, wall modeled large eddy simulation, and immersed boundary methods. His work addresses complex configurations relevant to turbomachinery, aerospace, and propulsion systems, with the aim of improving predictive accuracy while maintaining computational efficiency. He has been involved in several national and international research projects linking academic research with industrial needs and has authored numerous peer reviewed publications on advanced CFD methodologies and their verification.

Ernesto Benini

Ernesto Benini is Full professor in aircraft engines and turbomachinery at the Dipartiment di of Industrial Engineering of the Università degli Studi di Padova, Italy. His research activities span aircraft propulsion technologies, turbomachinery, and energy conversion systems, with a strong focus on numerical modeling and multidisciplinary design optimization. He has coordinated and participated in several competitive international research projects in collaboration with industry and research institutions, contributing to the development of innovative design methodologies. ASME and AIAA fellow Ernesto Benini is the author of a large body of scientific publications in top-leading engineering journals.

Preface

This Reprint collects a selection of peer reviewed articles that illustrate current advances in industrial computational fluid dynamics and fluid modeling in engineering. Its subject is the application of numerical simulation to complex industrial flow problems, with a scope that spans fundamental modeling developments, validated methodologies, and practical engineering applications across a wide range of sectors.

The aim of this Reprint is to provide readers with a coherent overview of how CFD is being used today to support analysis, design, and optimization in realistic industrial environments. While computational capabilities have grown significantly, industrial fluid mechanics continues to pose demanding challenges due to complex geometries, multiphysics interactions, unsteady phenomena, and the need for reliable predictions within reasonable computational times. The contributions gathered here address these challenges from complementary perspectives, combining methodological innovation with application-driven studies.

The motivation for assembling this Reprint is twofold. First, it seeks to document the maturity reached by many CFD approaches that are now routinely employed in industry, such as validated turbulence models, multiphase flow simulations, and hybrid modeling strategies. Second, it highlights emerging directions, including the coupling of CFD with machine learning, reduced order models, and multi scale simulation frameworks, which are opening new possibilities for faster and more informative engineering analyses. Together, these works demonstrate how CFD continues to evolve from a purely investigative tool into a central component of simulation-driven engineering practice.

We hope this can be encouraging for further research that bridges academic innovation and practical engineering needs.

Francesco De Vanna and Ernesto Benini
Guest Editors

fluids

MDPI

Editorial

Industrial CFD and Fluid Modeling in Engineering—2nd Edition

Francesco De Vanna

Dipartimento di Ingegneria Industriale, Università degli Studi di Padova, Via Venezia, 1, 35121 Padova, Italy; francesco.devanna@unipd.it

Following the success of the first edition of "Industrial CFD and Fluid Modeling in Engineering", this second edition continues to showcase cutting-edge research that pushes the boundaries of computational fluid dynamics (CFD) in addressing complex industrial flow problems. In recent decades, advances in CFD algorithms and computing power have enabled simulations that were previously infeasible. Yet, applying CFD to realistic industrial scenarios remains challenging due to complex geometries, multiphysics couplings, and the need for validated modeling assumptions. The first edition laid a strong foundation by bringing together leading ideas in industrial flow modeling, with a particular emphasis on RANS validation in practical geometries and emerging CFD strategies. In this second edition, we build upon that foundation with 13 new contributions that highlight innovative methodologies and diverse applications—from improved turbulence models and optimization techniques to multiphase flow simulations—reflecting both continuity with the prior collection and significant new advancements. Each paper in this Special Issue illustrates how researchers are addressing challenges in industrial fluid mechanics through novel CFD approaches, collectively advancing the state of the art in the field.

We begin with a comprehensive review that bridges past and present modeling techniques, illustrating the continuity of research in industrial CFD. Delvar et al. present an extensive literature review on single- and twin-screw extruders for polymerization processes, tracing the evolution from early analytical models to modern CFD-based simulations. This review highlights progress in reactive extrusion modeling while underscoring the current challenges and future directions for CFD in this domain. The authors particularly emphasize that reactive extrusion poses unique difficulties due to the interplay of mixing, heat transfer, chemical reactions, and non-Newtonian fluid behavior under high shear, which is further complicated by multiphase flows and complex extruder geometries.

Notably, the themes identified in the review—such as multiphase flow management, turbulence modeling, and the need for efficient computation—resonate throughout many papers in this edition. The continuity between editions is evident: while the first edition emphasized foundational CFD approaches and their applicability, this second edition expands into innovative hybrid methods, design optimization, and cross-disciplinary techniques (including machine learning integration) to address industrial challenges. Collectively, these works maintain a scholarly dialogue with earlier studies, extending their scope into new industries and increasingly complex phenomena. In doing so, this Special Issue reaffirms the transformative role of CFD in engineering while pushing it into new and dynamic arenas.

Several contributions in this edition focus on multiphase flows and mixing processes, which are critical in various industries, from construction to aquaculture. Ferrari et al. [1] introduce a new numerical mixing index designed to quantify mixing efficiency in concrete production. By applying a statistical methodology to a transient Eulerian–Eulerian CFD simulation of a concrete truck mixer, they assess how aggregate particles and cement paste homogenize during the initial revolutions of the mixer. The proposed cell-based

https://doi.org/10.3390/fluids10070164

mixing index, defined using statistical measures of concentration, allows for a quantitative evaluation of distribution uniformity as mixing progresses. Their results show that as the drum rotates, material distribution improves and vortical structures enhance mixing, with the computed index increasing accordingly. This method offers a practical tool for the predictive evaluation of new mixer designs, enabling engineers to estimate how design parameters—such as internal geometry, rotation speed, and mixture composition—affect mixing outcomes prior to prototype construction. This capability streamlines the development of more efficient concrete mixers and is also applicable to other types of mixing equipment.

In the context of viscoplastic fluids, Benmoussa and Páscoa explore how impeller geometry can improve mixing performance. Their study examines a chamfered anchor impeller design for agitating non-Newtonian (yield-stress) fluids, comparing various chamfer angles to a standard design. Through CFD simulations conducted in an unbaffled mixing vessel, they evaluate each impeller's ability to overcome yield stress, enhance flow circulation, and improve thermal homogeneity. The findings reveal that a 67.5° chamfered impeller significantly enhances flow distribution and reduces dead zones, particularly near vessel walls where stagnant regions often occur. Additionally, it promotes stronger vertical mixing and a more uniform temperature field, demonstrating that subtle geometric modifications at the blade tip can meaningfully impact the mixing of highly viscous fluids. These insights provide valuable guidance for designing more efficient mixers in industries such as food processing, polymer production, and chemical manufacturing, where handling viscoplastic fluids is common.

We examined the impact of the cut/sweep ratio of a "Lily" impeller on the distribution of dissolved oxygen in shrimp cultivation ponds. Aeration and oxygenation are crucial in aquaculture for maintaining water quality; in this case, five impeller designs with varying blade cut percentages were analyzed using CFD with a user-defined oxygen transport model. The simulations demonstrate that modifying the cut/sweep ratio substantially affects the pond's flow patterns and oxygen dispersion. Specifically, variations in the impeller's cut significantly influenced dissolved oxygen levels, dynamic pressure, and flow velocity throughout the pond. The "no-cut" impeller variant produced the highest average oxygen concentration and the strongest circulation, indicating superior aeration performance. However, the study also found that only the region immediately surrounding the aerator achieved sufficiently high oxygenation to sustain aquatic life, leaving distant areas under-oxygenated. These findings suggest that additional aerators or improved impeller designs are necessary to ensure a uniform oxygen distribution across large ponds. By quantifying how design modifications influence oxygen transfer efficiency, this work offers practical insights for developing better aeration systems in sustainable aquaculture.

Another key theme in this edition is the application of CFD for design optimization in aerodynamic and turbomachinery systems [2,3]. Yun et al. present a two-stage multi-objective optimization of a high-speed train's nose shape, aimed at improving performance across multiple criteria. In the first stage, CFD-based shape optimization yields a more pointed nose that reduces aerodynamic drag by approximately 8.7% relative to the baseline model. Building on this, the second stage introduces additional objectives: minimizing tunnel micro-pressure waves (to reduce pressure transients when entering tunnels) and enhancing stability under crosswinds, while ensuring that drag remains near its minimized value. By applying constraints from stage one, the authors obtain a Pareto-optimal set of nose geometries that simultaneously reduce micro-pressure wave intensity and improve crosswind safety, with only a minimal (<1.5%) trade-off in drag relative to the single-objective optimum. This two-stage CFD optimization strategy exemplifies an effective balance between competing aerodynamic considerations. It provides railway engineers with a refined design that meets the stringent requirements of next-generation high-speed trains—reducing drag to enhance energy efficiency while mitigating tunnel pressure effects and stability risks. This study underscores how CFD-based multi-objective design can deliver innovative and practical solutions in vehicle aerodynamics.

In the realm of fluid machinery, Avanzi et al. [4] introduce ARES, a novel mean-line design and analysis code for axial-flow pumps, specifically developed for an "Outboard Dynamic-Inlet Waterjet"—a marine pump system inspired by aero-engine technology. The solver enforces radial equilibrium through streamlines along the pump's meridional flow path, using empirical correlations to account for complex effects such as secondary flows, tip leakage, and end-wall losses. By comparing ARES predictions with experimental data from several test cases, including single-stage rotors and rotor–stator configurations, the authors assess the code's accuracy across a range of operating conditions. At design point flow rates, ARES closely predicts hydraulic efficiency compared to measurements. Minor discrepancies emerge at off-design conditions, where loss models are less accurate, leading to increased errors in efficiency predictions. Nonetheless, ARES effectively captures radial distributions of pressure and velocity, particularly at midspan, though some complex end-wall boundary layer effects remain difficult to reproduce. This work demonstrates that simplified 1D/2D CFD approaches, like ARES, can provide fast and reliable tools for preliminary waterjet pump design, enabling engineers to rapidly evaluate design iterations with reasonable accuracy.

Holtmann and Key [5] focus on enhancing turbulence modeling fidelity for unsteady diffuser flows in centrifugal compressors. Building upon an "isolated diffuser" CFD methodology—which models the diffuser passage with an unsteady inlet boundary condition abstracted from the impeller to reduce computational cost—they address discrepancies from full-stage simulations caused by oversimplified inlet turbulence assumptions. To improve accuracy, the authors introduce a non-uniform turbulence inflow model for the isolated diffuser and assess its impact across multiple operating conditions, from choke to near-surge. This enhanced approach prescribes realistic spatial variations in turbulence at the diffuser inlet, resulting in significantly better agreement with full-stage results. Across four loading conditions, the isolated diffuser with non-uniform inlet turbulence yields much closer predictions of flow structures and one-dimensional performance parameters, matching full 3D simulations far better than the uniform-turbulence model. Importantly, this improvement comes without additional computational costs, preserving the efficiency advantage of the isolated model. By refining turbulence modeling in reduced-order simulations, this work offers more reliable and efficient design evaluations for vaned diffuser compressors, benefiting turbomachinery designers seeking faster turnaround times without sacrificing accuracy [6,7].

Beyond specific design studies, several papers in this issue contribute to enhanced CFD methodologies and the integration of CFD with other modeling approaches, expanding the tools available for industrial fluid dynamics. McConnell et al. [8] examine the performance of an advanced turbulence simulation technique—an Improved Delayed Detached-Eddy Simulation (IDDES)—in predicting unsteady aerodynamic flows. Using the canonical case of turbulent vortex shedding behind a triangular bluff body, they assess IDDES against both experiments and a traditional Delayed Detached-Eddy Simulation (DDES). The IDDES model captures key flow features with high fidelity, including recirculation length, velocity profiles, and Reynolds stress distributions, all of which align well with experimental data. However, compared to the simpler DDES, the IDDES exhibits a slight overprediction (about 4%) of recirculation zone length and a modest (about 3%) increase in the computational cost. Encouragingly, the IDDES maintains good accuracy even on coarser meshes, demonstrating its robustness. The study offers valuable guidance on the trade-offs of using the IDDES in industrial applications: while more complex, it can adaptively transition between RANS and LES modes to handle a wide range of flow regimes, providing improved predictive power for challenging cases such as bluff body flows, with only a minor increase in cost. This work contributes to the ongoing refinement of hybrid RANS–LES techniques for industrial CFD applications.

A notable cross-disciplinary innovation is presented by Santos et al., who develop a framework that integrates conventional simulations with machine learning (ML) to predict structural damage in subsea structures exposed to ocean wave conditions. The authors train an ML model on a dataset generated from a combination of analytical formulations (Morison's equations for wave loading), finite-element structural analysis, and 2D/3D CFD simulations of wave–structure interactions [9–12]. This streamlined data generation strategy—combining fast analytical and reduced models with high-fidelity CFD—provides the ML model with diverse scenarios of wave parameters and corresponding structural responses. The resulting model can quickly predict key metrics of structural stress and damage under various ocean conditions. Validation against a six-month field experiment involving a benthic lander equipped with strain gauges confirms the accuracy of the predictions. The study demonstrates the feasibility of using trained ML surrogates to anticipate structural loads and potential damage in a fraction of the time required for full FSI simulations. This work exemplifies the potential of integrating CFD with machine learning and real-world data, signaling a future where AI-assisted CFD models support decision making in offshore engineering and other fields by delivering fast, reliable estimates of complex fluid–structure phenomena.

Ajmani et al. [13] address an important problem in public health and building engineering by developing an efficient co-simulation strategy for indoor airflow and pollutant dispersion. Focusing on a mechanically ventilated lecture hall, they combine a 1D fluid network model of the ventilation system with detailed 3D CFD and discrete particle modeling (DPM) to simulate the dispersion of exhaled aerosol particles representing respiratory droplets. This hybrid modeling approach significantly reduces computational costs by coupling a simpler network flow solver for ducts with high-fidelity CFD for room airflow. The study demonstrates that such a workflow can reliably predict local aerosol concentrations and identify poorly ventilated zones. The authors find that the existing ventilation system was imbalanced and inefficient at managing the aerosol distribution: despite a high overall air exchange rate, certain areas experienced prolonged aerosol residence and elevated concentrations. Through integrated 1D–3D simulations, they propose practical modifications (such as altering flow circulation paths via diffuser adjustments or adding fans) and show how these changes can be efficiently implemented within the co-simulation framework to markedly improve air quality. The findings underscore the value of CFD-based analysis in designing healthier indoor environments and highlight the benefits of coupling multi-scale models (network models for quick global insights with CFD for detailed local predictions) to optimize ventilation performance in large occupied spaces.

Caccavaro et al. [14] provide a valuable methodology-focused contribution by comparing mesh-based and meshless CFD approaches for simulating ship hydrodynamics. They examine a classic test case (a Wigley hull) and a full-scale 30 m ship hull using two open-source codes: OpenFOAM (an Eulerian, grid-based finite volume solver) and DualSPHysics (a Lagrangian, meshless smoothed particle hydrodynamic solver). Each approach is paired with an appropriate turbulence modeling strategy—RANS k–ϵ in OpenFOAM and a particle viscosity model in SPH—and validated against experimental data. The comparative study reveals that mesh-based simulations (OpenFOAM) provide more compact resistance predictions with shorter computational times, while the meshless SPH approach excels in capturing free-surface deformations, such as wave profiles around the hull, with comparable accuracy. In practice, OpenFOAM delivers reliable drag force estimates rapidly, making it advantageous for iterative design processes, whereas DualSPHysics offers a better resolution of complex free-surface phenomena, like bow and stern waves, making it particularly valuable when detailed surface behavior is of interest. The authors discuss the trade-offs: the efficiency and robustness of the Eulerian approach versus the strengths of the Lagrangian method in handling complex interfaces and large deformations. By outlining the benefits and limitations of each, this work guides naval architects in selecting suitable CFD tools or combining them (e.g., applying SPH for specific components) to achieve a comprehensive assessment of a vessel's hydrodynamic performance.

Taken together, the papers in this Special Issue underscore the diverse applications of CFD in contemporary engineering, demonstrating its remarkable versatility in tackling challenges across numerous sectors. From indoor air quality management to marine and agricultural engineering, CFD has become an indispensable tool for both analysis and innovation. For example, Hanspal and Cryer illustrate the widespread adoption of CFD within the agricultural industry, showcasing how a major agrochemical company employs CFD to address a broad spectrum of manufacturing and environmental challenges. Their article presents case studies ranging from optimizing pesticide production equipment to simulating pollutant dispersion in environmental scenarios—situations where experimental testing would be prohibitively costly, hazardous, or impractical. By digitally evaluating "what-if" scenarios—such as chemical plant explosion aftermaths or river contamination events—CFD enables engineers to predict outcomes and plan mitigation strategies in cases that would be impossible to fully test empirically. The inclusion of this study highlights CFD's expanding role beyond traditional industries and underscores its critical importance in both process optimization and risk assessment. Similarly, the role of CFD in process safety and loss prevention highlights its critical contribution to hazard mitigation in industrial settings [15].

A particularly noteworthy addition to this collection is the contribution by Gutiérrez et al. [16], which explores the temperature effects on fluid dynamics inside a continuous casting tundish. Their study rigorously compares isothermal and non-isothermal CFD simulations, challenging long-standing assumptions. Through detailed analysis of buoyancy and inertial forces, as well as inclusion removal rates, the authors demonstrate that temperature gradients induce substantial changes in flow patterns, particularly in zones where buoyancy forces dominate. The findings reveal that while isothermal models may suffice under certain conditions—especially when flow control devices are present—non-isothermal simulations become essential when these controls are absent or weak, as temperature gradients significantly affect flow behavior and inclusion removal efficiency. This nuanced understanding of tundish hydrodynamics not only enhances process control but also informs more precise design and operational strategies in steelmaking.

In conclusion, this second edition of "Industrial CFD and Fluid Modelling in Engineering" exemplifies how cutting-edge CFD research is advancing both the accuracy and the scope of fluid flow simulations in industry. The collected studies explore new turbulence modeling techniques, propose efficient co-simulation and optimization frameworks, validate CFD approaches through novel experimental comparisons, and integrate CFD with other disciplines such as structural mechanics and machine learning. Collectively, these works aim to make CFD more reliable and impactful for addressing real-world engineering challenges [17]. By improving modeling fidelity and computational efficiency, these studies collectively push the boundaries of what CFD can achieve, while also paving the way for future innovations in industrial fluid mechanics. As the Guest Editor of this Special Issue, I would like to extend my sincere gratitude to all the authors for their high-quality contributions and to the reviewers for their rigorous and constructive feedback. I also wish to thank the editorial team for their invaluable support in bringing this second edition to fruition. Together, these efforts have produced a rich and insightful collection of papers that will undoubtedly stimulate further research and development in industrial CFD and fluid modeling.

Conflicts of Interest: The author declare no conflicts of interest.

References

1. Ferrari, C.; Beccati, N.; Magri, L. Numerical Mixing Index: Definition and Application on Concrete Mixer. *Fluids* **2025**, *10*, 72. [CrossRef]
2. Necci, C.; Ceresola, N.; Guglieri, G.; Quagliotti, F. Industrial computational fluid dynamics tools for the evaluation of aerodynamic coefficients. *J. Aircr.* **2009**, *46*, 1973–1983. [CrossRef]
3. Karbasian, H.R.; Vermeire, B.C. Application of physics-constrained data-driven reduced-order models to shape optimization. *J. Fluid Mech.* **2022**, *934*, A32. [CrossRef]
4. Avanzi, F.; De Vanna, F.; Magrini, A.; Benini, E. ARES: A Meanline Code for Outboard Dynamic-Inlet Waterjet Axial-Flow Pumps Design. *Fluids* **2025**, *10*, 66. [CrossRef]
5. Holtmann, B.L.; Key, N.L. Non-Uniform Turbulence Modeling in Isolated Unsteady Diffuser Computational Models for a Vaned Centrifugal Compressor. *Fluids* **2024**, *9*, 270. [CrossRef]
6. Javiya, U.; Chew, J.; Hills, N.; Zhou, L.; Wilson, M.; Lock, G. CFD analysis of flow and heat transfer in a direct transfer pre-swirl system. In Proceedings of the Turbo Expo: Power for Land, Sea, and Air, Glasgow, UK, 14–18 June 2010; Volume 43994, pp. 1167–1178. [CrossRef]
7. Yan, C.; Wang, B.; He, X.; Zhao, F.; Zheng, X.; Vahdati, M.; Zheng, X. Extension and Validation of the Turbomachinery Capabilities of SU2 Open Source CFD Code. In Proceedings of the Turbo Expo: Power for Land, Sea, and Air. American Society of Mechanical Engineers, Boston, MA, USA, 26–30 June 2023; Volume 87103, p. V13CT32A034. [CrossRef]
8. McConnell, M.R.; Knight, J.; Buick, J. Improved Delayed Detached-Eddy Simulation of Turbulent Vortex Shedding in Inert Flow over a Triangular Bluff Body. *Fluids* **2024**, *9*, 246. [CrossRef]
9. De Vanna, F.; Picano, F.; Benini, E. An immersed boundary method for moving objects in compressible flows. In Proceedings of the Direct and Large Eddy Simulation XII, Madrid, Spain, 5–7 June 2019; Springer: Cham, Switzerland, 2020; pp. 291–296. [CrossRef]
10. De Vanna, F.; Baldan, G.; Picano, F.; Benini, E. On the coupling between wall-modeled LES and immersed boundary method towards applicative compressible flow simulations. *Comput. Fluids* **2023**, *266*, 106058. [CrossRef]
11. Silvestri, L. CFD modeling in Industry 4.0: New perspectives for smart factories. *Procedia Comput. Sci.* **2021**, *180*, 381–387. [CrossRef]
12. Davydzenka, T.; Tahmasebi, P. High-resolution fluid–particle interactions: A machine learning approach. *J. Fluid Mech.* **2022**, *938*, A20. [CrossRef]
13. Ajmani, A.; Kirchhof, L.; Rouhi, A.; Mehring, C. Assessing the Effect of Air Ventilation on the Dispersion of Exhaled Aerosol Particles in a Lecture Hall: Simulation Strategy and Streamlined Workflow. *Fluids* **2025**, *10*, 132. [CrossRef]
14. Caccavaro, D.; Tagliafierro, B.; Bilotta, G.; Domínguez, J.M.; Caravella, A.; Gaudio, R.; Cassano, A.; Altomare, C.; Lauria, A. An analysis and comparison of the hydrodynamic behavior of ships using mesh-based and meshless computational fluid dynamics simulations. *Fluids* **2024**, *9*, 266. [CrossRef]
15. Shen, R.; Jiao, Z.; Parker, T.; Sun, Y.; Wang, Q. Recent application of Computational Fluid Dynamics (CFD) in process safety and loss prevention: A review. *J. Loss Prev. Proc. Ind.* **2020**, *67*, 104252. [CrossRef]
16. Gutiérrez, E.; Garcia-Hernandez, S.; Davila, R.M.; de Jesus Barreto, J. Deeper Flow Behavior Explanation of Temperature Effects on the Fluid Dynamic inside a Tundish. *Fluids* **2024**, *9*, 21. [CrossRef]
17. Ourliac, M.; Wah, T.Y.; Rida, A. Computational fluid dynamics along the energy value chain: From natural gas processing to industrial end-use. *Energy Procedia* **2017**, *143*, 381–388. [CrossRef]

fluids

MDPI

Article

Assessing the Effect of Air Ventilation on the Dispersion of Exhaled Aerosol Particles in a Lecture Hall: Simulation Strategy and Streamlined Workflow

Arnav Ajmani *, Lars Kirchhof, Alireza Rouhi and Carsten Mehring

Institute for Mechanical Process Engineering, University of Stuttgart, 70199 Stuttgart, Germany; lars.kirchhof@imvt.uni-stuttgart.de (L.K.); alireza.rouhi@imvt.uni-stuttgart.de (A.R.); carsten.mehring@imvt.uni-stuttgart.de (C.M.)

* Correspondence: arnav.ajmani@imvt.uni-stuttgart.de

Abstract: An efficient solution strategy based on fluid network modeling, computational fluid dynamics (CFD) and discrete particle modeling (DPM) is presented in order to predict and improve air quality, specifically regarding breathing aerosol concentration, in a person-occupied mechanically ventilated room. The efficiency of the proposed workflow is evaluated for the specific case of a lecture hall. It is found that the actual vent system is imbalanced and inefficient in managing the aerosol concentration within the room. Despite a high volumetric exchange rate, aerosol residence times and local aerosol concentrations remain high over an extended period of time, without additional efforts to alter air flow circulation throughout the room. The proposed strategy illustrates how such changes can be efficiently implemented in the basic 1D/3D co-simulation workflow. Analysis of the lecture hall and vent system shows that the execution time for the overall process workflow can be optimized by the following: (1) CAD geometry generation of the room via 3D laser scanning, (2) mesh generation based on the anticipated air discharge behavior from the vent system and (3) by employing HPC resources. Additional simplifications such as the decoupling of vent air flow and room aerodynamics, as observed for the investigated test case, one-way coupling between air flow and aerosol dispersion at low aerosol concentrations and the successive solution of flow field equations can further reduce the problem's complexity and processing times.

Keywords: computational fluid dynamics; 1D flow network model; aerosol; ventilation; virtual reality; laser scanning; discrete phase model; ANSYS; FluidFlow

1. Introduction

Many studies on the transmission of pathogens, i.e., infectious microorganisms such as viruses, bacteria or fungi, in closed quarters such as hospitals or office spaces suggest that room ventilation is one of the major methods for reducing and thereby controlling their spread via the airborne route [1–3]. Although increasing the ventilation air flow rate does dilute concentrations better when the contaminant source is constant, it does, however, not necessarily increase ventilation's effectiveness [1].

The outbreak of highly contagious airborne diseases, such as the past COVID-19 pandemic, requires quick actions in implementing various social distancing and other safety measures. This also includes regulations and recommendations for occupancy limits and ventilation requirements for closed public spaces, including public transport, work spaces, doctors' offices, classrooms and lecture halls [4–6]. In the past, room ventilation regulations were limited to specifications for volume exchange rate (based on room volume) and occupancy (based on square-footage per person and independent of room height). Specifics of the room geometry, vent-in and vent-out configuration and the resulting room air flow field characteristics were only considered more recently, since they greatly influence pathogen or aerosol dispersion and local aerosol/pathogen concentrations within a room.

Clearly, the question of if a person's physical presence in a specific space is safe, with respect to infection with an airborne pathogen emitted from a source (e.g., human host(s)) also present within that room, will not only depend on regulatory parameters, such as the number of occupants, their personal protective measures (such as breathing masks) and room air exchange rates facilitated by the ventilation system, but also on the room's geometry, the position of the individual(s) in that room and the specifics of the employed venting system, including vent-in and vent-out components.

To answer the prescribed "safety question" for a specific space in a timely fashion under a crisis situation, on-site experimental measurements are impractical, too time-consuming or are simply not possible. On the other hand, virtual simulations of the room aerodynamics and aerosol or pathogen dispersion using computational fluid dynamics (CFD) can be effectively deployed to (a) determine actual aerosol/pathogen concentrations and distribution for a given setting and (b) optimize or modify the vent air inlet/discharge system and the positions of the room's available seating areas, as well as the positions of additional obstacles or mobile fan or filtrations systems, to control and reduce aerosol concentrations throughout the room [7,8]. In context of item (b), it is important to note that changes to the vent-in/vent-out system will generally also impact the air distribution in the overall vent system itself, thereby modifying the air flow rate through the vent-in/vent-out locations and thereby the room aerodynamics.

A comprehensive review of studies conducted using CFD methods to understand the airborne dispersion and transmission of the COVID-19 virus has been presented by Mohamadi and Fazeli [9]. Some of the recent works that are concerned with aerosol dispersion and air quality in a lecture hall or classroom were presented by Arpino et al. [10], Mahmoud et al. [11] and Zivelonghi and Lai [12]. The aerosol dispersion and effect of facemasks on indoor transmission have also been analyzed numerically by Aderbigbe et al. [13] and Aliyu et al. [14]. These studies analyzed room aerodynamics based on inflow and outflow boundary conditions imposed at the room's immediate periphery. However, as noted earlier, for actively ventilated rooms, two-way coupling between a room's aerodynamics and the flow within the vent system, the latter of which can efficiently be described by a 1D flow network model, might occur and has to be accounted for in the overall analysis. This is also relevant and practical in cases where vent inlets and/or vent outlets are not readily accessible for measurements such that boundary-condition values are not directly available for the CFD analysis of the room. In addition, incorporating a flow network model allows one to readily investigate different vent system operating conditions without having to perform measurements at a potentially large number of vent inlets and/or outlets.

While the coupling of 1D flow network models with comprehensive CFD analysis models (1D-3D co-simulation) can be found in the analysis of building energy systems (see for example Wei at al. [15]), as well as in other disciplines such as the modeling of respiratory flows [16] and gas turbine components [17], the authors are not aware of any published study wherein a network model for the ventilation system has been coupled with 3D-CFD simulations as part of the overall workflow to assess the effect of air ventilation on aerosol particle dispersion in a lecture hall.

In the present study, a streamlined process workflow for the coupled analysis of the aerodynamics inside an actively ventilated room and the flow inside the corresponding vent system is presented. The topics addressed include the problem set-up, modeling approach, solution strategy and data post-processing, allowing for an expedited assessment of risk for airborne droplet infection that depends, among other things (see below), on the predicted spatial distribution and concentration of breathing aerosols. See also Mahmoud et al. [18] and Tang et al. [19] in this context. It is noted here that no direct risk assessment was conducted as part of this study, as the primary focus was on analyzing the distribution of exhaled aerosols under forced air ventilation conditions and a suitable efficient workflow to do so. Aside from the distribution of breathing aerosols in the room, assessing a person's infection risk for pathogens such as viruses within the liquid phase requires additional considerations, including viral activity (which is affected by room temperature and humidity), the specific viral strain(s), exposure duration, virus decay rate(s) and interactions with deposition surfaces and individual immune system responses, excluding indirect transmission from touching contaminated surfaces.

The proposed workflow consists of eight major steps: (1) data collection, (2) vent system analysis, (3) CAD geometry generation of the room, (4) flow-field mesh generation and quality check, (5) CFD solver selection and settings, (6) selection of boundary and initial conditions, (7) stepwise solution process for room aerodynamics and aerosol dispersion and (8) data post-processing. Included in the workflow, also, are process steps relevant for the subsequent mitigation of excessively high local aerosol concentrations leading to higher exposure risks, even though those steps are not being pursued within the present case study. The specific case under consideration here is an actual student lecture hall at the University of Stuttgart, occupied by the maximum allowable number of individuals (based on 2021 COVID state regulations) and subject to air circulation imposed by an existing (unmodified) forced air ventilation system (without heating or cooling). The proposed workflow illustrates the tasks necessary to quickly ascertain both the flow through the vent system and the coupled aerodynamic flow field in the room, the latter of which also provides the distribution of breathing aerosols exhaled by the room's occupants, thereby allowing one to evaluate the effectiveness of the vent system and the usability of the room if occupied by individuals (at least in part) with a contagious airborne disease.

2. Room and Venting System—Data Collection

The lecture hall depicted in Figure 1 situated at the University of Stuttgart has been considered as a test case to illustrate the streamlined solution strategy presented in Section 3 for determining the airborne aerosol concentration in a room with an installed forced ventilation system.

The hall has a floor area of 417 m^2, with a normal seating capacity of 396 people. Since our initial university-internal study was conducted during the COVID-19 pandemic, the hall was assumed to be occupied by 32 people, corresponding to the maximum seating capacity based on COVID-19 regulations at the time (May 2021) in the State of Baden-Württemberg (BW), Germany [4]. The present work builds on that initial study and focuses on the workflow for jointly analyzing vent system air flow together with room aerodynamics and the breathing aerosol concentration therein. Hence, the original seating of occupants and room configuration has been retained here. Naturally, the proposed workflow can serve not only for improving an existing indoor air-managing system but also for the design of a room's interior geometry and vent system from the outset at the planning stage. In either case, this might include the positioning of additional fans/blowers (as analyzed by Rasam et al. [20] and Yang et al. [21]) or using partitioning walls/room dividers (see, for example, Tao et al. [22]). In addition, social distancing in confined spaces remains a valuable and relevant measure in a layered infection prevention strategy, especially where architectural, usage and/or cost constraints prevent the implementation of other measures.

Figure 1. Lecture hall with forced air ventilation system considered in the present study. View from back to front of the room. One vent-out channel (dark rectangle) located on the left side in front of the room, separated by a divider (gray) from an access door (brown) next to the blackboards (green) is visible.

In our test case, occupants were assumed to be seated in uniformly distributed way across the hall, irrespective of the ventilation arrangement. The torso and head of each occupant have been included in a simplified fashion as part of the boundary conditions for determining the room's aerodynamics. The mouths of the occupants as a source of the aerosol particles were modeled as circular orifices (see Section 5.4 in this context).

The ventilation system of the hall consists of two identical piping branches near the ceiling for air inflow, each comprised of ten swirl nozzles (or swirl diffusers), six jet (or long-throw) nozzles and four air inlet grills. Two outlet vent grills, each connected to an air outflow channel, are situated on either side in the front of the lecture hall. The outflow channels vent the air out of the building and into the atmosphere or alternatively circulate the room air back into the venting. The latter set-up is used during winter-time operation. In the present investigation, the former set-up is in place.

The three types of inflow vents installed in the ventilation system were manufactured by SCHAKO Ferdinand Schad KG, Kolbingen, Germany [23–25]. Each type of nozzle or vent produces an air stream with a distinctive flow characteristic and pressure loss. The air streams entering the lecture hall through the various nozzles and vents, and the location and geometry of the outflow vents, dominate the overall air flow pattern in the hall (forced ventilation). The specific characteristics of the flow entering the flow domain through the prescribed nozzles and vents have been captured by the inflow boundary conditions imposed in the room's simulation model at the respective component outflow cross-sections.

The geometric dimensions of the lecture hall and of the ventilation system were taken directly from the architectural drawings and the Heating, Ventilation and Air Conditioning (HVAC) layout available at the university's building management and maintenance office. Selected geometric dimensions of the room and the vent nozzles were verified by on-site measurements, using a ruler, calipers and a laser distance-measuring tool (and an electronic inclinometer). Some other equipment in the hall not featured on the available drawings (such as the furniture and some ventilation peripherals) were also measured on site. In a second step, the lecture hall's geometry was also captured by a laser scanning system (Leica RTC360 manufactured by Leica Geosystems AG, Heerbrugg, Switzerland), providing a 3D point cloud of surface points for further processing. Time requirements for generating the overall room geometry and computational mesh using the two prescribed methods were compared.

The specifics of the venting system geometry not exposed within the room could not be verified directly. The flow or operating conditions of the venting system were provided by the system operators on site in the form of real-time flow rate measurements taken continuously at a specified location within the venting system.

3. Process Workflow

3.1. Overall Flow Field Solution—Room/Vent Coupling (Inner Loop)

The overall flow problem to be solved consists of the air flow within the ducting/venting system and the air circulation within the room. Both flow domains are coupled with each other at the inflow and outflow planes of the vent-in and vent-out components. The characteristic lengths of these components and their geometric features are substantially smaller than the characteristic length of the entire hall. A simultaneous, spatially fully resolved solution of the entire flow domain, comprising the room and venting system, would require a substantial amount of computational resources and computing time. On the other hand, if there is limited (or no) two-way interaction between the air flow within the venting system and the air flow inside the lecture hall, then an iterative but decoupled solution of the two flow domains is possible. Here, coupling only occurs at the flow interfaces, i.e., the vent-in and vent-out planes. Accordingly, the solution process for determining the overall flow field in the present study has been separated into three sequential steps using two different modeling approaches, i.e., 1D-3D co-simulation. The three steps are as follows:

- Step 1: calculation of the flow distribution in the venting system based on a reduced-order flow network model;
- Step 2: analysis of the air flow inside the room/lecture hall by means of computational fluid dynamics (CFD) analysis;
- Step 3: extracting the boundary information from Step 2 and repeating Step 1 with the updated information if the old and new boundary information differ by a pre-defined margin.

In the present test case, it was found that the iterative step (Step 3) was not necessary, since a meaningful two-way coupling between room aerodynamics and venting system air flow does not exist. Specifically, in the investigated case, the outflow air vents and air inflow vents are spatially separated from each other (and amongst themselves) far enough, such that their local flow fields do not interact directly, thereby influencing the overall flow distribution in the venting system. Also, areas in front of the vent locations were not geometrically constrained or blocked by room installations, which might result in flow field fluctuations at these boundaries. Note that as part of Step 3 above, only information on the boundaries common to both the network model and the CFD model is passed between the two solvers. This information exchange can be handled by using a script or a code coupler such as preCICE v2 [26]. Clearly, the network solver and CFD solver solve for the flow in their respective domains independently of each other in Step 1 and 2.

The prescribed use of a network model for the vent system does not only provide for the better convergence behavior of the overall solution in comparison to a fully resolved simulation of both room interior and vent system together; it also reduces the total computational cost and resource requirements, while retaining a detailed flow-field resolution where necessary. Clearly, using the network model eliminates the need for volumetric 3D meshing of the complex vent system geometry, swirl nozzles and air vents. By bypassing the complex flow topology in the vent system and its components and the associated meshing challenges, the overall simulation model achieves better convergence, while at the same time, the omission of spatially resolving the vent components has only a minor impact on the overall room aerodynamics.

In the absence of a temperature difference between room air and vent air and without the warm air exhaled by the room's occupants, the prescribed (iterative) procedure provides a complete picture of the flow-field aerodynamics within the room as a consequence of the air forced through the room by the venting system. However, since the aim of the investigation is the determination of the breathing aerosol distribution in the room, at least two additional steps, i.e., Steps 4 and 5, have to be taken to determine the latter:

- Step 4: Solution of the energy equation, alongside mass and momentum conservation equations and based on an initialization using the already calculated isothermal flow field from Step 2. This is necessary in order to account for buoyancy effects due to the higher temperature of the exhaled air and thermal plumes as a consequence of heat transfer from the occupants' bodies;
- Step 5: Tracking of aerosol trajectories, allowing for the determination of the aerosol concentration over time within the room (and on surfaces). Here, it is assumed that the aerosol volume concentration in the room is small, such that the room aerodynamics is not influenced by the presence of the aerosol phase; i.e., a one-way coupling between aerosol phase and gas phase is assumed.

3.1.1. Flow Field in the Venting System

The flow-field solution inside the ventilation system is determined subject to the given operating conditions and ambient/outflow conditions at the vent discharge locations, i.e., at the vents that discharge the vent system air into the lecture hall. A venting or HVAC system consists, in general, of an air handler, heating and cooling components, a thermostat, various valves and the piping system moving the air through the vent "inlets" into the room and managing the air leaving the room through the vent "outlets". Since in the present case, the vent-out air, i.e., the air leaving the room, is not recirculated (not even in parts) but directly exhausted into the atmosphere, only the portion of the ventilation system delivering the air through the ducting system into the lecture hall needs to be considered. Also, in the investigated configuration, the air admitted to the room is assumed to have a constant temperature. Temperature variations of the air in the lecture hall are assumed to be solely a result of the warm air exhaled by the room's occupants. The HVAC system is controlled to provide a pre-set total volume flow rate through a pair of identical (open-loop) ventilation branches, subsequently referred to as legs, and the associated periphery. Each leg consists of air ducts, branching components and various reducing and diversion components, along with the discharge nozzles.

The fluid flow through one of the two identical legs is analyzed using a 1D network model and by imposing one-half of the total air volume flow rate at its upstream boundary and atmospheric pressure conditions as the outflow condition at each of the discharge nozzles. Analysis results for the volume or mass flow rates at the various discharge nozzles obtained from the 1D model analysis are then mapped on the corresponding nozzles of the second leg and subsequently used to specify the inflow boundary conditions for the CFD analysis of the lecture hall, i.e., the second step in the prescribed solution process.

When interpreting the results for the room aerodynamics and the spatial aerosol concentration within the room, it should be recalled that the simplified 1D model analysis of the venting system will result in certain inaccuracies with respect to the distribution of the vent air across the various inlets, compared to a detailed CFD analysis of the venting system, thereby affecting also the analysis results for the room flow field and aerosol concentration to a certain extend.

3.1.2. Flow Field Within the Lecture Hall

A steady-state flow-field solution within the lecture hall is computed prior to accounting for the distribution of the airborne aerosol concentration in the room. A commercial computational fluid dynamics (CFD) solver is employed to solve the Navier–Stokes and mass conservation equations for isothermal turbulent flow (as part of Step 2) and the energy equation for non-isothermal flow (as part of Step 4) after spatially discretizing the flow domain. After having determined the steady-state flow-field solution, including turbulence effects and buoyancy effects due to the higher temperature of the air exhaled by the occupants, aerosol particles are injected into the air flow exhaled by the occupants of the room. Lagrangian tracking of the particles in the room is performed using a DPM algorithm and a Random Walk model for turbulent dispersion, from which the aerosol distribution and its concentration can be determined. Droplet vaporization was not considered in the present study, even though it can impact droplet trajectories. It is noted that the vaporization effect could be considered with or without accounting for the generated water vapor by means of an additional transport equation and considering an air–vapor mixture phase in lieu of air.

The flow in both the venting system and in the lecture hall was considered to be incompressible, since the Mach number Ma everywhere in the flow field was substantially lower than 0.3. The maximum fluid velocity was found at the outlet of one of the jet or long-throw nozzles, with a value of 8.58 m/s. To model buoyancy effects, the well-known Boussinesq model has been used.

3.2. System Modifications—Optimization (Outer Loop)

The solution process described in the previous section provides information on the breathing aerosol distribution in the lecture hall based on a selected seating configuration, assumed occupants' breathing behavior (e.g., taken from the literature) and for a given vent system operating condition. Based on the analysis results, modifications to the overall set-up might have to be considered in order to reduce the aerosol concentration in the room below acceptable levels and permit continued usage of the room based on given health regulations, for example. There are various modifications that can be considered to achieve this goal, each of which will require changes at different points within the prescribed workflow and different levels of additional modeling efforts to assess their efficacy, following the renewed execution of Steps 1–5 described in the previous section.

Vent System Changes

One possibility to better manage the aerosol concentration within the room is to modify the room's aerodynamics via changes to the vent system. This concerns changes in the volumetric or mass flow rate of the vent system blower (for example, by increasing the blower speed) or changes to vent system component geometries, e.g., the swirl angle of swirl nozzles, direction of jet nozzles or opening area of air grids.

Modifications of this type will require changes to the network model and, in the latter case, also changes to the inflow boundary conditions within the CFD analysis model for room aerodynamics.

Changes to Room Interior and/or Occupants

This relates to changes in the room conditions in general, in order to change the room's aerodynamics and thereby also the aerosol concentration throughout the room. These modifications include the addition of room dividers or drop ceiling features, the removal or rearrangement of furniture, the addition of secondary air movers (fans, blowers) with or without filtration features and a change in room occupancy or the position of occupants. Naturally, the prescribed changes are constrained by the need to have a minimal or no impact on the usability of the room for the desired purpose.

Hence, any proposed changes to the room's architecture (which passively or actively impact the room's aerodynamics), and their impact on the accessibility of seating areas, visibility and comfort level for occupants or functionality of the room in general, should be evaluated before conducting computationally intensive meshing procedures and CFD simulations. To do so, immersive reality such as a VR headset or a VR Powerwall/CAVE can be useful.

3.3. *Workflow Summary*

The overall workflow for determining (a) the flow behavior within the room (via CFD modeling) and the venting system (via a 1D network model), (b) the dynamics of aerosol distribution within the room and (c) the process for implementing modifications to the system in order to achieve the desired requirements with respect to aerosol loading within the room is illustrated in Figure 2.

There are four major activity or task blocks, which will be described subsequently in more detail: (1) a data collection block, (2) a network modeling block for the venting system, (3) a CFD modeling block for the room aerodynamics and aerosol distribution and (4) a post-processing block focusing on a detailed analysis of the room's aerodynamics and the resulting aerosol distribution, based on which changes to the room/vent system are selected if operational requirements are not met. The diagram also illustrates the two feedback loops described previously.

Block 1: Data Collection

As part of the "data collection" activity, Block 1, the geometric and operational data for the room and vent system are collected. The operational data also entail the location and distribution of individuals within the room based on user needs or regulations. While geometric data for a room are in general available from construction or architectural drawings, geometric data for room installations might have to be collected directly on site. One method to collect all the required data might be by 3D laser scanning of the room. For the present investigation, both methods have been employed. Figure 3 shows a picture of the point cloud data collected from laser scanning at one position using a Leica 3D Laser Scanner (model: RTC360).

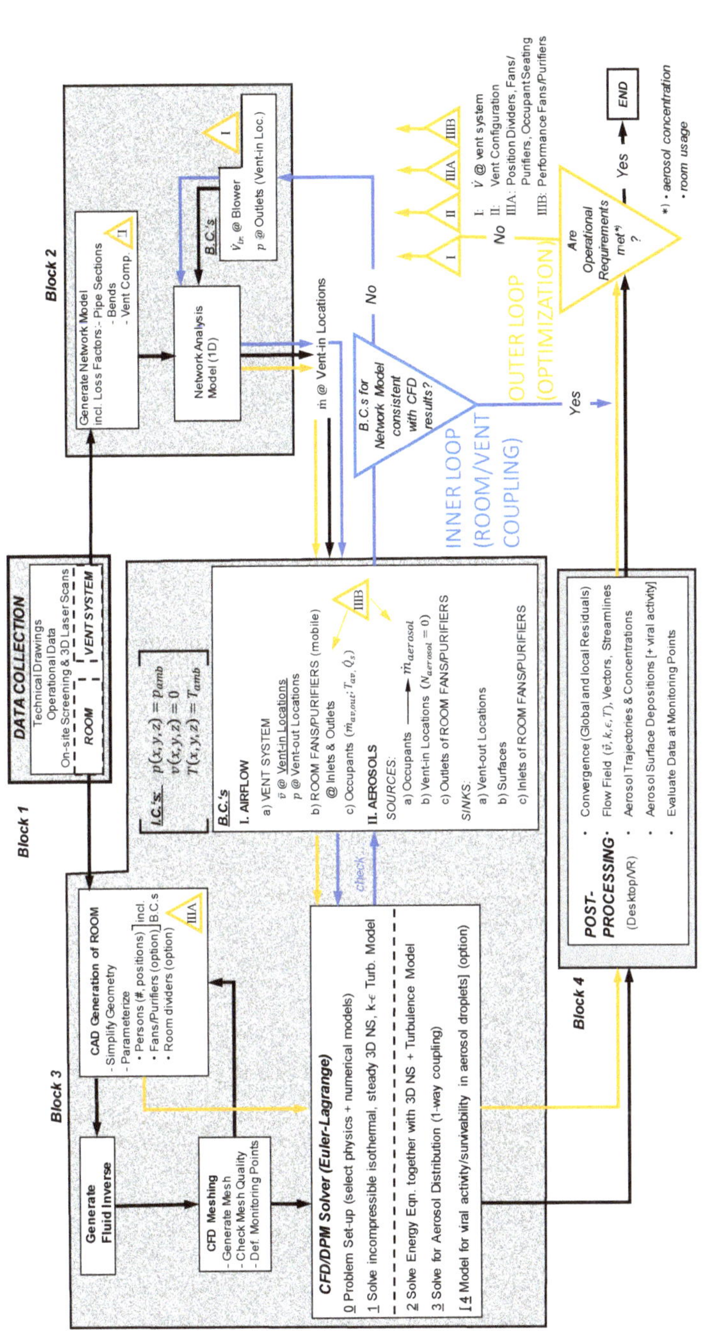

Figure 2. Overall workflow for the analysis of the breathing aerosol concentration in an occupied room with forced air ventilation and its vent system air flow, including measures for improving system performance. Shown are the four major activity blocks, with two iteration loops connecting a flow network block for the vent system (Block 2) with a CFD-modeling block for room aerodynamics (Block 3); i.e., an "inner loop" (blue) solving for the coupled 1D/3D flow fields of a given set-up and an "outer loop" (orange) to iterate the system's performance based on room/vent system modifications.

Figure 3. Laser scanning data (point cloud) from a selected measuring location (large red dot) inside the lecture hall. Smaller red dots indicate other (prior) scan positions. Leica 3D Laser Scanner (model: RTC360).

Block 2: 1D Network Model/Vent System Flow Distribution

The network modeling block (Block 2) determines the quasi one-dimensional flow throughout the vent system and includes a set-up task, defining the vent system network geometry, including vent diameters and lengths, the geometry of bends, contractions and vent outlets (together with relevant wall roughness parameters and loss coefficients) and system operating conditions, including fluid properties and (pressure) boundary conditions at the vent system outlets, corresponding to the vent-in locations of the lecture hall. Assuming the air vented out of the room is not recirculated in the venting system, the room's vent-in locations provide the only coupling between the network model and the subsequently described CFD-modeling block. After the network model has been set up, it is solved numerically, even though for the cases considered in the present study, an analytical solution would also have been possible. Aside from mean flow velocities within the vent system network, the solution provides mass flow or volume flow rates, leaving the various vent nozzles "feeding" the room with fresh air.

Block 3: CFD Modeling/Room Aerodynamics and Aerosol Distribution

The CFD-modeling block (Block 3) begins with a CAD task in which the room geometry is modeled, including furniture and (to a certain degree of abstraction) the bodies of the occupants seated in the room. The latter should be modeled in a parametric fashion, so that their position can be placed arbitrarily in the room. This also applies to any number of room dividers or secondary air movers potentially placed in the room at a later time (as part of the system optimization pursued in the "outer loop" described earlier). It might be beneficial to make provision for the addition of room dividers as part of the initial room geometry generation process but to consider them to be open to flow initially. The feasibility of the latter will depend on the employed CFD solver and the ability to prescribe "internal face boundaries". The subsequent CFD set-up tasks follow the best practice guidelines available for incompressible single-phase fluid flows; see, for example, [27]. They consist of suitable geometry simplifications, the generation of a fluid inverse, the generation of an adequate computational mesh and the setting of proper boundary conditions at inlet vents, exit flow vents and solid surfaces. In addition, monitoring points within the room can be set here for monitoring the convergence of the overall room/vent simulation and for post-processing purposes. Once a converged solution has been obtained, best practices requires a mesh convergence study to be conducted as part of the CFD analysis. This can be performed for a fixed set of boundary conditions (provided their changes remain within certain limits as part of the optimization loop) and without coupling with the network model.

As described earlier, the CFD analysis of the room and the prediction of aerosol dispersion can be carried out sequentially, due to the low volume concentration of aerosol. As part of the weakly coupled isothermal room/vent simulation, the steady RANS (Reynolds–Averaged Navier–Stokes) equations for incompressible flow (together with a suitable turbulence model) and the conservation of mass equation are solved within the room. Upon convergence of the overall room/vent air flow, the energy equation is solved, in addition to the other equations and in order to capture the buoyant behavior of the warm air exhaled by the occupants. Upon the renewed convergence of the flow field, now modified by buoyancy effects, aerosol inflow conditions are activated, and the trajectories of aerosol particles/droplets are calculated within the flow field, assuming a one-way coupling (i.e., the effect of the aerosol particles on the flow field is neglected). Tracing the trajectories of the droplet aerosols can be augmented by a model for droplet vaporization and a model for the survivability of pathogens contained in the aerosol droplets once being released by a host/occupant and potentially being deposited onto a surface within the lecture hall.

Block 4: Post-Processing Simulation Data

Aerosol admission to the flow is carried out for a set physical time or until steady-state conditions with respect to the concentration field of suspended aerosols in the room is reached. In the former case, the set physical time might be determined by the usage cycle of the room or set events, upon which changes in the boundary conditions change the flow field within the room, such as the opening of doors or windows. Typical quantities to be analyzed during post-processing are those relevant for determining if usage requirements are met, as well as those needed for validating the computations and those providing guidance for needed flow-field modifications in case the requirements are not met. These include the following: iso-surfaces of aerosol concentration throughout the room, ideally breaking down which local concentration can be attributed to which occupant; aerosol trajectories; aerosol surface deposition rates (potentially with viral activity over time); streamline and vorticity field plots; contour plots of velocity components; and turbulent quantities, as well as various quantities at predefined monitoring points for the purpose of experimental validation of the simulation results. Commonly, post-processing is conducted at a standard desktop; alternatively, post-processing can also be completed in a virtual environment using a VR headset or VR Powerwall or CAVE. Immersive post-processing is especially useful, if in the course of the optimization (outer) loop, room dividers or other obstacles are placed in the room, potentially interfering with its usage, such as visibility to the front of the lecture hall, impacting the well-being of students (e.g., due to air drafts) or the execution of emergency protocols, e.g., the ability to exit the room within a certain amount from any room position. Novel augmented functionality in a VR environment, for example through “wind devices” for skin sensory input of local air velocity, can in the future provide valuable additional information during post-processing [28,29].

Specific details regarding the physical and numerical models employed for the major simulation blocks (CFD and network modeling) as part of the present case study are presented in the following section.

4. Simulation Methods and Models for Test Case

4.1. 1D Network Model—Ventilation System

The ventilation system was analyzed using the commercial software FluidFlow v3.47. The software features a built-in database for fluid properties and loss coefficients for a number of piping system components.

The conservation of mass equation and the extended Bernoulli equation are the underlying governing equations of the model. In the framework of a steady-state fluid network model, this implies that (a) at any junction or node in the network, the incoming and outgoing mass flow has to sum up to zero and (b) the pressure loss between any junction or node is independent of the path taken between the nodes. FluidFlow software calculates the total pressure losses, with the assumption that no energy is dissipated; hence, the total energy needs to be tracked [30]. Mass conservation equations for the various nodes and pressure loss conditions between the nodes based on the prevailing mass flow rates are solved simultaneously and iterated upon. The convergence criteria for pressure and mass flow rate are set at a 0.01% and 0.001% change between two consecutive iterations, respectively.

In the model, air duct pressure loss Δp_D [Pa] is calculated using the Darcy–Weisbach equation, i.e.,

$$\Delta p_D = f_D \frac{L}{D} \frac{\rho V^2}{2} \tag{1}$$

where f_D $[-]$ denotes the friction coefficient from the Moody diagram [31], V $[\mathrm{m/s}]$ is the mean fluid velocity in the duct, L [m] the length of the duct section, D [m] the hydraulic diameter of the duct section and g $[\mathrm{m/s^2}]$ the acceleration due to gravity.

Constant pressure loss factors K_N $[-]$ for swirl diffusers, jet nozzles and ventilation grills (air grids) were extracted from manufacturers' data sheets and had to be defined manually in the network model. Pressure loss Δp_N [Pa] across these components is given by

$$\Delta p_N = K_N \frac{\rho V^2}{2} \tag{2}$$

Cross-junctions were modeled using the coefficients by Idelchik [32], elbows and various other components according to Crane [33]. At the duct inlet, the volumetric or mass flow rate of air was set, and at the outflow vent components, the static pressure was prescribed. See also Section 5.4 in this context. The solutions from the network model analysis are the volumetric or mass flow rates at the various vent outlets, i.e., nozzles and ventilation grills. Using that information, together with available geometric information for these components, velocity vectors were determined in the respective outflow planes.

4.2. 3D CFD Model—Room Aerodynamics and Aerosol Dispersion

Computational fluid dynamics (CFD) was employed to solve for the flow field and the aerosol concentration inside the lecture hall. The turbulent flow field governing aerosol dispersion was assumed to be steady, thereby excluding large-scale flow unsteadiness, which can potentially lead to bifurcations in flow patterns, for example. Because of the small aerosol particle size and the low aerosol concentration in the flow domain, only one-way coupling was considered for the interaction between fluid flow and the aerosol particle phase within the flow domain. In other words, while the aerosol dispersion in the room is governed by the air flow in the room, the aerosol particles do not impact the fluid velocity nor its turbulent components. In addition, and due to the same reason, particle–particle or rather droplet–droplet interaction has not been considered in this study.

Gas-Phase Governing Equations

The differential forms of the conservation laws for mass, momentum and energy used to solve for the turbulent flow field of an incompressible fluid of a density ρ flow can be written compactly in a generalized form, i.e.,

$$\rho \frac{\partial \Phi}{\partial \tau} + \rho \, div(\vec{V}\Phi) - div\left(\Gamma_{\Phi,\, eff} \; grad \; \Phi\right) = S_{\Phi} \tag{3}$$

with the velocity vector $\vec{V} = (U, V, W)$ and $\Phi = 1$ for mass conservation, $\Phi = U,\ V,\ W$ for conservation of momentum in the x, y and z direction and $\Phi = T$ for conservation of energy. Here, mass, momentum and energy equations have been time-averaged over the turbulent time-scale of the fluid flow, and U, V, W and T denote Reynolds-averaged velocity components (in the x, y and z direction) and Reynolds-averaged temperature, respectively. $\Gamma_{\Phi,eff}$ represents an effective diffusivity due to molecular and turbulent diffusion and S_{Φ} denotes additional (source) terms pertaining to a specific Φ-equation. Their respective values are listed in Table 1. Note that in the present investigation, a steady-state flow-field solution is sought. The unsteady term on the left-hand side of the generalized governing equation is employed in a pseudo time-stepping approach to obtain the steady-state solution and vanishes in that limit. It is noted here that the validity of the steady-state assumption was evaluated by performing a time-accurate unsteady simulation after steady-state conditions were reached with the pseudo time-stepping approach. Doing so, the flow field was found to be quasi-steady, with low frequent limited amplitude flow asymmetries along the centerline of the symmetric lecture hall.

Table 1. Effective diffusivities $\Gamma_{\Phi,eff}$ and source terms S_{Φ} pertaining to mass, momentum and energy conservation equations according to Equation (3).

Equation	Φ	$\Gamma_{\Phi,eff}$	S_{Φ}
Continuity	1	0	0
x-momentum	U	$\mu + \mu_t$	$-\frac{\partial p}{\partial x} + \rho g_x + S_x$
y-momentum	V	$\mu + \mu_t$	$-\frac{\partial p}{\partial y} + \rho g_y + S_y$
z-momentum	W	$\mu + \mu_t$	$-\frac{\partial p}{\partial z} + \rho g_z + S_z$
Energy	T	$\mu/\sigma_l + \mu_t/\sigma_t$	S_T

Here, $p[Pa]$ represents the local fluid pressure; $g_x\ [\mathrm{m/s^2}]$, $g_y\ [\mathrm{m/s^2}]$ and $g_z\,[\mathrm{m/s^2}]$ are the components of gravitational acceleration; $S_x\,[\mathrm{kg/m^2s^2}]$, $S_y\,[\mathrm{kg/m^2s^2}]$ and $S_z\,[\mathrm{kg/m^2s^2}]$ denote additional external forces acting on the fluid; and $S_T\,[\mathrm{J/m^3s}]$ is due to heat sources within the fluid domain. Note that in the present analysis $S_{x,y,z,T} = 0$. In the $\Gamma_{\Phi,eff}$ column of Table 1, $\mu[\mathrm{kg/ms}]$ and $\mu_t[\mathrm{kg/ms}]$ are the molecular and turbulent dynamic viscosity and $\sigma_l[-]$ and $\sigma_t[-]$ represent laminar and turbulent Prandtl numbers, respectively. Since the overall temperature differences are small, buoyancy effects are considered by using the well-known Boussinesq model. The latter accounts for buoyancy forces by expressing the density in gravitational force terms by

$$\rho = \rho_0[1 - \beta(T - T_0)] \tag{4}$$

where $\rho_0\ [\mathrm{kg/m^3}]$ denotes the density at the operating temperature T_0 [K] (i.e., the temperature with which the vent air enters the room in the present case) and β = 0.00338 [1/K] is the thermal expansion coefficient of air. In all other governing equations, the density is assumed to be $\rho_0\ [\mathrm{kg/m^3}]$.

In the present study, the well-known two-equation k-ε turbulence model by Launder and Spalding [34] has been employed to calculate the local values of the turbulent or eddy viscosity μ_t [kg/ms], which is defined as, $\mu_t = \rho\, C_\mu\, k^2/\varepsilon$, where k $\left[m^2/s^2\right]$ and ε $\left[m^2/s^3\right]$ represent local turbulent kinetic energy and its dissipation rate, and C_μ $[-]$ is a model coefficient. Values for k and ε are obtained by solving two additional transport equations, one for k and one for ε. These equations can also be written in the form of Equation (3); the respective values for $\Gamma_{\phi,eff}$ and S_Φ are listed in Table 2.

Table 2. Effective diffusivities $\Gamma_{\Phi,eff}$ and source terms S_Φ pertaining to transport equations for turbulent kinetic energy k and turbulent kinetic energy dissipation rate ε according to Equation (3).

Equation	Φ	$\Gamma_{\Phi,eff}$	S_Φ
k-equation	k	$\mu + \mu_t/\sigma_k$	$G - \rho\varepsilon + G_B$
ε-equation	ε	$\mu + \mu_t/\sigma_\varepsilon$	$[\varepsilon(C_{\varepsilon 1}G - C_{\varepsilon 2}\rho\varepsilon)/k] + C_{\varepsilon 3}C_{\varepsilon 1}G_B(\varepsilon/k)$
$\mu_t = \rho C_\mu k^2/\varepsilon$, $G = \mu_t(\partial U_i/\partial x_j + \partial U_j/\partial x_i)\partial U_i/\partial x_j$, $G_B = g_i\beta(\mu_t/\sigma_t)\,\partial T/\partial x_i$			

Here, G $\left[kg/ms^3\right]$ represents the generation of turbulence kinetic energy due to the mean velocity gradients, and G_B $\left[kg/ms^3\right]$ is the generation of turbulence kinetic energy due to buoyancy. $C_{\varepsilon 1}$ $[-]$, $C_{\varepsilon 2}$ $[-]$ and $C_{\varepsilon 3}$ $[-]$ are constants. σ_k and σ_ε are the turbulent Prandtl numbers for k and ε, respectively.

The model constants have been assigned the default values as per [34] and are as follows:

$$C_{1\varepsilon} = 1.44,\ C_{2\varepsilon} = 1.92,\ C_\mu = 0.09,\ \sigma_\varepsilon = 1.3 \text{ and } \sigma_k = 1.0.$$

The prescribed governing equations are discretized in ANSYS® Fluent® v19.1 based on the finite-volume method with the steady-state flow-field solution being computed via the pseudo-transient solution method, as noted earlier, and subject to the boundary conditions described in Section 5.4. The SIMPLE scheme has been used for pressure–velocity coupling. In Table 3, the numerical schemes for spatial discretization in the present study have been summarized. Scaled residuals were used according to [35], and threshold values were set to 1×10^{-3} for the conservation equations of mass, momentum and turbulent quantities and 1×10^{-6} for the energy equation.

Table 3. Numerical schemes used for spatial discretization in the present study.

Spatial Discretization	Numerical Scheme
Gradient	Least Squares Cell-Based
Pressure	Second-Order
Momentum	Second-Order Upwind
Turbulent Kinetic Energy	Second-Order Upwind
Turbulent Dissipation Rate	Second-Order Upwind
Energy	Second-Order Upwind

Disperse-Phase Model

In general, a dispersed droplet phase such as an aerosol moving in a surrounding gas or fluid phase can exchange mass, momentum and energy with the continuous phase. For the purpose of the present study, this phase coupling is modeled by the so-called discrete phase model (DPM) available in ANSYS® Fluent® v19.1. The trajectories of the discrete-phase particles/droplets are tracked within the flow domain in a Lagrangian fashion by integrating the particles' equation of motion [35]

$$m_p \frac{d\vec{u}_p}{dt} = m_p \frac{\vec{u} - \vec{u}_p}{\tau_r} + m_p \frac{\vec{g}(\rho_p - \rho)}{\rho_p} + \vec{F} \tag{5}$$

together with the kinematic condition

$$\frac{d\vec{x}_p}{dt} = \vec{u}_p \tag{6}$$

where m_p [kg] is the particle mass; $\vec{u}$ [m/s] and $\vec{u}_p$ [m/s] denote the local fluid-phase velocity and particle velocity, respectively; ρ [kg/m^3] and ρ_p [kg/m^3] are the fluid and particle densities; $\vec{F}$ [kg m/s^2] denotes additional forces acting on the particle aside from gravity and buoyancy; and $m_p \frac{\vec{u} - \vec{u}_p}{\tau_r}$ represents the drag force acting from the surrounding flow on the particle, where τ_r [s] is the droplet or particle relaxation time under the assumption of creeping flow around the particle or droplet (i.e., $Re < 1$) [36]:

$$\tau_r = \frac{\rho_p d_p^2}{18\mu} \frac{24}{C_D Re} \tag{7}$$

Here, μ [kg/ms] is the molecular viscosity of the fluid, d_p [m] is the particle diameter, and Re [$-$] is the particle's Reynolds number based on the relative velocity between particle and surrounding fluid flow,

$$Re = \frac{\rho d_p \left| \vec{u} - \vec{u}_p \right|}{\mu} \tag{8}$$

and $C_D = 24/Re$ [$-$] denotes the drag coefficient for creeping flow around spherical droplets.

As the density of the continuous fluid phase (air) is considerably lower than that of the dispersed aerosol particles, the so-called "virtual mass" force, i.e., the force required to accelerate the fluid surrounding an aerosol particle, has been neglected. On the other hand, the shear lift or Saffman lift force $\vec{F}_{L,Saff}$ [kg m/s^2] acting on the particle has been considered, i.e.,

$$\vec{F}_{L,Saff} = 1.615 \rho d_p^2 \sqrt{\nu} \frac{\frac{\partial u}{\partial z}}{\sqrt{\left|\frac{\partial u}{\partial z}\right|}} (\vec{u} - \vec{u}_p) \tag{9}$$

The effect of turbulence on particle dispersion has been modeled using the well-known Discrete Random Walk (DRW) model, which describes the interaction of a particle with a turbulent eddy. Here, a randomized fluctuating velocity is imposed onto the mean velocity over a time interval associated with the respective turbulent eddy lifetimes of the turbulent eddies [35]. Since the aerosol particle size considered here is larger than 0.3 μm, Brownian dispersion has been neglected. Note that in the present investigation, droplet evaporation (affecting droplet trajectory) is not considered and $m_p = constant$ in the above equations.

5. Simulation Set-Up

The following section summarizes the model parameters, geometric parameters and boundary conditions specific to the present study, leading to the CFD and network model results presented in Section 6.

5.1. Material Properties

The ventilation air was assumed to be incompressible ideal gas having a temperature of 20 °C, a density of 1.1894 kg/m^3 and a dynamic viscosity μ of 1.8232×10^{-5} Pa·s. The same constant properties were also assumed for the room air at the point of the problem's initialization. Upon changing the temperature of the air exhaled by the room's occupants in the course of solving the energy equation within the room, the air density appearing in the gravitational force terms of the momentum equations has been modified according to Equation (4).

Respiratory droplets released by exhaled breathing air consist of saliva and other materials found in the respiratory tract, also including various cell types and, potentially, pathogens such as SARS-CoV-2. As respiratory droplets evaporate, the concentration of non-volatile compounds increases, thereby reducing the droplet evaporation rate. The equilibrium diameter at which no further vaporization takes place is reported to be about 20–40% of its initial value [37]. However, in the present analysis, droplet vaporization has been neglected; the exhaled aerosol particles or droplets were assigned the material properties of water at 34 °C [38] with a density of 998.2 kg/m^3.

5.2. Vent System Geometry and Boundary Conditions

The flow through the air venting system has to be modeled in order to determine the air mass flow rates through the various vent components, which discharge the vent air into the lecture hall. These mass flow rates, together with the specific outlet geometries of the employed long-throw jet nozzles, swirl nozzles and air grids/grills, served to define boundary conditions within the subsequent 3D CFD analysis of the air movement inside the lecture hall.

Since the lecture hall's ventilation system is an open-loop system (i.e., none of the air leaving the room via the vent-out ports is recirculated), and since the air delivery portion of the venting system consists of two identical branches, only one such branch or "leg" is simulated. The network pertaining to this leg is shown in Figure 4 (right) and is comprised of various air discharge components (i.e., vent inlets with respect to the room), whose pressure loss coefficients have been taken from the manufacturers' technical specifications. Although the pressure loss coefficients are not constant for different flow regimes, the variations were found to be negligible within the considered operating range.

Hence, average values for the pressure loss coefficients of the employed discharge nozzles and ventilation/air grids were considered. Their respective values are listed in Table 4.

Table 4. Loss coefficients for vent discharge nozzles and vent grids used in the present venting system and employed in the network model.

No.	Component	K Factor
1	Ventilation Air Grid or Grill	1.40
2	Swirl Nozzle/Diffuser	20.00
3	Jet/Long-Throw Nozzle	1.14

To limit acoustic noise in the lecture hall to acceptable levels, the ventilation system is run at 70% peak capacity, which corresponds to a volume flow rate of 4800 m^3/h or 1.586 kg/s of air. Half of this flow rate has been imposed at the 1D vent system branch illustrated in Figure 4. The vent outlets, i.e., nozzles and air grids, have been defined as fixed static pressure outlets, the values of which have been initially set to 101,325 Pa.

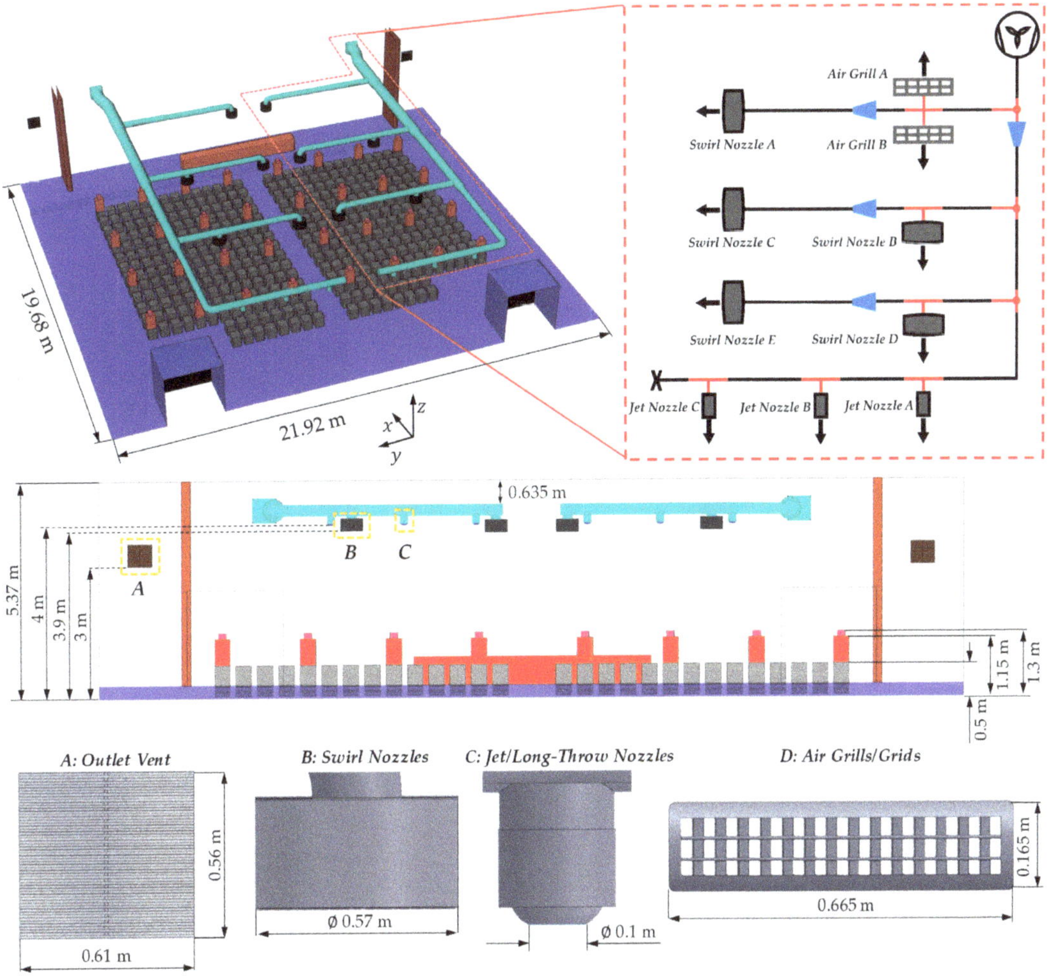

Figure 4. Top left: CAD model of the simulated lecture hall (with side walls removed), including seating area (gray), occupants (red), vent system channels with vent inlets (cyan) and divider walls in front of the room (brown) next to vent outlets (black). Top right: Schematic of one half of the overall vent system, featuring two ventilation grids in the front of the room, five swirl nozzles/diffusers distributed across the room and three throw nozzles in the back of the room. See also Figure 1. The "fan" symbol illustrates that the incoming air mass flow rate is driven by an air blower. Bottom: Detailed dimensional information on the room and vent geometry and relative positions.

Room Geometry/CAD Model

Figure 4 (top left) illustrates the CAD model of the lecture hall with its side walls omitted but including seating areas (gray), the shapes of occupants (red) and the part of the venting system that is located inside the lecture hall. The hall has a total width of 21.92 m, a total length of 19.32 m and maximum height of 5.37 m. The vent pipes are located approximately 0.635 m below the ceiling, and the vent-in boundaries of the swirl nozzles and long-throw nozzle have a distance of 3.9 m and 4 m from the floor. The two cubic sections in the back of the room are the foyers by which the students access and leave the hall. The doors have been assumed to be closed for the simulations. In the front of the hall, there is a slightly elevated stage area occupied by a long rectangular front desk.

There are only two square-shaped vent outlets (black) located in the front of the hall, each one situated between a front divider wall and the neighboring side wall at a height of 3 m above the floor and with a cross-sectional area of 0.3416 m^2. Dimensional information can also be found on the bottom diagram in Figure 4.

In order to keep the size of the computational mesh needed for the CFD analysis of the room manageable, only geometric features essential to the room aerodynamics have been retained. This includes the exact positions and cross-sections of air inflow (vent-in) and outflow (vent-out) boundaries, the walls of the foyer entrances in the back of the room, divider walls near emergency exit doors (which have not been modeled) in front of the room and all seating rows, as well as 32 occupants positioned in every fourth row behind each other and four seats apart from each other within each row. Care has been taken to include the slope of the floor and to precisely position the ventilation duct system within the room; see Figure 4. Geometric features that have been simplified to reduce the meshing time and mesh size include the complexity of the human bodies representing the occupants and details of the seating rows. Each seat within a seating row has been modeled as a separate solid block, and for an occupied chair, the torso and head of the occupant have been approximated as rectangular cuboids sitting atop that block. This has resulted in a considerable reduction in the number of computational mesh elements and a reduction in the complexity of the fluid mesh, the latter of which also had a positive effect on the stability and convergence behavior of the CFD analysis.

Computational Mesh/Discretization of Flow Domain

The geometric or CAD model of the previous section has been used to generate the fluid domain subject to the air flow within the lecture hall. This has been accomplished by using ANSYS® DesignModeler™ after importing the CAD model from SOLIDWORKS® 2018. The computational mesh for the fluid domain has been generated using ANSYS® ICEM-CFD™ 19.1.

Local mesh resolution was set based on anticipated gradients in flow velocity, which are largest near the jet flows originating from the long-throw/jet nozzles, the swirl diffusers/nozzles, the vent air grids and the occupants' mouths. For solid surfaces such as the room's floor, walls and seating positions, wall functions were used to properly and efficiently capture the flow physics in the "near wall" regions. Mesh generation had to be performed with care, since no adaptive re-meshing was employed within the present CFD analysis.

The computational mesh used to obtain the numerical results presented in Section 6.2 initially consisted of approximately 476 million tetrahedral cells and was subsequently transformed into a polyhedral-type mesh. The latter provides better computational efficiency, as well as better accuracy, stability and convergence properties with respect to the numerical solution, in comparison to tetrahedral elements [39]. Special care had to been taken to refine the mesh in the vicinity of the mouths of the occupants, swirl nozzle and jet nozzles; see Figures 5 and 6. After transformation, the total number of polyhedral cells was approximately 85 million.

Cell size dimensions varied from 0.25 mm (in the vicinity of air flow inlets such as swirl and jet nozzles and the mouths of occupants) to 0.1 m. A mesh quality check was conducted. The so-called Cell Squish Index and Face Squish Index are used to assess the mesh quality of polyhedral cells [40,41]. For the exact definition of the Cell Squish Index and Face Squish Index, see [35]. Around 90% of the cells had a squish index below 0.1 and an orthogonal quality index above 0.9, which is considered sufficient for the present study.

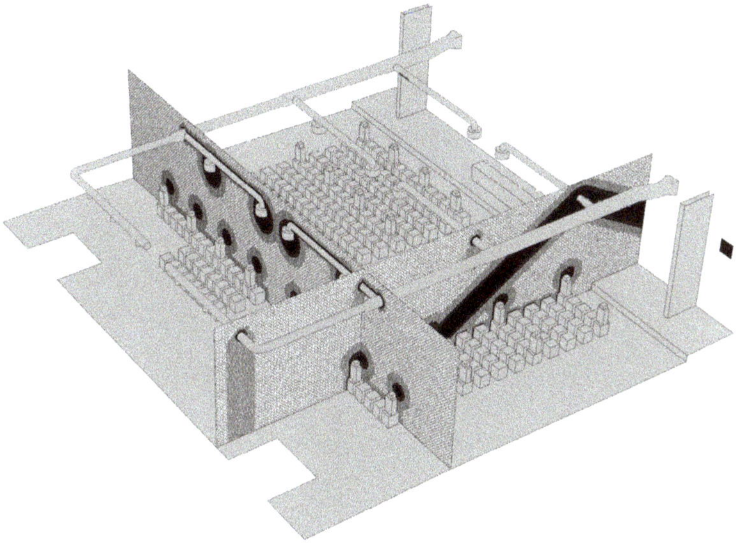

Figure 5. Two perpendicular planes cutting the polyhedral mesh of the fluid domain in the lecture hall. One cut-plane (top left to bottom right) was centered through a set of swirl nozzles/diffusers; the other cut-plane (bottom left to top right) was centered through the exhaling boundary of a set of occupants seated behind each other and the air grills of the ventilation. [No jet nozzle boundary is being cut].

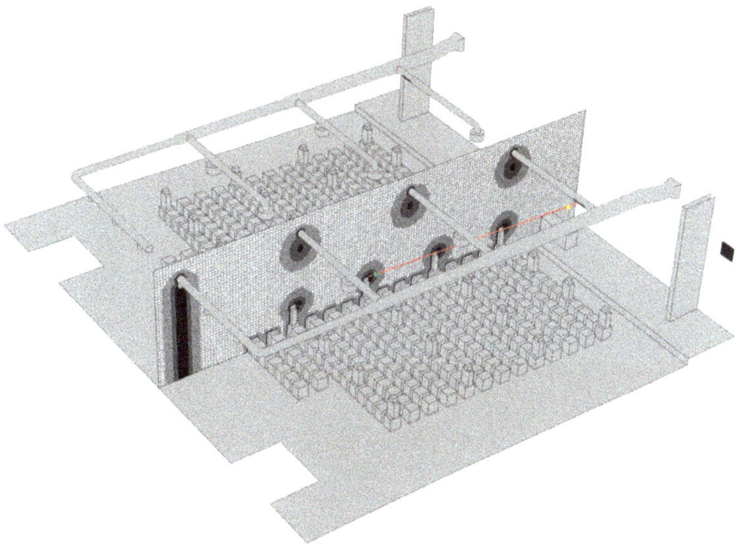

Figure 6. Horizontal cut-plane showing increased mesh resolution around the seating rows and near the floor below the jet nozzles. Vertical cut-plane (from bottom left to top right) through polyhedral mesh showing mesh refinement to resolve the air jet(s) discharging from the jet nozzles in the back of the room and mesh refinement around occupants' mouths and next to swirl nozzles/diffusers near the ceiling. It can be noted that the mesh in low-velocity areas and away from geometric features has been kept coarser. The red line marks the path along which the predicted velocity magnitude has been evaluated as part of the mesh study; see Section 5.3. Furthermore, the green and yellow markers along the red line correspond to Points #1 and #2, respectively, which were selected as reference locations for the grid convergence study.

5.3. Mesh Sensitivity Study

To ensure that the numerical results for the flow field in the lecture hall are mesh-insensitive, a mesh-independence study was conducted for the non-isothermal case and employing three meshes with a mesh count of 32, 66 and 99 million polyhedral cells, corresponding to a mesh count refinement factor of 2 and 1.5, respectively. Specifically, a GCI (Grid Convergence Index) study was undertaken, as already described in [42–44], while considering the velocity magnitude at two meaningful locations within the lecture hall as values of interest or parameters indicative of grid convergence. A GCI value is computed for pairs of grids with successive sizes. For the mesh pair having the two highest resolutions, the GCI is defined as

$$GCI = \frac{F_s|\varepsilon|}{r^p - 1} \tag{10}$$

where F_s denotes a safety factor, which is commonly assigned a value of 1.25 when three (or more) grids are considered as per Roache [44]. The grid refinement ratio r is defined as the ratio between cell sizes of the two grids being compared, i.e., $r = h_2/h_1$, where h_1 denotes the grid spacing of the finer and h_2 the grid spacing of the prior (coarser) mesh. According to the NPARC (National Project for Application-oriented Research in CFD) Verification and Validation Guide [43], $r \geq 1.1$ is required in order to assess the error due to mesh resolution.

For our unstructured meshes and following Celik et al. [42], we used the volume-equivalent mesh size h_{vol} obtained from the average cell volume for each mesh in order to evaluate r, whereby

$$h_{vol} = \left(\frac{1}{N}\sum_{i=1}^{N}\Delta V_i\right)^{\frac{1}{3}} \tag{11}$$

with N being the total cell count of the mesh and ΔV_i denoting the volume of the ith cell.

The parameter ε in Equation (10) represents the relative difference (regarding the predicted value of interest (in our case the velocity magnitude at specific points)) between the two meshes. Accordingly, for the mesh pair with the highest resolution

$$\varepsilon = \frac{f_2 - f_1}{f_1} \tag{12}$$

Finally, p in Equation (10) represents the order of convergence and is calculated as

$$p = \frac{\ln\left(\frac{f_3 - f_2}{f_2 - f_1}\right)}{\ln(r)} \tag{13}$$

where f_1, f_2 and f_3 represent the value of the velocity's magnitude predicted by the finest, medium and coarsest mesh, respectively, at a chosen point of interest in the simulation domain, i.e., the lecture hall. For the present study, two such points, positioned on the line reaching across the lecture hall as depicted in Figure 6, have been chosen. The first point (#P1, indicated by green marking in Figure 6) is located about 20 cm (off center) in front of an occupant's mouth and within the range of the exhaled air plume. At this location, the local velocity profile plays a critical role in aerosol dispersion. The second point (#P2, indicated by the yellow marking in Figure 6) is situated in the shear layer of one of the air jets emanating from one of the air grids. The accurate prediction of shear layers is of great importance for accurately predicting the overall room aerodynamics.

According to [43], when the ratio $(GCI_{23}/(GCI_{12} \cdot r^p))$ is approximately one, it indicates that the solutions are well within the asymptotic range of convergence. The GCI values for the respective mesh pairs, order of convergence and the ratio indicating asymptotic convergence for both these monitoring points are listed in Table 5.

Table 5. Summary of the grid convergence study.

Grid	Elements [in Millions]	Mesh Size (h_{vol}) [m]	U [m/s]		GCI [%]		p [-]		$\frac{GCI_{23}}{(GCI_{12} \cdot r^p)}$	
			#P1	#P2	#P1	#P2	#P1	#P2	#P1	#P2
G1	99	0.0279	0.15	1.79	1.47	0.57				
G2	66	0.0319	0.144	1.77			10.97	9.18	1.04	1.01
G3	32	0.0406	0.042	1.54	6.73	1.99				

Figure 7 illustrates an exemplary comparison of results obtained for the three meshes (including an additional result for a mesh with 85 million polyhedral cells). Shown are predictions for the velocity's magnitude along an exemplary straight line inside the lecture hall, which passes through low- and high-speed areas. The chosen line extends from the vicinity of an occupant's mouth to the close proximity of one of the air grid inlets, as shown in Figure 6. Figure 8 and similar flow-field comparisons throughout the computational domain (not illustrated) show that the mesh with 85 million cells provides a good accuracy at a reasonable computational expense. Hence, this mesh has been used to produce the flow-field results presented in Section 6.2.

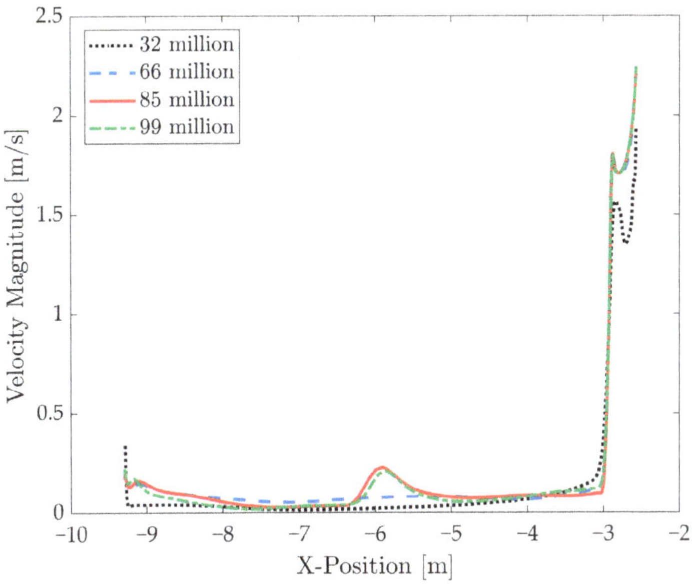

Figure 7. Comparison of predicted magnitude of velocity along the straight line shown in Figure 6 for four different mesh resolutions.

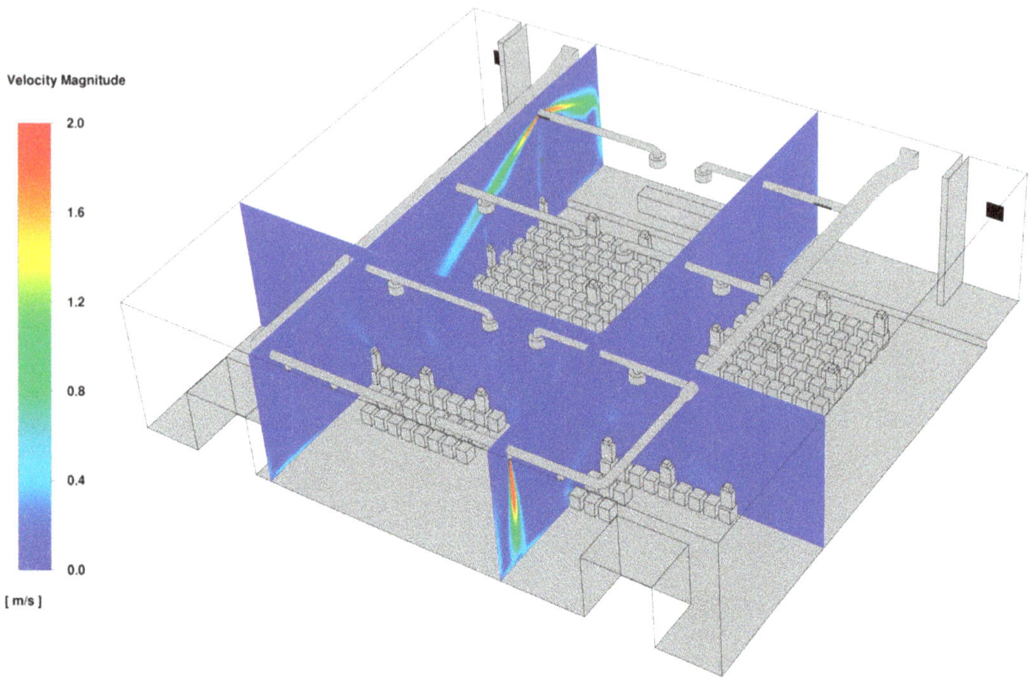

Figure 8. Contour plots of magnitude of velocity in three planes through the lecture hall, cutting a pair of air grids (left plane), a long-throw nozzle (right plane) and a set of swirl diffusers (cross plane).

5.4. Boundary and Initial Flow-Field Conditions

Inflow conditions related to vent system components, i.e., swirl nozzles, throw nozzles and vent grids, were set according to the locally discharged mass flows predicted by the vent system network model solution. Specifically, velocity inlet conditions were specified to components' exit flow boundaries (i.e., the respective room inflow boundaries) based on mass flow and the specific component geometry given by the manufacturers' technical drawings. The calculated velocity vectors were assumed to be constant throughout the boundary surfaces. In the case of the swirl nozzles/diffusers, the exit flow boundary consists of a number of radial slit openings uniformly spaced in the circumferential direction, whereby two sets of internal blades are responsible for the air swirl effect and determine the direction of the velocity vectors at the slit openings. Each slit opening was assigned a specific direction for the air velocity vector, i.e., at 0° or 30° towards the local normal direction. For the present investigation, an angle of 0° was chosen for all swirl diffusers. Defining the boundary conditions for the throw or jet nozzles and vent grids was straightforward.

There are only two vent outlets through which air leaves the lecture hall. These are located at the front wall of the room at a height of approximately 3 m. They are separated by a short divider wall from the area where the blackboards are mounted (see Figure 1). At these boundaries, pressure outlet conditions have been specified, with a static pressure of 101,325 Pa.

To model the occupants' breathing activity and the release of aerosol particles with the exhaled air, the mouth of each occupant is modeled as a circular orifice having a surface area of 20 cm^2. In the present study, dynamic breathing, i.e., an inhalation cycle followed by an exhalation cycle, or coughing, is not modeled. Instead, continuous exhalation at a volume flow rate of 0.27 L/s is assumed. The latter value corresponds to the mean flow rate of a real exhalation cycle under normal breathing conditions; see [45]. Since the mouth opening is specified as a "velocity inlet" boundary condition, a uniform velocity of 0.135 m/s has been imposed here.

No-slip boundary conditions were imposed for solid or impenetrable surfaces; this includes the side walls, floor and ceiling of the room, outer airduct surfaces within the room seating areas and the surfaces of the occupants' bodies (except for their open mouth surfaces, across which breathing air is exhaled).

A summary of the boundary conditions for mean flow quantities is given in Tables 6 and 7. Based on the calculated exit velocities of the swirl diffusers, air grids and long-throw nozzles, a medium turbulence level (with turbulence intensity = 5%) and a viscosity ratio of 10 were assumed for all inflow components. Accordingly, boundary values for k and ε were derived using these assumptions. While this approach is suitable for the swirl diffusers, it may underestimate turbulence for the long-throw nozzles. Nevertheless, at the present Reynolds number, free jet spreading is predominantly influenced by jet–ambient air interactions rather than inlet turbulence, as discussed in [46,47]. Although using varying viscosity ratios for different inflow boundaries could enhance the accuracy, a uniform value was adopted in the present study.

Table 6. Boundary conditions imposed as part of the CFD analysis of the lecture hall and their respective mean values. Turbulent quantities (k, ε) at velocity inlets obtained from imposed turbulence intensity and viscosity ratios, i.e., 5% and 10, respectively.

Boundary Location	Boundary Type	Value
Occupants' Mouths	Velocity Inlet	$V_n = 0.135$ m/s
Inlet Vents	Velocity Inlet	see Table 7
Impermeable Surfaces	No-Slip Wall	$V = 0$ m/s
Outlet Air Grids	Pressure Outlet	$p_{out} = 101{,}325$ Pa

Table 7. Discharge flow rates and respective mean velocities through various vent system components for the vent network of Section 5.2. Component specification according to the network diagram in Figure 4 (right).

Serial No.	Component Name	Flowrate [m^3/h]	Mean Inlet Velocity [m/s]
1	Air Grid A	568.47	2.52
2	Air Grid B	568.47	2.52
3	Swirl Diffuser/Nozzle A	126.91	0.27
4	Swirl Diffuser/Nozzle B	164.49	0.35
5	Swirl Diffuser/Nozzle C	133.93	0.28
6	Swirl Diffuser/Nozzle D	121.56	0.26
7	Swirl Diffuser/Nozzle E	123.33	0.26
8	Jet/Long-Throw Nozzle A	157.21	5.56
9	Jet/Long-Throw Nozzle B	173.14	6.12
10	Jet/Long-Throw Nozzle C	242.49	8.58

Using conservation of mass and momentum equations (i.e., Equation (3) for $\Phi = (1, U, V, W, k, \varepsilon)$), the steady-state isothermal velocity field within the lecture hall has been determined based on the prescribed boundary conditions and using a pseudo time-stepping approach towards the steady-state solution. The initial velocity, pressure and turbulence field within the room was assumed to be uniform, with $(U, V, W, P, k, \varepsilon) = (0, 0, 0, 101{,}325\ \text{Pa}, 0, 0)$. After a converged isothermal flow field solution was obtained, the temperature at all vent system inflow boundaries was set to 293.15 K and that of the air exhaled by the occupants was set to 307.15 K (corresponding to the temperature in the human oral cavity). In addition, zero heat flux was imposed at all impermeable surfaces, except for the occupants' bodies, where a heat flux of $\dot{q} = 58\ \text{W}/\text{m}^2$ was assumed, following DIN EN ISO 7730:2023-04 [48]. Also, before the renewed flow-field solution together with the energy equation ($\Phi = T$ in Equation (3)), the initial temperature on all internal cells was set to 293.15 K.

If the vent or HVAC system generates a temperature difference between the room air and/or room walls, the room aerodynamics can change considerably and the CFD analysis within the room/vent loop workflow now also has to include the solution of the energy equation. If in addition, the room exhaust air is recirculated into the HVAC system, the convergence of the overall flow field, i.e., the room aerodynamics and air flow in the vent network, might be impacted considerably.

Boundary Conditions Aerosol Phase/DPM Set-Up

The distribution of breathing aerosols within the lecture hall has been determined based on the steady-state non-isothermal flow field described earlier and based on the assumption that the aerosol phase does not alter the flow field (i.e., a one-way coupling between the aerosol and gas phase has been assumed).

The calculation of particle trajectories has been performed as described in Section 4.2. According to Hartmann et al. [49], the number of particles emitted through the mouth varies between 134 particles/s and 1018 particles/s. For the present study, we have tracked 10,186 aerosol droplets per person, corresponding to the particles released during a four-second breathing cycle (based on an average release rate of 849 particles per second). Note that for a coughing event, the latter authors measured a mean particle release rate of 13,709 particles/s and a maximum rate of 287,697 particles/s. Particle release positions at the open-mouth boundary were equally distributed across the mouth surface. Based on measured particle size distributions in exhaled air [49–52], a uniform aerosol droplet diameter of 0.5 μm was assumed.

An "escape" boundary condition was set at both air outlet vents, i.e., aerosol particles reaching those outflow boundaries are discarded within the simulation. Aerosol particles that reach an impenetrable surface are assumed to be trapped or deposited at that position.

6. Results and Discussion

This section describes the analysis results from the combined analyses of vent system, room aerodynamics and aerosol dispersion within the room for the specific configuration presented in Section 5 and using the models described in Section 4. Flow-field coupling between the vent system and the room air flow, as covered by the process workflow of Section 3 (see also Figure 2), was not observed; i.e., the static pressure of 101,325 Pa, initially imposed at the vent system outlets within the network model, did not meaningfully vary from those predicted within the subsequent CFD analysis.

6.1. Flow Within the Vent System

Simulation results from the network model, subject to the imposed boundary conditions and fluid properties, provide the air flow rate at every vent opening in the ventilation system. These results are summarized in the third column of Table 7. The listed volume flow rates have subsequently been used to determine the velocity inlet conditions at the various vent components (see Table 7) needed for the CFD analysis of the room's aerodynamics. The velocity magnitudes are listed in column 4 of Table 7.

The calculated mean velocity values were validated by spot-checking the discharge velocity across the outlets of various jet nozzles, air grids and swirl nozzles on site, using a portable vane anemometer. The maximum error between the predicted mean velocities and those calculated from experimental data was 5%.

6.2. Flow Field and Aerosol Dispersion Within the Lecture Hall

Figure 8 shows contour lines of magnitude of velocity (in m/s) in three planes of the fluid domain, cutting either through a set of inlet air grids, a jet/long-throw nozzle or a set of swirl diffusers/nozzles. It can be observed that the air being discharged from the air grids has a major influence on the air circulation in the room. In the front of the room, the two grids directed towards the front wall induce a strong air roller or strong circulation pattern, with air being pushed down the front wall, then back along the elevated stage section towards the front seating rows and from there again upwards; see also Figure 13 (left) in this context. Evidence of the latter is given, also, by the temperature plots shown in Figure 9 and specifically the strongly upwards-directed warm air exhaled by the occupants seated in the front rows. In addition, streamline plots illustrated in Figure 10 show that the left and right sides of the room, down to about half of the room's length from the front wall, are dominated by air flow originating from the air grids directed towards the front wall.

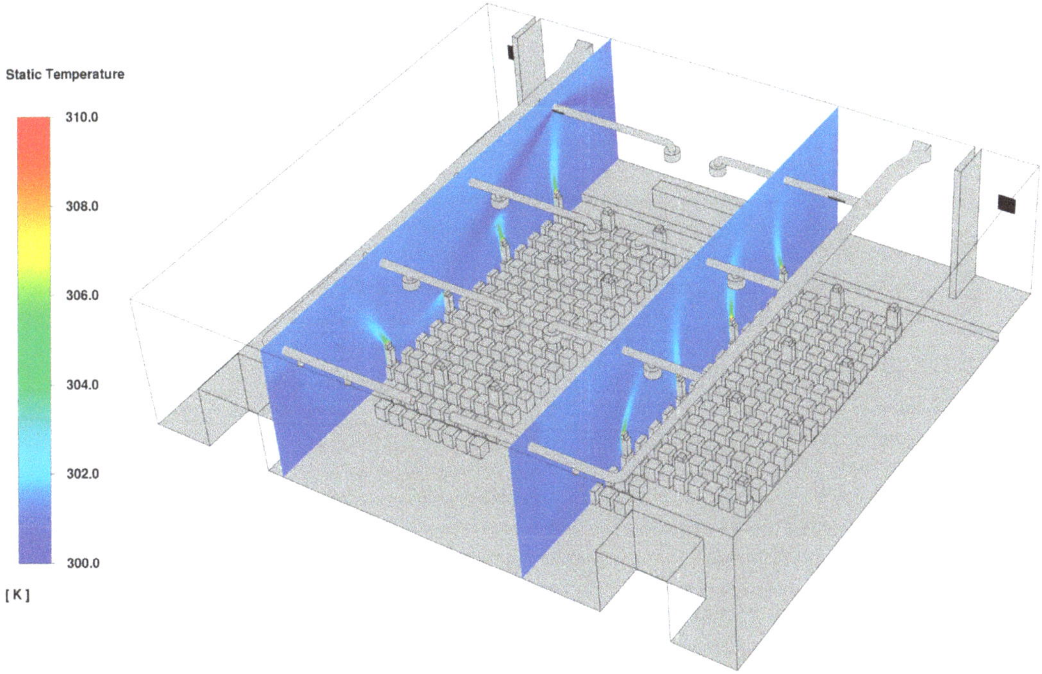

Figure 9. Contour plots of temperature in two planes throughout the lecture hall, cutting through the circular open-mouth boundary of a series of occupants aligned behind each other in different rows of the lecture hall. Temperature of the vent air is 20 °C (=293.15 K), and that of the exhaled breathing air is assumed to be 34 °C (=307.15 K).

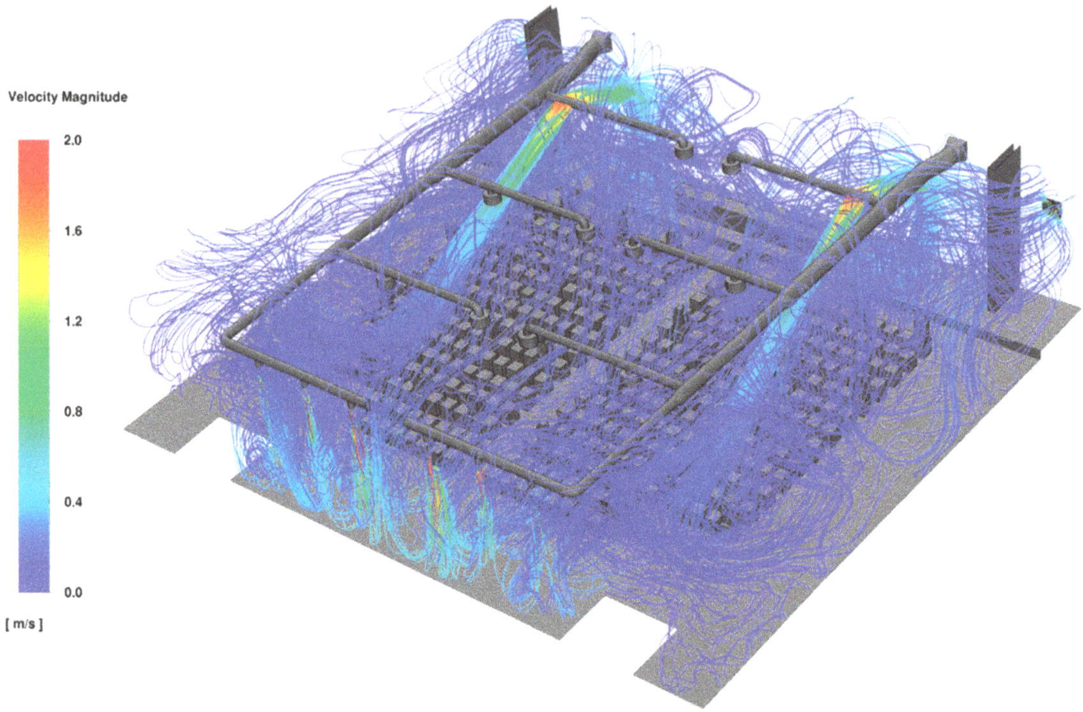

Figure 10. Streamlines for air flow discharged from the air grids (two directed towards the front wall and two towards the center region of the room), colored by magnitude of velocity.

The air being discharged from the two air grids that are directed towards the center region of the room (i.e., down towards the seating rows) reaches far across the entire room, including the last row of seats with occupants. Air from these two vent grids, together with the air being ejected through six jet nozzles in between the two box-shaped foyers, determines the flow field dynamics in the back of the lecture hall (side wall to side wall), as well as in the center region of the room. Air introduced via the swirl diffusers/nozzles mainly affects the air flow in the center region of the room near the center aisle. Here, flow behavior is also influenced by body plumes, as well as by the surrounding air streams originating from the two inward-directed air grids and by the strong air currents induced by the long-throw nozzles at the back end of the room.

Figure 9 shows contour lines of temperature (in kelvin) in two parallel planes along the length of the lecture hall. Each plane cuts through the circular open-mouth boundary of a series of occupants aligned behind each other in different rows. Due to the higher temperature of the exhaled air, that air experiences a buoyancy force in the upward direction. As a consequence of different surrounding air flow conditions at the various seating positions, the body plumes take different shapes, as can be observed from the temperature contours. The occupants seated in the front row experience the upward flow effect of the air roller discussed earlier; here, the temperature plumes are pointed sharply upwards. The occupants in the second occupied row of seats find themselves below the air current generated by the inward-directed air grids. Here, the flow near the occupants is primarily driven by their body's thermal plumes. The occupants seated in the last two occupied rows and in line with the air current discharged from the inward-directed air grids experience the strong air flow of this current now approaching the lower area of the room. Here, the temperature plume has nearly disappeared (second to last row) or is washed upwards behind the heads of the occupants towards the back of the room; see Figure 8 (left contour plane) in this context.

Figure 10 illustrates streamlines originating from the air grids, colored by the local magnitude of velocity. As noted earlier, the vent air discharged towards the inner region of the room reaches well across the lecture hall, approaching the room floor at about the last row where occupants are seated. One also observes that a significant portion of that air enters into the left and right back region of the hall without seats. Some of that air first mixes with the air discharged from the downward-directed jet nozzles in the back of the room. Figure 11 shows streamlines originating from the jet nozzles in the back of the room colored by the residence time of virtual massless tracer particles released at time 0 s on the respective starting points of the streamlines. One observes that, after mixing with air from the vent air grids and impinging on the floor, the air from the jet nozzles creates a strong forward- and subsequently upward-directed air current in the central region of the room. This air current also induces a forward air motion above the backward-directed air streams from the vent air grids before being entrapped by these, only to be pushed upwards and forward again along the center aisle later on. Some of the air discharged from the jet nozzles also reaches the side regions in the back of the room. As a consequence of the various prescribed air currents, the air flow surrounding occupants seated in the center and outer region of the second row is mainly driven by natural convection from thermal plumes.

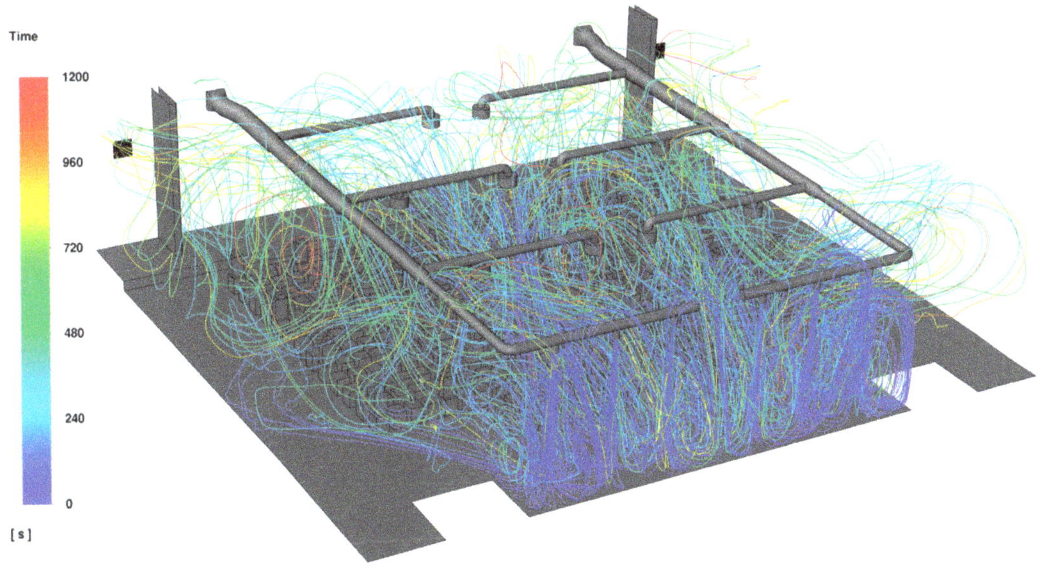

Figure 11. Streamlines for air flow discharged from the six vents or long-throw nozzles in the back of the lecture hall, colored by the residence time of massless tracer particles (w/o inertia) released at time $t = 0$ s at the respective inflow boundaries.

As already noted earlier, much of the air discharged from the air grids towards the front wall of the lecture hall remains confined to the front, i.e., the stage area, where it either drives a rolling air motion between the front wall and the first seating row or is diverted to the right-side and left-side areas between the walls and the seating rows, taking up about one-half of the length of the hall. From there, it can leave the room through the exit flow vents. Alternatively, upon its upward motion along the side walls, that air can enter the central region of the room, where it is again pushed towards the front of the room and then discharged through the exit flow vent or entrained by the air streams and pushed towards the back of the lecture hall.

Figure 12 illustrates streamlines for the air flow discharged from the ten swirl diffusers/nozzles of the vent air distribution system. The air discharged from the four swirl diffusers over the center aisle towards the back of the room (nozzles C and E in Figure 4 for one vent system branch) is pushed downwards and then redirected forwards, due to the influence of the air currents generated by the jet nozzles. The air from the two swirl nozzles in the front of the room (nozzle A in Figure 4 for one vent system branch) is diverted away from the center aisle. The air from the two neighboring swirl diffusers in the center of the room on either side (nozzle B in Figure 4) is entrained by the downward-directed air flow originating from the air grids facing the center of the room and subsequently diverted towards the front of the room.

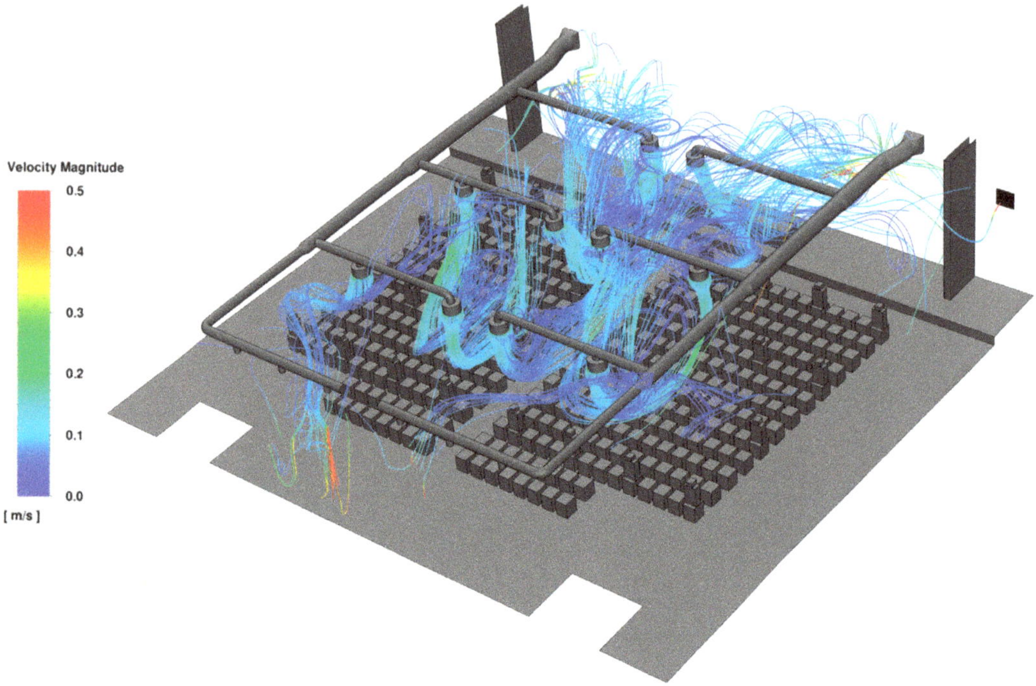

Figure 12. Streamlines for air flow discharged from the 10 swirl nozzles/diffusers near the ceiling of the lecture hall, colored by magnitude of velocity.

The air from the outer swirl diffusers in the back of the room (nozzle D in Figure 4) initially moves downwards towards the side walls of the room, followed by an upward motion, at which point the flow splits into a backward-facing (directed towards the jet nozzles) and a forward-facing stream. Note that the upward-directed thermal air currents or body plumes and the downward air currents generated by the swirl diffusers support each other, due to the relative offset between swirl diffuser positions and occupants' seating positions (in the x-y plane). As a consequence, the breathing aerosols initially rise fairly quickly after being exhaled by the occupants; see Figure 15 (top picture) in this context.

The streamline plots in Figures 10–12 provide an insight into the air distribution patterns, which cannot be quickly obtained by cut-plane plots of magnitude of velocity, velocity components or in- or out-of-plane velocity vector plots, for example. An even deeper understanding of room aerodynamics (and aerosol distribution patterns) can be gained from flow field and particle trajectory visualization in a virtual reality (VR) environment. Here, an interactive tracer particle release at arbitrary positions quickly provides an insight into the complex aerodynamics of the room. Unlike VR desktops or headsets, VR Powerwalls or a VR CAVE (as used in the present work; see Figure 13) provide an additional opportunity for working collaboratively in an immersive environment, which can accelerate discussions or decisions on flow-field optimization or the assessment of a room's usability. For example, flow dividers might be considered for optimizing aerodynamics, but they potentially impact visibility across the room, as well as access to escape routes in case of emergency evacuation. The latter assessment can be more readily be made in VR. In addition, future sensing solutions for immersive VR systems, such as electronic skins [53], will further enhance information concerning an occupant's well-being, such as the sensation of air drafts when being exposed to a strong air current within the room—information that is otherwise difficult to ascertain.

Figure 13. Three-dimensional visualization of room aerodynamics in fully immersive VR CAVE. Left: streamlines (blue) of air roller in front corner of lecture hall and velocity vectors (white) in horizontal plane near overhead vent air grids. Right: streamlines across the entire lecture hall (view towards the back of the room; high velocities in rose color, low velocities in blue).

In the following, aerosol dispersion within the lecture hall is discussed. In this context, it is instructive to remember that, here, only one size of aerosols is considered, all having a diameter of 0.5 μm, and that considering other aerosol particle sizes [51,52] and tracking them within the computed quasi-steady flow field would provide a more realistic picture of the overall aerosol load in the room. Figures 14 and 15 illustrate the trajectories of a selected number of aerosol particles (d = 0.5 μm) exhaled by the occupants and colored by their residence time within the room after being released from the occupants' mouths. One observes that the aerodynamics within the room results in a broadly scattered distribution of aerosol droplets throughout the room, even though the breathing aerosols initially rise fairly quickly towards the ceiling (see Figure 15, top picture). Also, even though the occupants are seated at the same positions on the left and right side of the symmetrically designed lecture hall (including the venting system), with the symmetry axis along its center aisle, the aerosol trajectories differ notably on both sides. This is partly due to turbulent dispersion effects (and the employed Random Walk model) and to a certain extent due to the slight asymmetry of the mean flow velocity distribution with respect to the center aisle.

To describe the aerosol dispersion observed for different seating positions, it is useful to subdivide the occupant pool into (square) blocks of four, i.e., R12C12, representing the two four-person blocks next to the center aisle in the front of the room, and R12S12, representing the two four-person blocks near the front but to the sides of the lecture hall. The blocks R34C34 and R34S34 correspondingly represent the four-person blocks near the center aisle and to the sides in the back of the room; see Figure 15 in this context. Based on this classification, Figure 15 shows that aerosols released from block R34C34 will in general rise upwards and travel towards the front of the room, where they will either be entrained by the air circulation near the front of the room (air roller) or be entrained and "thrown back" towards the room exits (in the back) by the air discharged from the air grids whose openings point inwards into the room. In contrast aerosols released in block R34S34 initially travel mostly backwards (towards the room exits) and are then directly entrained by the long-throw nozzles in the back of the room, or they (initially) might find themselves in the back corners and sides of the room. Aerosols released in block R12C12 are taken up by the air roller established by the air grids discharging their air towards the front of the room. The same is true for aerosols released from the occupants near the center aisle in the front row of block R12S12. Aerosols released by the occupants of block R12S12, which are seated in the second occupied row nearer to the center aisle, can be found (at least initially) within the air circulating in the sides of the room, after being entrained by the air currents originating from the air grids directed towards the inside of the room. Aerosols released from occupants seated in block R12S12 near the wall are either exhausted directly through the vent outlets or indirectly after participating in the air circulation at the side of the room. Overall, based on the actual flow field within the room, only a limited amount of aerosols is directly exhausted through the two exit vents in the front of the room. Evaluation of the residence times and trajectories of aerosol particles shows that after a period of 20 min (starting at $t = 0$ s), 39% of the aerosol particles released during a four-second period after $t = 0$ s will leave the lecture hall over the exit vents, while in the same time period the vent air volume admitted to or exhausted from the room corresponds to 1.45 times the total volume of the lecture hall. Figure 16 shows particles suspended in the room 505 s and 1205 s after release from the occupant's mouth at $t = 0$ s (for a time period of 4 s). One observes that after 1205 s, the particles are more or less uniformly distributed across the room, while after 505 s streaks of locally higher aerosol concentration can be observed.

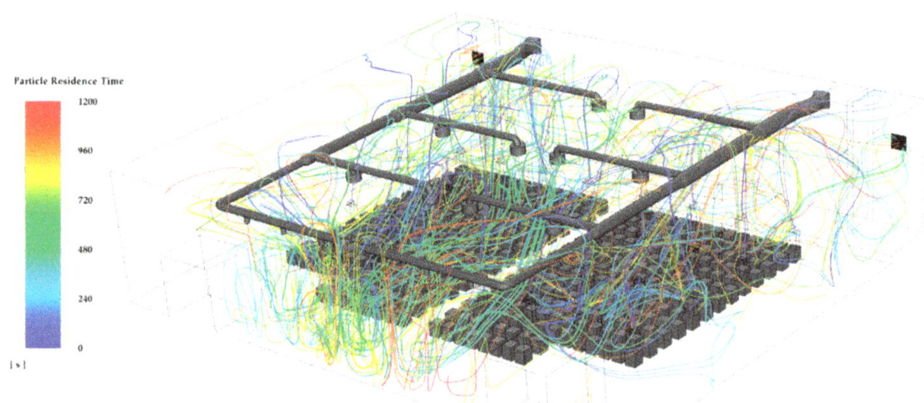

Figure 14. Three-dimensional view of aerosol particle trajectories released from the occupants' mouth openings over a time period of 4 s according to the boundary conditions of Section 5.4. Illustrated are aerosol particle trajectories over a tracking time of 20 min after being released (at $t = 0$ s). Trajectories colored according to their respective residence times.

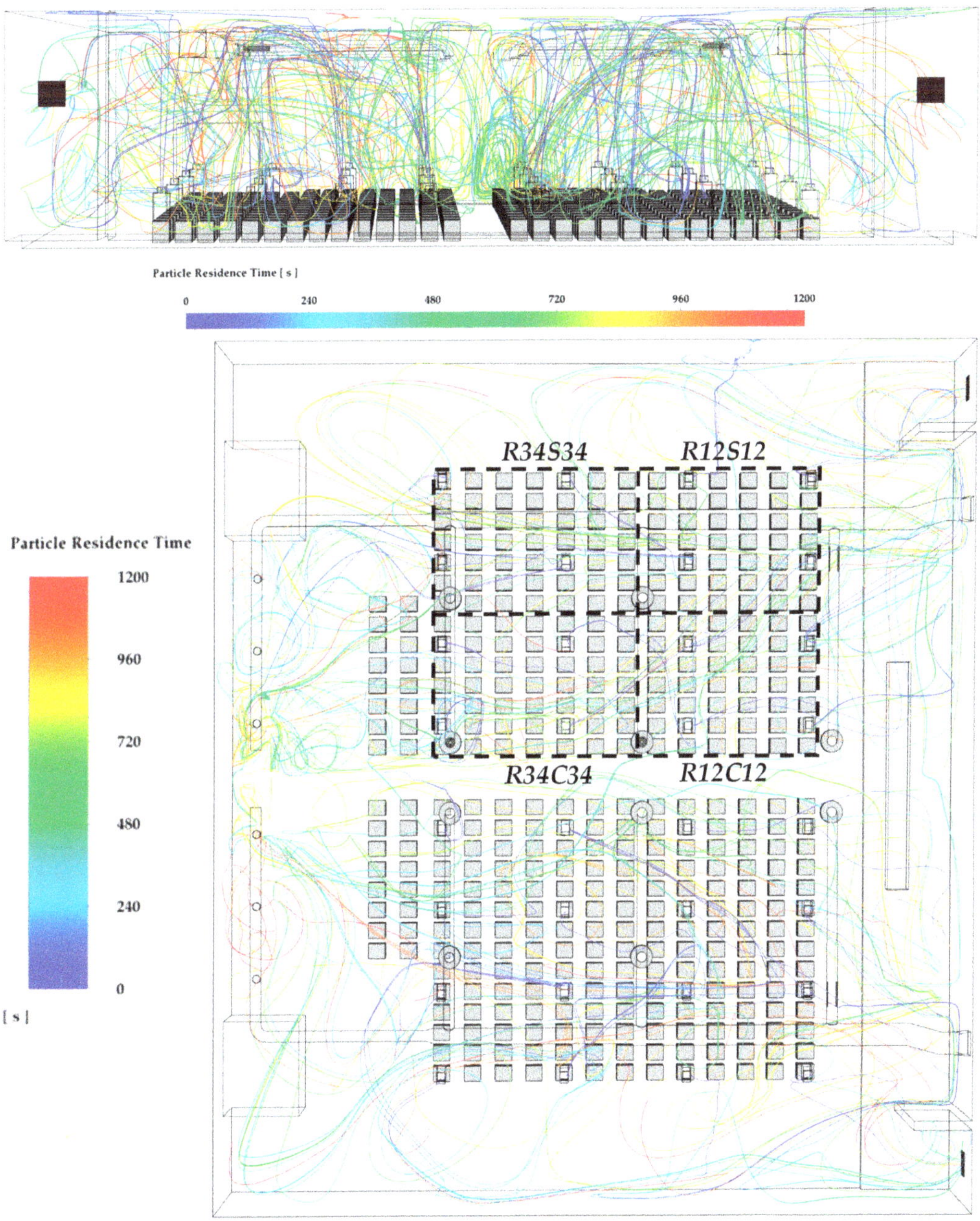

Figure 15. Front-to-back view (top picture) and top view (bottom picture) of the 3D aerosol particle tracks within the lecture hall illustrated in Figure 14.

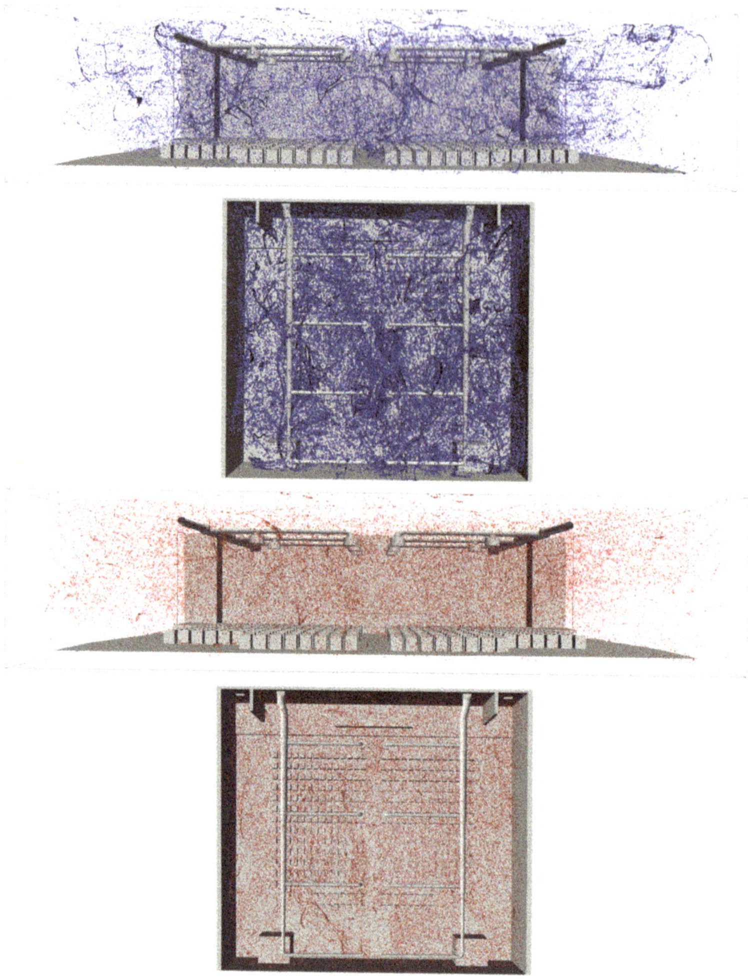

Figure 16. Locations of suspended aerosol particles within the lecture hall, 505 s (top; particles in blue color) and 1205 s (bottom; particles in red color) after starting release from the occupants' mouth openings at t = 0 s for a time period of 4 s (each occupant releasing particles at a rate of 849 particles/s). Number of particles suspended after 505 s is 98,500 particles and after 1205 s, 66,200 particles, corresponding to 90% and 61% of the total particles released.

Table 8 summarizes the time requirements for the various work packages of the process workflow shown in Figure 2, specific to the present investigation. Note that, here, no system optimization ("outer loop") has been included, triggering changes in room or vent geometry, seating positions, vent system volumetric flow rate or other boundary conditions. Also, the pressure conditions predicted by the CFD analysis were consistent with the pressure imposed in the vent system network model; accordingly, no "inner loop" iterations were necessary as part of the present test case.

Table 8. Wall-clock times for the tasks according to the workflow presented in Figure 2 for the present analysis case of a lecture hall with a forced air vent system.

Task	Time Consumption [h]
Network model	**3.5014**
Set-up	3.5
Calculation	0.0014
Alternative 1: CAD Generation for Room	**26**
In situ measurements in lecture hall	4
Measurements of flow conditions	2
HVAC/architecture plan measuring (AutoCAD)	4
CAD generation in SOLIDWORKS®	16
Alternative 2: CAD Generation with 3D-Laserscanner Equipment	**13**
In situ measurements in lecture hall	3
Measurements of flow conditions	2
Post-processing of point cloud data	8
Mesh Generation	**48**
Geometry preparation for pre-processing (DesignModeler)	2
Meshing set-up (sizing parameters, refinement)	6
Mesh execution	10
Check mesh quality and improve mesh (3 iterations)	30
CFD Simulation + DPM	**16**
Conversion of tetrahedral mesh to polyhedral mesh	2
Definition of boundary conditions	1
Selection of numerical models and schemes	1
Run steady simulation without energy equation (5000 iterations); 192 CPU in-house cluster Hydra	4
Coupling flow with energy equation and run simulation (1000 iterations); 192 CPU in-house cluster Hydra	1
DPM set-up	1
Calculation of particle trajectories	6
Post-Processing	**12**
CFD and DPM results	10
Visualization of results (VR via CAVE)	2
Total Time Consumption	**118.5014**

Regarding the CPU time reported in Table 8 for the CFD analysis, it is noted that the CFD simulations were carried out on a local computing cluster featuring 192 CPU cores (processor model: Intel® Xeon® CPU E5-2690 manufactured by Intel Corporation headquartered in Santa Clara, California, USA). The post-processing of the results was performed using ANSYS Fluent and ParaView on a desktop PC, as well as by using COVISE (Collaborative Visualization and Simulation Environment) and a CAVE system available at the HLRS High Performance Computing Center Stuttgart, Germany.

7. Validation of Model Predictions

While not included in the process workflow diagram of Figure 2, any flow-field prediction should naturally be validated against experimental data. To this extent and as noted in Section 6.1, mean velocity values at the vent-in boundaries obtained as part of the network model analysis (and imposed as boundary conditions for the CFD analysis of the lecture hall) were checked by measuring the discharge velocities at selected jet and swirl nozzles and vent air grids using a Ø 3 mm hot ball (thermal) anemometer (manufacturer: Testo SE & Co. KGaA, Titisee-Neustadt, Germany) with a measuring range from 0 to 10 m/s flow velocity and an accuracy of ±0.03 m/s. In addition, predicted flow velocities were compared with measured values at singular positions in the jets discharging from the long-throw nozzles and at vent outlets, with errors ranging from 4 to 5% (see Table 9 in this context). Local measurements within the room were not conducted, as the measurements were taken without the presence of occupants. Also, measurements at swirl nozzles could not be performed as they were not safely accessible. Overall, a good agreement between numerical and experimental data was found at the sampled positions.

Table 9. Comparison of calculated velocities (via 1D network model or 3D CFD analysis) with experimentally measured velocities at various positions.

S. No.	Location	Simulation [m/s]	Experiment	Percentage Error
1	Jet/Long-Throw Exit Nozzle C	8.58 (1D network)	8.06–8.3 m/s (±0.03 m/s + 5% of measured value)	4.89%
2	Jet/Long-Throw Nozzle B (measured 1.5 m below the nozzle outlet)	2.72 (3D CFD)	2.74–2.99 m/s (±0.03 m/s + 5% of measured value)	5.06%
3	Outlet 1 (right side of the lecture hall)	3.754 (3D CFD)	3.6–4.2 m/s (±0.03 m/s + 5% of measured value)	3.74%

For a more comprehensive validation of the predicted flow field, laser-based methods like particle image velocimetry and/or laser doppler anemometry are preferred. While these methods are more accurate, their set-up and calibration is elaborate and tedious to execute in a timely manner in a public place. Hence, they have not been employed here. Another method to gage the accuracy of the numerical solution, or rather the predicted major flow patterns, is (a) smoke testing, i.e., the admission of tracers such as smoke or fog, possibly in conjunction with laser-sheet illumination or (b) the timed admission of gases to the vent air and their measurement at various selected positions over time, using suitable sensors.

8. Conclusions

Room ventilation is, aside from personal protection, essential to prevent the spread of any airborne pathogen in closed public spaces such as doctor's offices, hallways, subways or lecture halls.

While increasing the ACH (air change per hour) is a good approach to decreasing harmful particle concentration in a ventilated room, it is well known that the aerodynamic flow-field pattern within a room, driven by the vent system flow rate (at inlets/outlets), buoyancy effects as consequence of temperature gradients and occupants' positions (or movement) and their breathing behavior, can have a dominant effect on each individuals' exposure risk for harmful pathogens released into or entering the room [54–56].

While there have been attempts to consider air flow effects by a single lumped parameter for design purposes [57], resolved CFD simulations are needed in order to accurately assess rooms' air quality and potential exposure risks. Intermediate models, such as RTD (Residence Time Distribution) models [58], multizone air flow network models or, similarly, RN (Reactor Network) models known from gas-turbine design (which can be viewed as network models for a single room), can provide reliable information at a low computational expense, as long as changes in boundary conditions do not result in a change in the overall flow structure, bifurcating flow or bi-/multi-stable flow behavior. However, these models rely on simulations or experimental data to identify characteristic flow-field zones or network nodes.

The present study provides a detailed process workflow for the numerical analysis of a mechanically vented room and its connected vent system, with the emphasis on simulation efficiency. Actual data with regards to the execution/processing time of the various basic workflow tasks is provided for the case of a large lecture hall with a selected student seating arrangement and under normal breathing conditions.

For the considered test case, flow coupling between vent air flow and room aerodynamics did not occur. Steady turbulent flow conditions were assumed for the flow field within the room and modeled by the RANS equations, together with the original k-ε turbulence closure model by Launder and Spalding. Governing equations were solved using the pseudo-transient method in the commercial CFD solver ANSYS Fluent v19.1. Mesh refinement in areas where high-velocity gradients could be expected, and a mesh sensitivity study, resulted in a suitable mesh consisting of 85 million polyhedral cells. Given the low aerosol concentration in the air exhaled by the occupants, and the lack of vent air heating/cooling or heat transfer through walls or windows, a one-way coupled approach between the continuous and disperse phase and the general dominance of forced vs. natural convection could be assumed. Accordingly, the CFD analysis of the room aerodynamics was efficiently performed by a stepwise increasing of the complexity of the flow problem, i.e., (1) solution of mass and momentum conservation equations subject to the given boundary conditions, (2) additional solution of the conservation equations for turbulent quantities k and ε (not necessarily up to full convergence) and (3) additional solution of the energy equation to account for the rising of warm breathing air and thermal plumes generated by the occupants. These successive steps have been performed on a local computing cluster featuring 192 CPUs (Intel® Xeon® CPU E5-2690) for a total of 10,000 iterations or pseudo time steps, requiring five hours' total wall clock time. The Lagrangian tracking of a total of 326,000 monodisperse aerosol particles (d = 0.5 μm) within the calculated turbulent flow field was performed on a local workstation. The results indicate that the vent system air flow entering the lecture hall is imbalanced and inefficiently used. The room aerodynamics is driven by strong, slightly downward-directed air jets discharged from four air grids in the front area of the room. Two of the four jets reach far across the lecture hall and interact with a series of downward-directed throw nozzles at the other end of the hall, driving an air current along the center isle back towards the front of the room and directed slightly upwards. A series of ten swirl nozzles accounts for approx. 28% of the air flow and is distributed across the center region of the room at the height of the air grids. Their positions are generally offset from the seating positions of the occupants below and generate locally downward air currents that support the upward-directed thermal body plumes of the occupants; such breathing aerosols initially rise fairly quickly after being exhaled by the occupants.

Twenty minutes after being released with the air exhaled by the occupants, 61% of the released aerosols are still within the room and mostly suspended in the room's air. Only 39% have discharged with the vent-out air, while during the same time period the overall volume of air in the room has been changed 1.45 times.

For the considered open (flow-through) vent air system, building the network model accounted for about 3% of the overall process workflow, with a negligible amount of model run time. Control measurements of flow rates and of the basic room geometry, comparison with architectural drawings and CAD data generation using SOLIDWORKS® 2018 took about 22% of the overall process workflow. Alternatively, performing room measurements with a laser scanning device and using an existing workflow and tools for point cloud processing was considerably more efficient and reduced that time down to 12%. Mesh generation and mesh quality checking accounted for a significant amount of the time, i.e., 40% (not including a mesh sensitivity study), and posed a bottleneck in the overall process workflow. While the mesh preparation step also included estimates for the spreading behavior of freely discharging jet and air streams originating from the vent system (thereby allowing for the selection of a suitable mesh), it shows that streamlining this process will greatly improve overall process turn-around times, provided that enough computational resources are available for the subsequent computing step, especially in the case of large computational meshes. In this context, the use of HPC and Cloud Computing services that also provide CFD software licenses can yield cost-effective and timely solutions. Using the computing cluster available at our institute, the CFD run time for room aerodynamics amounted to 9%, and calculating particle trajectories on a desktop PC to an additional 6%, of the total process time. Data post-processing in order to understand the flow aerodynamics and resulting aerosol concentration accounted for 10%, including both traditional desktop post-processing and post-processing in a VR (Cube) environment. In the latter case, data management and rendering are significantly more efficient if suitable tools for data importing and pre- and post-processing are available and if the user has been properly trained. In fact, developers of high-performing VR systems frequently employ more powerful tools than those used for "traditional" desktop CFD post-processing (e.g., with respect to the cache and buffer optimization). In addition, VR processing allows for the evaluation of features that are not readily accessible otherwise, such as visibility across the room if dividers are placed for the purpose of improving a room's aerodynamics, for example.

From an engineering perspective, the effective numerical analysis of room aerodynamics and its connected vent flow system requires a careful evaluation/determination of the dominant flow physics, flow coupling and initial mesh generation based on estimates of the expected overall aerodynamics and specifically the extent of forced air streams/jets (and their shear layers) discharged from the vent system. Flow-field meshing retains the largest potential for reducing the overall workflow process time, if sufficient computing resources are available for the problem at hand.

Author Contributions: Conceptualization, A.A., L.K., A.R. and C.M.; methodology, A.A., L.K., A.R. and C.M.; software, A.A., L.K. and A.R.; validation, A.A., L.K. and A.R.; formal analysis, A.A., L.K., A.R. and C.M.; investigation, A.A., L.K., A.R. and C.M.; resources, A.A., L.K., A.R. and C.M.; data curation, A.A., L.K. and A.R.; writing—original draft preparation, A.A., L.K. and C.M.; writing—review and editing, A.A., L.K., A.R. and C.M.; visualization, A.A., L.K. and A.R.; supervision, C.M.; project administration, C.M.; funding acquisition, C.M. All authors have read and agreed to the published version of the manuscript.

Funding: This research received no external funding.

Data Availability Statement: Further data can be provided after contacting the corresponding author.

In order to evaluate the distribution of concentrations, the statistical parameters used are the mean and the standard deviation, with the latter calculated as the square root of the variance. This method is versatile, as it makes it possible to analyze the mixing either experimentally, by taking an arbitrary number of samples from the mixture, or numerically, by analyzing the distribution of components in certain points in the simulated fluid domain.

In statistics, according to Equation (1), the sample mean of a quantity X over a sample number n is defined as follows:

$$\overline{X} = \frac{\sum_{j=1}^{n} X_j}{n} \tag{1}$$

The sample standard deviation S is defined according to Equation (2):

$$S = \sqrt{\frac{\sum_{j=1}^{n} \left(X_j - \overline{X}\right)^2}{n-1}} \tag{2}$$

The two parameters, when applied to the concentration of a component on n samples, allow us to assess whether the mixing has occurred homogeneously or if the components have been segregated.

For a mixture, the Lacey mixing index can be defined [18] on the basis of Equation (3):

$$M_L = \frac{{S_0}^2 - S^2}{{S_0}^2 - {S_R}^2} \tag{3}$$

where S_0 is the standard deviation for complete segregation and S_R is the standard deviation for optimal mixing. As it is defined, Lacey's index is a number between 0 and 1 and can be understood as the ratio between the actual extent of the mixing and the ideal optimal limit to which the mixing can tend [19].

Similarly, by omitting the segregation term S_0 from Equation (3), the index can be simplified, obtaining Equation (4):

$$M_L = \frac{S^2}{{S_R}^2} \tag{4}$$

In this definition, mixing can be called optimal to the extent that the value of M is close to one, admitting values greater than one as well.

The standard deviation method has been successfully employed by various authors for the study of simulated liquid–solid suspensions by a multiphase CFD approach [20–22]. The increased homogenization (improved suspension quality) results in a reduction in the standard deviation. Ref. [23] presented a method based on the distribution, mean, and standard deviation of the velocity magnitude, a method which can be easily applied in the comparison of various velocity profiles from CFD simulations. In other words, the Lacey mixing index and standard deviation are the most commonly used methods and are well suited to the problem. These methods are not only widely recognized but also simple to apply. CFD simulation can be an excellent and cost-effective tool to calculate mixing efficiency using CFD simulators, even with complex impeller geometries [24]. Ref. [25] defined a new mixing index for measuring the efficiency of a Twin Screw Extruder; the CFD simulation was validated by a comparison with the experimental data.

Regarding the study of concrete mixing by numerical means, CFD models can be used for the simulation of a truck mixer [26], a paddle and screw mixer for cement [27], and a cement slurry mixer [28]. However, in these studies, the mixing index and overall quality of mixing were not calculated, and the fluid was considered homogeneous. Instead, in the current study, the mixing of the phases—water, cement, sand, and gravel—is taken into account by calculating the volumetric fractions of each component. As shown above, the calculation of the mixing index by means of a statistical method has not yet been applied to concrete mixing that has been investigated using an Eulerian computational approach. There are few works in the literature on the simulation of concrete mixing, and the DEM approach is used in all of them. This approach can be difficult to apply, especially in dense flows with non-Newtonian dynamics where the motion is governed by viscous effects, which occur in cement paste and concrete dynamics. The advantage of the DEM approach is that the mixing index is easily calculated. On the contrary, the study of mixers for concrete with an Eulerian multiphase CFD approach can lead to a more accurate result on the viscous and non-Newtonian dynamics of the mixture (cement paste or concrete); however, but it would be complicated to apply the concept of a mixing index. In other words, it is difficult to estimate how well the flow is mixed with an index.

In recent years, we have focused on the study of fresh concrete dynamics in a concrete mixer via multiphase CFD methods. In the work by [29], a CFD model was presented that simulated the discharge of fresh concrete from a concrete mixer; the flow characteristics of the model were calibrated with experimental tests in the field. In the work by the same authors [30] issued within the following year, however, the same model was used to predict the capacity level of a concrete mixer as a function of the consistency class of the concrete. In both of these works, fresh concrete was simulated as the Bingham fluid, and they focused on studying the dynamics of the fresh concrete just produced, while the mixing process was not addressed. Instead, in [31], a multiphase Eulerian model was developed in a transient regime, which was useful for studying the concrete mixing process inside a truck concrete drum mixer. For the first time, the model simulated the distribution of the four phases (air, cement paste, gravel, and sand) inside the mixer as the number of revolutions of the rotating drum increased. In this work, the model was validated and some fluid dynamic parameters characteristic of the mixing process, which were previously unknown, were calculated. However, the work lacked a method for the quantitative calculation of the mixing of the various phases.

In this paper, we illustrate a methodology focusing on the definition and calculation of the numerical mixing index through the implementation of some statistical variables in the commercial fluid dynamics code used for the computation. Through the model and the application of the mixing index calculation, as carried out in this work, it is possible to optimize the geometry and operating parameters such as rotation speed and inclination.

Before this work, the mixing index had never been applied to the study of concrete mixing simulated with a multi-phase CFD approach. The mixing efficiency of a truck mixer is currently evaluated indirectly, through the measurement of the absorbed torque or with other experimental methods; numerically, it is evaluated when the truck mixer is simulated with DEM approaches. The definition and application of the mixing index allow for the evaluation of the mixing process and its efficiency using a numerical Eulerian approach. When the mixer is studied numerically with an Eulerian method, without the cell mixing index, the evaluation of mixing quality would be impossible, necessitating a Lagrangian approach. The cell mixing index is the variable that enables the calculation of a multiphase mixing process when studied with a numerical Eulerian method.

This work makes it possible to quantify the efficiency of a concrete mixer through a universally adopted index, even when studied with a multiphase CFD approach, and can be applied to other types of solid–liquid mixers present in other fields such as process engineering or the chemical industry, which can be studied with the same approach.

2. Analyzed Machine

Off-road concrete mixer trucks are designed to produce homogeneous concrete mixtures directly at building sites. The inclination angle of the drum can vary from 15° to 20°; during the mixing process, this figure is kept at 18°. The maximum angular speed is 22 rpm, which is reached during the mixing phase. As the components are loaded into the drum, it starts revolving in one direction to permit mixing; once the mixing is complete, the mixture is discharged by reversing the speed direction. The drum has two double truncated cone sections, with different cone angles in the upper and lower parts, which are divided by a middle cylindrical section (Figure 1).

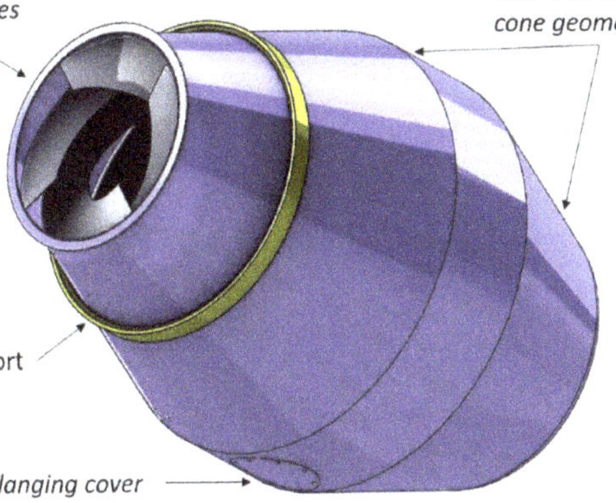

Figure 1. Drum geometry.

The raw materials are loaded using a water pump and a charging hopper for the solid aggregates. Generally, in the concrete production with batch mixers, it is possible to identify three different phases (Table 1). The first phase is the loading phase, which starts with a first hopper of gravel, followed by the addition of cement and water; afterwards, alternate loads of gravel and sand are added until the required mixture is obtained. Finally, an additional quantity of water is loaded in order to achieve the final concrete mix design. The second phase is the mixing phase, in which the cement paste, formed by water and cement, is absorbed by the inert materials, i.e., sand and gravel. This phase leads to the formation of agglomerates which subsequently dissolve, homogenizing the mixture. The third phase of the process is the discharge phase, which takes place once mixing has been completed. The mixture is discharged by reversing the direction, and the rotational velocity in this phase must be chosen carefully in order to speed up the cycle phase without causing segregation of the mixture, which would result in low-quality concrete. The mix design has to be chosen before loading the material, as it will determine the properties of the concrete.

Table 1. Working cycle of a self-loading truck concrete mixer.

Phase	Time [s]	Volume [m^3]
Charge material	30–320	0–1.35
Mixing	320–620	1.35
Discharge	620–700	1.35–0

The concrete mix design and the material loading order are presented in the following tables (Tables 2 and 3). The presented mix design is the most commonly used one for the main applications of the machine under investigation.

Table 2. Analyzed Mix Design.

Material	Mix Design [Weight %]
Water	7
Cement	13
Sand	38
Gravel	42

Table 3. Drum Loading Cycle.

Loading Order	Material	Weight [kg]
1st	Gravel	758
2nd	Cement	476
3rd	Water	149
4th	Gravel	758
5th	Sand	683
6th	Sand	683
7th	Water	99

3. Multiphase Model of the Mixing

3.1. Governing Equations

The numerical simulations were carried out with the commercial CFD code ANSYS CFX 19.2 [32]. This code solved the 3D Reynolds-averaged form of the Navier–Stokes equations by using a finite element approach based on a finite volume method. A second-order high-resolution advection scheme was adopted to calculate the advection terms in the discrete finite volume equations. An Eulerian inhomogeneous model with an interphase transfer particle approach was used for resolving the multiphase simulation. The governing equations are listed below:

$$\frac{\partial}{\partial t}(\rho_\alpha r_\alpha) + \nabla\cdot(\rho_\alpha r_\alpha U_\alpha) = 0 \tag{5}$$

$$\frac{\partial}{\partial t}(\rho_\alpha r_\alpha U_\alpha) + \nabla\cdot(r_\alpha \rho_\alpha U_\alpha \bigotimes U_\alpha) = -r_\alpha \nabla p_\alpha + \nabla\cdot(r_\alpha \mu_\alpha(\nabla U_\alpha + [U_\alpha]^T)) + \sum_{\beta=1}^{N_P} M_{\alpha\beta} \tag{6}$$

$$\sum r_\alpha = 1 \tag{7}$$

These 17 (12 momentum, 4 phasic continuity, 1 volume fraction) nonlinear partial differential equations must be solved for the 17 dependent variables $[(U_\alpha, V_\alpha, W_\alpha), r_\alpha, p]$, where subscript index $\alpha = 1, \ldots, 4$ identifies the phase. For the other characteristics of the multiphase model (i.e., Continuous–Dispersed, Continuous–Continuous, and Dispersed–Dispersed iterations), see [31].

3.2. Geometry and Mesh

The 3D CAD geometry was imported into a commercial meshing software program, ANSYS ICEM CFD 19.2 [33]. The discretization of the fluid domain was obtained using a mesh made up of hexahedral elements in order to increase the numerical accuracy and decrease the time and effort needed for the simulations. The generation of hexahedral elements needed a specific blocking construction, associating the vertex, edges, and faces of the blocks to the points, curves, and surfaces inside the geometrical domain. A mesh with uniform element density must be generated to achieve a consistent element size distribution throughout the entire internal drum geometry.

For the grid sensitivity test, the following three tetrahedral element grids were generated: Mesh 1, comprising 3.8×10^6 elements; Mesh 2, comprising 4.8×10^6 elements; and Mesh 3, comprising 6.8×10^6 elements. Examples of the meshes obtained are shown in Figure 2a,b.

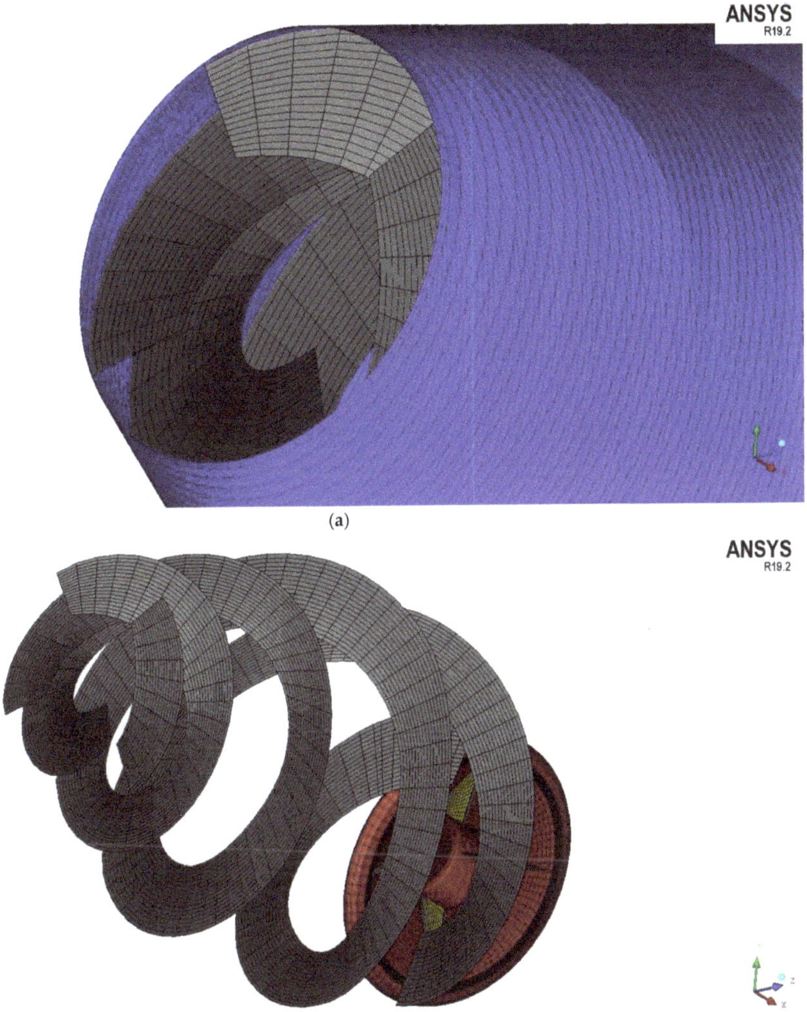

(**a**)

(**b**)

Figure 2. Examples of block layers with O-grids applied to the drum.

3.3. Boundary Conditions

All the phases are completely segregated and stratified at the start of the simulation. Gravel is placed on the bottom, sand is placed between gravel and cement paste, and air is set to fill the remaining volume. Inside the domain, the buoyancy model is used by applying the gravitational force. In order to stabilize the initialization of the simulation, the static pressure gradient is set along the height of the drum, following Stevino's law, as follows:

$$p(y) = \rho g y + p_{atm} \tag{8}$$

The parameters imposed for each species are shown in Table 4. Some additional parameters are the air at atmospheric pressure, 22 rpm for a constant drum rotational speed, and a free slip wall boundary condition for every wall boundary set.

Table 4. Material parameters used.

	Air	Cement Paste	Sand	Gravel
fluid scheme	continuous fluid	continuous fluid	dispersed solid	dispersed solid
ρ [kg/m^3]	1.2	1283	2560	2650
viscosity model	Newtonian	Bingham	-	-
particle diameter [mm]	-	-	2	20
maximum packing			0.62	0.62
volume fraction	-	0.2	0.38	0.42
turbulence model	laminar	laminar	-	-

According to the literature [34], cement paste can be considered a Bingham fluid with the appropriate yield stress and plastic viscosity, i.e., τ_0 and K, respectively.

$$\eta = \frac{\tau_0}{\dot{\gamma}} + K \tag{9}$$

where η is the apparent viscosity and γ is the shear rate. The plastic viscosity depends largely on the concentration of cement and the yield stress as a function of hydration time. From the mix design used (Table 2), the water/cement ratio is about 0.5; at this value, the yield stress is equal to 0.7 Ncm and the plastic viscosity is set to 0.5. Portland Cement with ρ_c = 1506 kg/m^3 was used; consequently, with the weights reported in Table 3, it is possible to calculate the density of cement paste.

A surface tension coefficient σ between air and concrete is set to 66 mN/m [35]. This value is applied in the continuum surface force model of [36], which is used in the CFX 19.2 software to calculate the effect of surface tension in the Cauchy moment equation. Every wall is modeled as an ideal No Slip smooth wall in order to avoid the effect of wall roughness on the results.

The transient analysis convergence is controlled by setting an adaptive time step around 10^{-4}:10^{-3} s, regulated by the RMS Courant number, which is set to 0.5. The convergence criteria on the equations' residual values are set to 10^{-4}. A second-order high-resolution advection scheme is implemented to compute the advection terms within the discrete finite volume equations. A trilinear interpolation technique is used to interpolate the values of pressure and velocity between the nodes of a three-dimensional grid. With these settings for the transient simulations, the global imbalances are normally kept below ±10%.

3.4. Validation

The model presented above was validated in several steps. The first validation was performed by comparing an experimental truck emptying over time with a computational drum emptying (Figures 3 and 4). This made it possible to validate the cement paste viscosity model and the aggregates' solid particle model used for numerical simulations. A grid independence analysis was conducted, and convergence was monitored both by keeping the imbalances within 10 percent, as mentioned above, and by analyzing the superficial velocity, which is a characteristic velocity used in multiphase flows [37].

Figure 3. Experimental test of discharge process with sand.

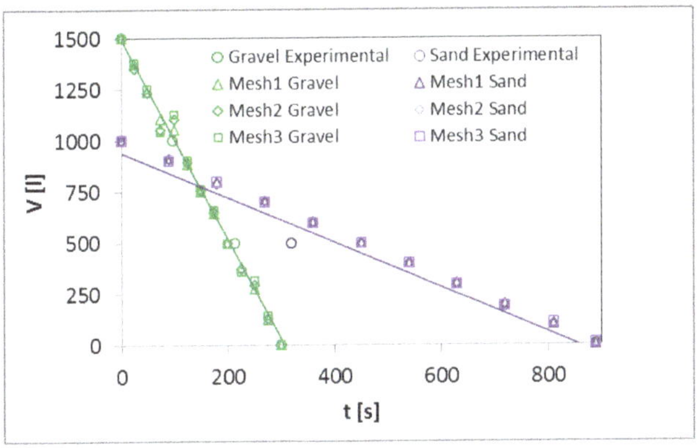

Figure 4. Comparison of the numerical testing discharge curves [31].

This showed that the flow velocity could be considered convergent only after eight drum revolutions.

Finally, the torque absorbed by the machine during the actual mixing process was measured during an experimental test and later compared with the calculated torque from the full numerical model simulation (Figure 5).

$$P_{mec} = M_t \, \omega \quad (10)$$

where M_t is the torque acting on the internal drum and blade surfaces, and ω is the rotational velocity.

The comparison is shown in Figure 5; as the model represents the first few minutes of the actual physical process, the torque curves are only compared for the first few revolutions.

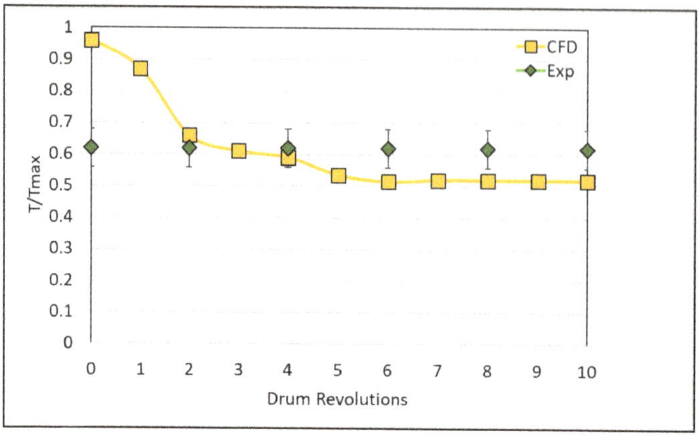

Figure 5. Dimensionless torque curve of the truck drum measured during the mixing phase (green), comparison with simulated curve (yellow), 20% error bar.

4. Numerical Mixing Index

Application of the Mixing Index at the Concrete Mixing Multiphase Simulation

An Eulerian–Eulerian model was adopted for multiphase simulations, which reproduces the problem by defining the volume fractions of the various phases present in the domain elements.

Following this approach, the mixing index adopted for the mixing analyses is defined by Equation (11):

$$R_{i,n} = \frac{V_{i,n}}{V_{i,ref}} \tag{11}$$

where $R_{i,n}$ is defined as the mixing index of the i-th phase in the n-th element; $V_{i,n}$ denotes the volume fraction of the i-th phase in the n-th element; and $V_{i,ref}$ is the mix design volume fraction of the analyzed concrete. The mix design values for each phase are given in Table 5.

Table 5. Calculation of design mix design.

Phase	Weight [kg]	Density [kg/m^3]	Volume [m^3]	Mix Design [%]
Cement paste	724	1700	0.426	28
Sand	1516	2660	0.570	37.4
Gravel	1366	2590	0.527	34.6

The R ratio calculated, as explained above, makes it possible to quantify if the volume fraction of each phase approximates the optimal design value over time for each domain element, as follows:

- $R < 1$—insufficient concentration of the phase in the element;
- $R = 1$—optimal concentration of the phase in the element;
- $R > 1$—excessive concentration of the phase in the element.

The study of this parameter by static methods therefore allows us to evaluate the distribution of phases in the domain during drum rotation, under the assumption that the optimal design is given by a homogeneous distribution of the various phases according to the mix design. The two statistical parameters adopted for the study are mean and variance of R.

- Mean of the mixing index:

Once the values of R in the domain have been calculated, the mean is averaged over the volume. The mix design calculation does not include the influence of air, which is present in the simulations performed. For this reason, to study the mean of R, the elements in the domain with a volume fraction of air equal to 1, i.e., all elements having only air inside, are excluded from the calculation.

For each i-th phase, at each instant of time, we can use Equation (12):

$$R_{m,i} = \frac{\sum_{j=1}^{n} R_{i,j}}{n} \tag{12}$$

where n is the number of elements in the analyzed domain, considering the assumptions about the air phase distribution just highlighted. The value of R_m has the same trend as R, so for an optimal simulation, a value as close to one as possible should be aimed for with Equation (12).

- Variance of the mixing index:

The variance of R is a parameter that indicates the dispersion of R around its mean value. For each i-th stage, at a given instant of time, the variance S^2 is calculated according to Equation (13):

$$S_i^2 = \frac{\sum_{j=1}^{n} \left(R_{i,j} - \overline{R_i}\right)^2}{n-1} \tag{13}$$

In the ideal case of perfect mixing, S^2 should be equal to zero, indicating a configuration in which all the elements in the domain have a value of R equal to R_m.

The absolute value of S^2 is affected by the number of grid elements used; therefore, it is necessary to apply grid-independent conditions for the S^2 parameter as well. Figure 6 shows the trends of S^2 as the number of simulated revolutions increases for different numerical grids; two meshes have been tested with 1.5×10^6 hexa elements and 5.0×10^6 hexa elements. As the number of elements increases (more uniform grid), it can be observed that the variance curve is more linear.

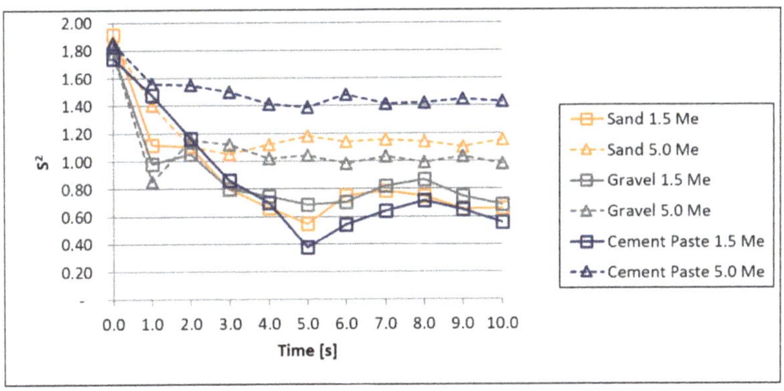

Figure 6. Variance as a function of the number of grid elements.

- Probability density function:

Once the steady-state values of R_m and S^2 are found for each phase, the resulting Gaussian probability distributions can be compared.

For the Gaussian distribution of a quantity with mean μ and variance σ^2, the probability density function $f_{(x)}$ is calculated according to Equation (14):

$$f_{(x)} = \frac{1}{\sigma\sqrt{2\pi}} e^{-\frac{(x-\mu)^2}{2\sigma^2}} \tag{14}$$

In an ideal case, in which Rm of each phase tends to 1 and S^2 tends to 0, the resulting Gaussian tends to a degenerate random variable in 1; meaning that, as the distribution of each phase improves, the Gaussian curve associated with the distribution of each phase in the domain should be increasingly closer and centered to 1.

- Statistical definition implemented in the code:

Equations (11)–(13) are implemented via CEL (Cfx Expression Languages) in the fluid dynamics code. In particular, the expressions are as follows:

```
R<phase>=<phase>.Volume Fraction/MixDesign<phase>
Rm<phase>=ave(R<phase>)@VolumeComponent
S2<phase>=(sum((R<phase>-Rm<phase>)^2) @VolumeComponent)/((count()@VolumeComponent)-1)
```

where the quantity MixDesign<phase> indicates the reference volume fraction values given by Table 5, and the VolumeComponent is the computational domain in which the air volume fraction is less than 1.

5. Results

5.1. *Motions of Mixing*

In terms of the direction of flow motion, the streamlines of the various phases are shown in Figure 7. The motions of mixing are caused by two combined effects, with the first one being the mechanical action given by the rotation of the drum and simultaneously by the movement of the helixes, as well as the second one being given by the difference in density and species of the simulated phases. The combination of these two aspects results in a series of vortices that facilitate the dispersion of the phases.

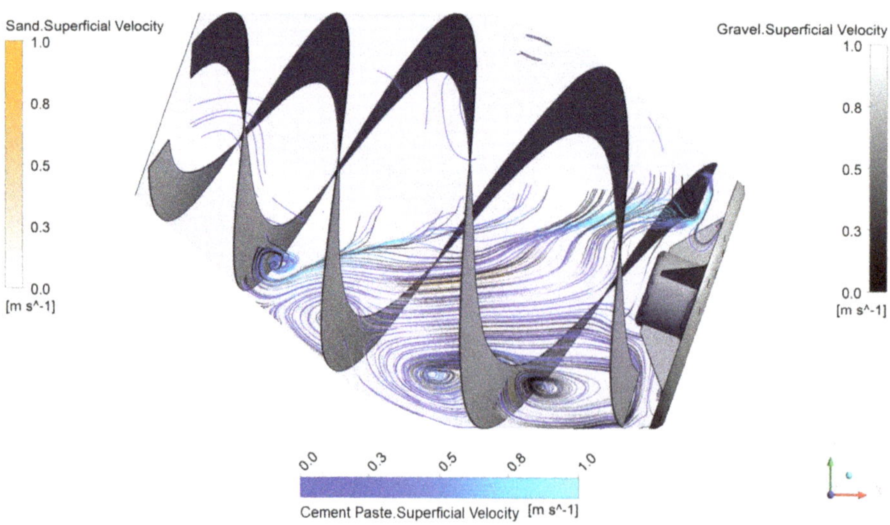

Figure 7. Surface streamlines of the flow.

5.2. *Type of Mixing*

Figure 8a,b show the phase concentrations in the initial condition and after 10 revolutions of the drum, respectively.

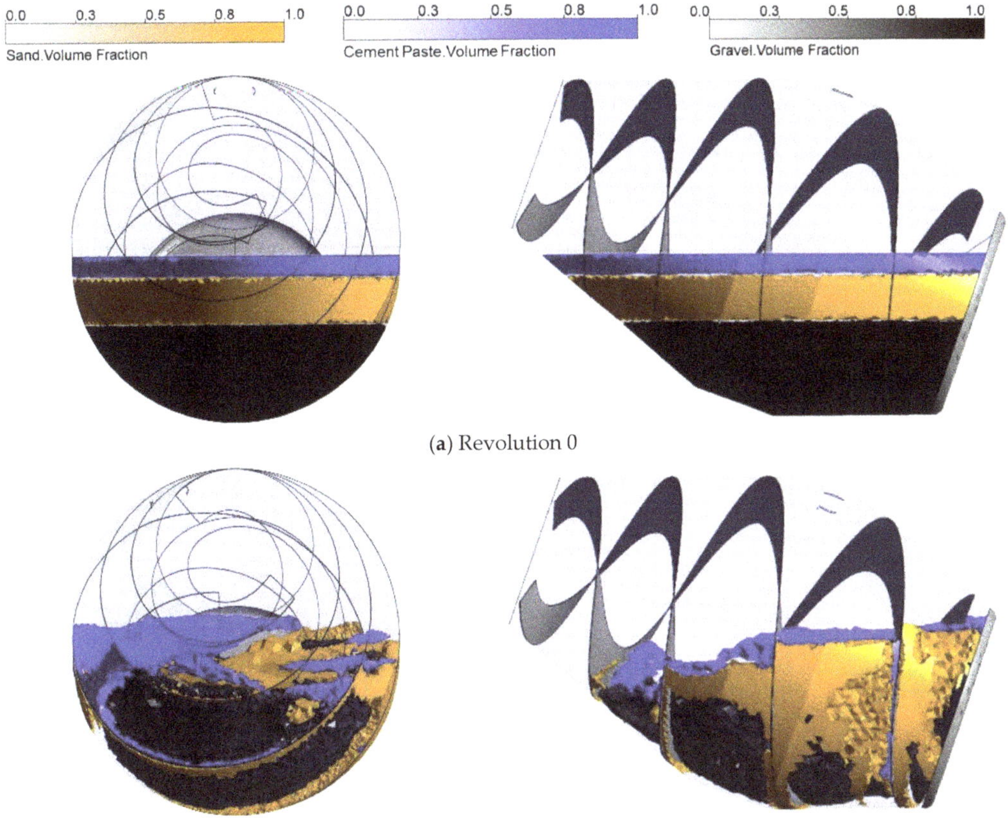

(**a**) Revolution 0

(**b**) Revolution 10

Figure 8. Volume fraction contour plots at different revolutions of the drum.

In Figure 8a, in the initial condition, the phases are fully segregated; the gravel is placed at the bottom of the mixer, the sand is placed in the middle, and the water is placed on top, in agreement with the initial condition set-up. After 10 revolutions (Figure 8b), the sand and gravel areas are approximately the same size, while the water areas are fewer and smaller, and most of the water remains on top. This is due to the transport and trail effects caused by the motion of the two-blade motion; in fact, their action, combined with the fact that the solid aggregates have a similar rheology, causes a mixing action along the helix profiles. This behavior is typical of the laminar mixing under high laminar stress [19,38]. The described behavior is justified by the quantity ratio (4:1) between the aggregates and the water.

5.3. *Computation of Numerical Mixing Index*

The values of the logarithmic tendency curves associated with the trend lines, calculated after 10 drum revolutions, were taken as the reference for R_m and S^2 for each phase. Once unique values of R_m and S^2 were found for each phase, the resulting Gaussian probability distributions could be compared.

By implementing the mean and variance values as the number of drum revolutions changes, the following trends can be derived (Figures 9 and 10).

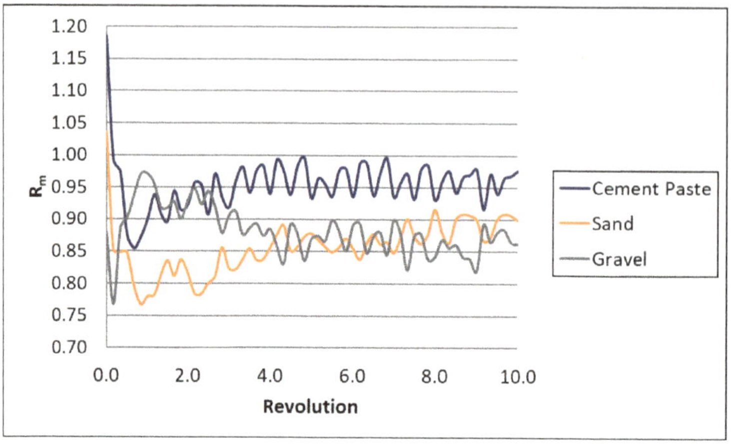

Figure 9. R_m of cement paste, sand, and gravel as the drum revolutions increase.

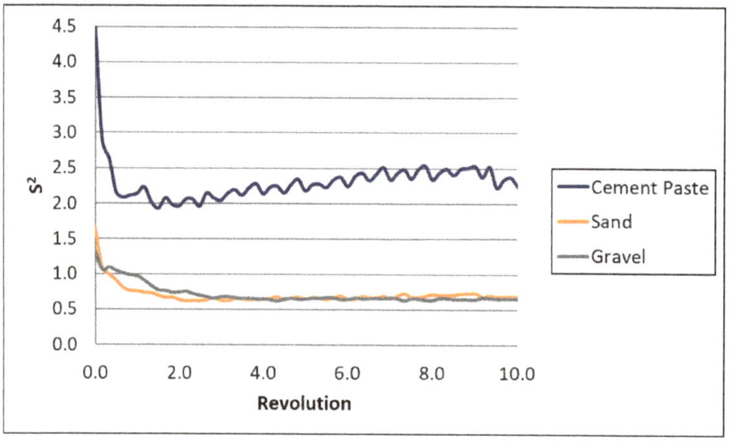

Figure 10. S^2 of cement paste, sand, and gravel as the drum revolutions increase.

The following features are evident from the graphs:

- The curves referring to R_m tend to stabilize on a value less than unity. The reason is given by the presence of air in some points of the domain, which is absent in the mix design calculation thus reducing the effective volume fraction of the various phases in the considered elements.
- The S^2 curve of the cement paste stabilizes at much higher values than the solid phases, indicating less dispersion of the continuous phase within the analyzed domain.

The main problem resulting from this methodology pertains to the formulation of a correlation that provides a quality index of the distribution of the various phases within the geometry. In fact, for each simulation, there is a total of six time-varying parameters (R_m and S^2 for each individual phase of cement paste, sand, and gravel) that need to be compared in order to establish whether one geometry or dynamic parameter of the drum results in a better or worse distribution of the different phases than the others.

To do so, it was decided to take the values of the logarithmic trend curves associated with the trends as reference values for R_m and S^2 of each phase, calculated after 10 drum revolutions.

For certain simulations, it was decided to remove the numerical values given by the initial instants, as they were not indicative of the full convergence, so as to reduce the logarithmic trend error. The obtained logarithmic trend curves are shown in Figures 11 and 12. From the logarithmic curves of S^2 and R_m, it is possible to obtain a distinct value at the considered revolution for each phase. At this point, it is possible, by means of the Gaussian distribution, to determine how the phases are distributed in the drum for each revolution. From the application of Equation (14), the curve shown in Figure 13 is obtained.

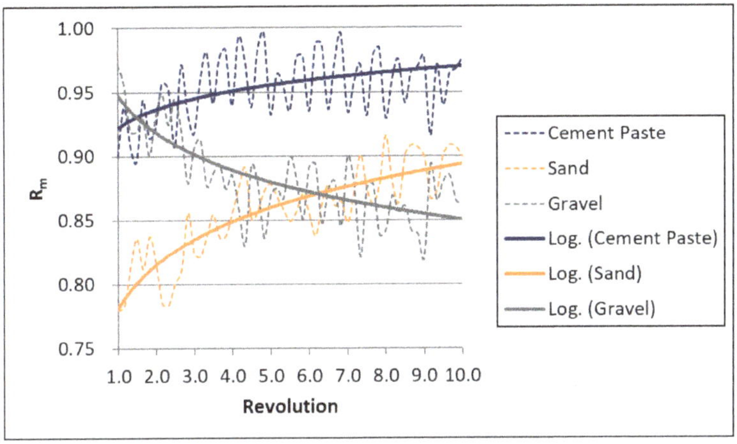

Figure 11. R_m logarithmic trend for cement paste, sand, and gravel.

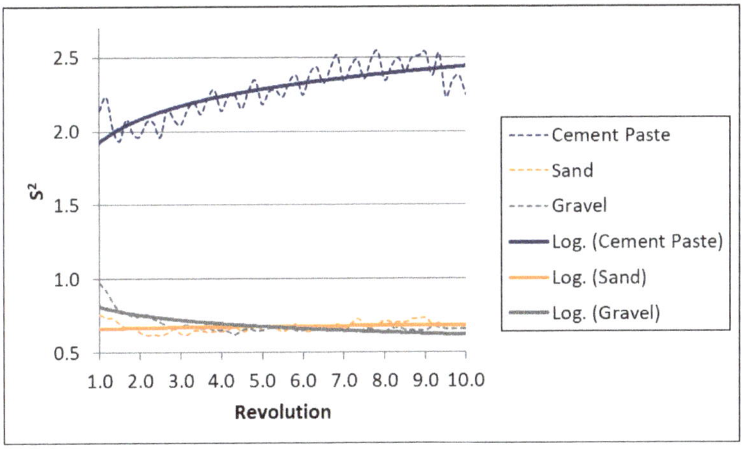

Figure 12. S^2 logarithmic trend for cement past, sand, and gravel.

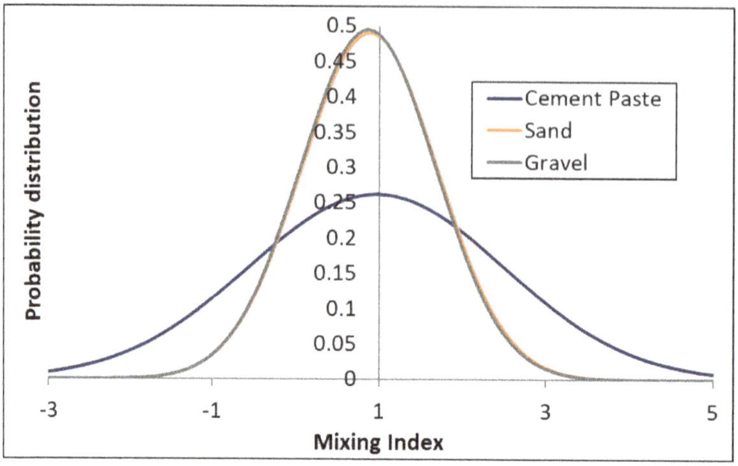

Figure 13. Probability density of the mixing index.

6. Conclusions and Future Perspectives

In a previous work [31], the authors developed a transient multiphase CFD model that simulated the distribution of aggregates in terms of gravel and sand of different grain sizes within cement paste in a concrete truck mixer. This work aimed to study the early stages of the concrete mixing process through an Eulerian CFD approach. Although the dispersion of various materials under the mechanical action of the drum blades was investigated, there was no quantitative method for calculating the mixing efficiency during drum rotation.

In this work, the statistical method for calculating the mixing index, defined as the ratio of the concentration of the considered phase (gravel, sand, or cement paste) in the i-th element to the optimal target concentration, was applied to the multiphase CFD model developed in the previous work. At the mixing index so defined, the terms of mean, variance, and probability density were calculated, which allowed the index to be extended from the cell (local) to the entire computational domain (concrete mixer). The establishment and implementation of the mixing index provided an effective means to evaluate the mixing process and its performance using a numerical Eulerian approach. This method ensured the precise assessment of mixing quality in the numerical study of mixers. Without the cell mixing index, such evaluations would be challenging, necessitating a shift to a Lagrangian approach. The cell mixing index was a crucial variable that enabled the analysis of multiphase mixing processes within the Eulerian numerical approach. By utilizing the model and applying the mixing index calculation as demonstrated in this study, it was possible to optimize the geometry and operational parameters, such as rotation speed and inclination.

The results showed that, after 10 revolutions, the gravel, sand, and cement paste reached the maximum distribution in the concrete truck mixer, equivalent to a steady-state value of the calculated mixing index. Furthermore, while the cement paste reached about 95% of the optimal distribution, the aggregates (sand and gravel), on the other hand, stopped between 85 and 90%. The values were intended to be averaged over the internal volume of the concrete mixer analyzed.

In conclusion, the following clarifications are presented:

The application of the mixing index as defined in this paper can only be applied if the target concentrations of the various species (mix design) are known.

- The cell mixing index, as well as its extension to the whole computational domain by statistical functions, is a method not yet applied to the simulation of concrete mixing with an Eulerian–Eulerian approach.
- The method can also be applied to other types of mixers.
- The application of the mixing index to the CFD model makes it possible to predict the mixing efficiency of a mixer from the CAD of the geometry, so the method allows an optimized mixer design to be defined by reducing the number of prototypes.
- The models for the simulation of concrete mixers are mainly CFD or DEM, in which the various phases are simulated through an Eulerian or Lagrangian approach, respectively. While for DEM-based models, the application of statistics to calculate the mixing index is sufficiently easy, precisely because they are based on the dynamics of an identifiable and well-defined number of particles, the application of the mixing index to the simulation of concrete mixing based on the Eulerian approach has not yet been defined and applied in this research area.

Author Contributions: Conceptualization, C.F.; Software, N.B.; Investigation, N.B.; Data curation, N.B. and L.M.; Writing—original draft, C.F.; Writing—review & editing, C.F. and L.M.; Visualization, L.M.; Supervision, C.F. All authors have read and agreed to the published version of the manuscript.

Funding: This research received no external funding.

Data Availability Statement: All data generated or analyzed during this study are included in this published article.

Conflicts of Interest: The authors declare no conflicts of interest.

References

1. El-Emam, M.A.; Zhou, L.; Shi, W.; Han, C.; Bai, L.; Agarwal, R. Theories and Applications of CFD–DEM Coupling Approach for Granular Flow: A Review. *Arch. Comput. Methods Eng.* **2021**, *28*, 4979–5020. [CrossRef]
2. Hanley, K.J.; O'Sullivan, C.; Oliveira, J.C.; Cronin, K.; Byrne, E.P. Application of Taguchi methods to DEM calibration of bonded agglomerates. *Powder Technol.* **2011**, *210*, 230–240. [CrossRef]
3. Xu, J.; Wang, F.; Abegaz, R. State of the Art of CFD-DEM Coupled Modeling and Its Application in Turbulent Flow-Induced Soil Erosion. *Geosciences* **2025**, *15*, 21. [CrossRef]
4. Zhao, Z.; Zhou, L.; Bai, L.; Wang, B.; Agarwal, R. Recent Advances and Perspectives of CFD–DEM Simulation in Fluidized Bed. *Arch. Comput. Methods Eng.* **2024**, *31*, 871–918. [CrossRef]
5. Delmas, H.; Barthe, L.; Cleary, R. Chapter 29—Ultrasonic mixing, homogenization, and emulsification in food processing and other applications. In *Power Ultrasonics*, 2nd ed.; Woodhead Publishing Series in Electronic and Optical Materials; Woodhead Publishing: Cambridge, UK, 2023; pp. 665–685; ISBN 9780128202548. [CrossRef]
6. Gall, V.; Runde, M.; Schuchmann, H.P. Extending Applications of High-Pressure Homogenization by Using Simultaneous Emulsification and Mixing (SEM)—An Overview. *Processes* **2016**, *4*, 46. [CrossRef]
7. Hlosta, J.; Jezerská, L.; Rozbroj, J.; Žurovec, D.; Nečas, J.; Zegzulka, J. DEM Investigation of the Influence of Particulate Properties and Operating Conditions on the Mixing Process in Rotary Drums: Part 2—Process Validation and Experimental Study. *Processes* **2020**, *8*, 184. [CrossRef]
8. Jadidi, B.; Ebrahimi, M.; Ein-Mozaffari, F.; Lohi, A. A comprehensive review of the application of DEM in the investigation of batch solid mixers. *Rev. Chem. Eng.* **2023**, *39*, 729–764. [CrossRef]
9. Chen, Y.; Chang, X.; Zhang, B.; Huang, H. Efficient and effective calibration of numerical model outputs using hierarchical dynamic models. *Ann. Appl. Stat.* **2024**, *18*, 1064–1089. [CrossRef]
10. Coetzee, C.J. Review: Calibration of the discrete element method. *Powder Technol.* **2017**, *310*, 104–142. [CrossRef]
11. Hesse, R.; Krull, F.; Antonyuk, S. Experimentally calibrated CFD-DEM study of air impairment during powder discharge for varying hopper configurations. *Powder Technol.* **2020**, *372*, 404–419. [CrossRef]
12. Magri, L.; Sequino, L.; Ferrari, C. Simulating the Electrochemical-Thermal Behavior of a Prismatic Lithium-Ion Battery on the Market under Various Discharge Cycles. *Batteries* **2023**, *9*, 397. [CrossRef]
13. Wasserfall, J.G.; Coetzee, C.J.; Meyer, C.J. A submerged draw down test calibration method for fully-coupled CFD-DEM modelling. *Front. Chem. Eng.* **2024**, *6*, 1376974. [CrossRef]

14. Hema, V.; Savithri, S. Mathematical Modelling of the Dynamics of Granular Materials in a Rotating Cylinder. Ph.D. Thesis, Cochin University of Science and Technology, Kochi, India, 2003. Available online: http://ir.niist.res.in:8080/jspui/handle/123456789/1173 (accessed on 21 August 2024).
15. Panneerselvam, R.; Savithri, S.; Surender, G.D. CFD modeling of gas– liquid–solid mechanically agitated. *Chem. Eng. Res. Des.* **2008**, *86*, 1331–1344. [CrossRef]
16. Poux, M.; Fayolle, P.; Bertrand, J.; Bridoux, D.; Bousquet, J. Powder mixing: Some practical rules applied to agitated systems. *Powder Technol.* **1991**, *86*, 213–234. [CrossRef]
17. Rhodes, M.J.; Wang, X.S.; Nguyen, M.; Stewart, P.; Liffman, K. Study of mixing in gas-fluidized beds using a DEM model. *Chem. Eng. Sci.* **2001**, *56*, 2859–2866. [CrossRef]
18. Lacey, P.M.C. Developments in the theory of particle mixing. *J. Appl. Chem.* **1954**, *4*, 257–268. [CrossRef]
19. Harnby, N.; Edwards, M.F.; Nienow, A. *Mixing in the Process Industries*; Butterworth-Hinemann: Oxford, UK, 1997.
20. Khopkar, A.R.; Kasat, G.R.; Pandit, A.B.; Ranade, V.V. CFD simulation of mixing in tall gas–liquid stirred vessel: Role of local flow patterns. *Chem. Eng. Sci.* **2006**, *61*, 2921–2929. [CrossRef]
21. Khopkar, A.R.; Kasat, G.R.; Pandit, A.B.; Ranade, V.V. Computational fluid dynamics simulation of the solid suspension in a stirred slurry reactor. *Ind. Eng. Chem. Res.* **2006**, *45*, 4416–4428. [CrossRef]
22. Murthy, B.N.; Ghadge, R.S.; Joshi, J.B. CFD simulations of gas–liquid–solid stirred reactor: Prediction of critical impeller speed for solid suspension. *Chem. Eng. Sci.* **2007**, *62*, 7184–7195. [CrossRef]
23. Till, Z.; Molnàr, B.; Egedy, A.; Varga, T. CFD Based Qualification of Mixing Efficiency of Stirred Vessels. *Period. Polytech. Chem. Eng.* **2019**, *63*, 226–238. [CrossRef]
24. Ameur, H. Energy efficiency of different impellers in stirred tank reactors. *Energy* **2015**, *93*, 1980–1988. [CrossRef]
25. Park, C.; Lee, K.; Hong, S.; Lee, J.; Cho, S.; Moon, I. Novel index for evaluating continuous mixing process with pulse injection of bimodal tracer particles. *Powder Technol.* **2019**, *355*, 309–319. [CrossRef]
26. Wallewik, J.E.; Wallewik, O.H. Analysis of shear rate inside a concrete truck mixer. *Cem. Concr. Res.* **2017**, *95*, 9–17. [CrossRef]
27. Stepanov, V.; Kireev, S. Efficiency Analysis for Mechanical Mixing Systems of Cementing Units. In *Networked Control Systems for Connected and Automated Vehicles. NN 2022*; Guda, A., Ed.; Lecture Notes in Networks and Systems; Springer: Cham, Switzerland, 2023; Volume 510. [CrossRef]
28. Li, S.P.; Yuan, Y.L.; Shi, L.G. Research on CFD Simulation of the Cement Slurry Mixer. *Adv. Mater. Res.* **2012**, *621*, 196–199. [CrossRef]
29. Beccati, N.; Ferrari, C.; Bonanno, A.; Balestra, M. Calibration of a CFD discharge process model of an off-road self-loading concrete mixer. *J. Braz. Soc. Mech. Sci. Eng.* **2019**, *41*, 76. [CrossRef]
30. Beccati, N.; Ferrari, C. Predicting the capacity of an off-road self-loading drum mixer with different concrete consistencies through three-dimensional fluid dynamics analysis. *J. Braz. Soc. Mech. Sci. Eng.* **2020**, *42*, 379. [CrossRef]
31. Ferrari, C.; Beccati, N. Mixing Phase Study of a Concrete Truck Mixer via CFD Multiphase Approach. *J. Eng. Mech.* **2022**, *148*, 04022002. [CrossRef]
32. Ansys. *CFX-Solver Theory Guide R19.2*; Ansys: Canonsburg, PA, USA, 2019.
33. Ansys. *Ansys Icem CFD User Manual*; Ansys: Canonsburg, PA, USA, 2019.
34. Wierig, H.J. *Properties of Fresh Concrete*; Chapman and Hall: New York, NY, USA, 1990.
35. Sahin, Y.; Akkaya, Y.; Boylu, F.; Tasdemir, M.A. Characterization of air entraining admixtures in concrete using surface tension measurements. *Cem. Concr. Compos.* **2017**, *82*, 95–104. [CrossRef]
36. Brackbill, J.U.; Kothe, D.B.; Zemach, C.A. Continuum Method for Modeling Surface Tension. *J. Comput. Phys.* **1992**, *100*, 335–354. [CrossRef]
37. Paul, E.L.; Atiemo-Obeng, V.A.; Kresta, S.M. *Handbook of Industrial Mixing: Science and Practice*; Wiley: New York, NY, USA, 2004.
38. Crowe, C.T.; Sommerfield, M.; Tsuj, Y. *Multiphase Flows with Droplets and Particles*; CRC Press: Boca Raton, FL, USA, 1998. [CrossRef]

fluids MDPI

Article

ARES: A Meanline Code for Outboard Dynamic-Inlet Waterjet Axial-Flow Pumps Design

Filippo Avanzi *, Francesco De Vanna †, Andrea Magrini † and Ernesto Benini †

Department of Industrial Engineering, Università degli Studi di Padova, 35131 Padova, Italy; francesco.devanna@unipd.it (F.D.V.); andrea.magrini@unipd.it (A.M.); ernesto.benini@unipd.it (E.B.)
* Correspondence: filippo.avanzi@unipd.it
† These authors contributed equally to this work.

Abstract: We introduce the solver ARES: Axial-flow pump Radial Equilibrium through Streamlines. The code implements a meanline method, enforcing the conservation of flow momentum and continuity across a set of discrete streamlines in the axial-flow pump's meridional channel. Real flow effects are modeled with empirical correlations, including off-design deviation and losses due to profile shape, secondary flows, tip leakage, and the end-wall boundary layer (EWBL). Inspired by aeronautical fan and compressor methods, this implementation is specifically tailored for the analysis of the Outboard Dynamic-inlet Waterjet (ODW), the latest aero-engine-derived innovation in marine engineering. To ensure the reliable application of ARES for the systematic designs of ODW pumps, the present investigation focuses on prediction accuracy. Global and local statistics are compared between numerical estimates and available measurements of three test cases: two single rotors and a rotor–stator waterjet configuration. At mass flow rates near the design point, hydraulic efficiency is predicted within 1% discrepancy to tests. Differently, as the flow coefficient increases, the loss prediction accuracy degrades, incrementing the error for off-design estimates. Spanwise velocity and pressure distributions exhibit good alignment with experiments near midspan, especially at the rotor exit, while end-wall boundary layer complex dynamics are hardly recovered by the present implementation.

Keywords: axial-flow pump design; meanline; blade element method; marine propulsion; outboard dynamic-inlet waterjet

1. Introduction

Outboard Dynamic-inlet Waterjets (ODW) have recently emerged as a real alternative in marine propulsion [1]. Advanced by industrial technology, this machine consists of an axisymmetric hydrodynamic nacelle shrouding a propulsive axial- or mixed-flow pump [2]. Submerged operation offers several advantages over existing solutions, including a uniform mass flow rate to the pump, no hull disturbance, no flow obstruction by the shaft, and alignment among the capture stream, thrust, and direction of advance—all addressing known limitations in naval propulsion [3–8].

To efficiently exploit the benefits of this configuration, the machinery requires a dedicated design. To this end, Blade Elements Methods (BEMs) have traditionally proved valuable techniques, enabling fast though accurate preliminary predictions of machines' fluid dynamics [9]. Their advantage relies on the possibility of solving simplified flow equations while iterating through to introduce real flow effects as empirical external correlations. These models can be both experimental and numerical, with the work from Lieblein [10] setting the milestone reference in this field. In that study, planar blade cascade

 https://doi.org/10.3390/fluids10030066

systematic measurements have served as the definition of several losses and deviation correlations against the incidence angle. Henceforth, the majority of the meanline methods for both compressible and incompressible applications has relied on those models [11–17], and many other theories have been developed over the years [18–22]. Although the estimates of BEM codes may reach significant accuracy levels, verification with high-fidelity Computational Fluid Dynamics (CFD) [23,24] and validation through experiments [25–27] are still mandatory steps.

To address the lack of existing methods for comprehensive ODW design, this study introduces ARES (Axial-flow pump Radial Equilibrium through Streamlines), a novel meanline solver specifically developed to predict ODW propulsive performance by modeling the interaction between the nacelle and the machine. The study focuses on the accuracy of the pump design module, which implements a BEM by enforcing flow momentum and continuity conservation and integrating these laws across streamlines. These axisymmetric surfaces discretize the machine's meridional section spanwise and are designed as Blade Elements (BEs) using the 2D planar cascades assumption. This approach leverages well-established correlations to model in- and off-design deviations, losses, and other real flow effects. Three axial-flow pump test cases from the literature are selected to compare global and local measurements with the ARES solution.

The work is organized as follows: Section 2 thoroughly details the methodology for ARES implementation. Section 3 discusses the numerical results compared with the available experiments. Finally, in Section 4, conclusions are drawn.

2. Methods

The present method solves the axial-flow pump meanline flow, relying on the satisfaction of continuity and momentum conservation laws, with the latter often labeled as the Radial Equilibrium (RE) equation. The differential laws integrals are computed across a finite number of BEs or streamlines adopted to discretize the machine meridional plane. The iterative routine marches while updating the primary variables, introducing empirical and numerical correlations to model specific flow effects. Under these hypotheses, the present implementation is effectively a lumped-element model, where flow conditions are evaluated at specific radial locations along streamwise stations. The empirical correlations are leveraged to model real-world fluid phenomena without solving their evolution. Specifically, the model assumes the geometry designed from 2D planar cascades allows for the adoption of several well-established approaches capable of modeling the primary 2D and 3D axial-flow machinery flow effects. Finally, consistency across the entire channel is ensured by iterating through the fluid dynamics fundamental laws. The complete implementation was independently carried out in Python 3.12.

2.1. Assumptions

The assumptions adopted to solve the governing equations are listed below:

1. The flow is incompressible, where $\rho = const$;
2. The flow is stationary, i.e., $\partial(\cdot)/\partial t = 0$;
3. The solution holds at planes perpendicular to the axis, which are placed downstream of each blade row (red stations in Figure 1);
4. At each plane, the RE is integrated through N_s streamlines, or BE (Figure 1), which are conceived as independent, axisymmetric surfaces of revolution about the machine axis. The corresponding spanwise location is iterated through to impose a mass flow balance, rather than employing constant radial increments, to enforce continuity within streamtubes;

5. The flow is inviscid: viscous effects are modelled using empirical correlations and introduced upon the equation's solution.

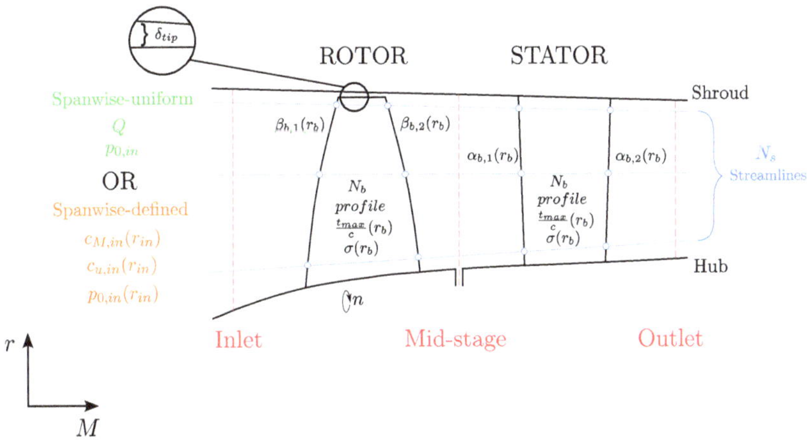

Figure 1. Schematic of a pump stage meridional view, including the domain discretization, input variables, and parameters required by the solver.

The streamtube flow is regulated by cascade shapes stacked to generate the blade's entire span. Thus, hub-to-shroud distributions of the key geometrical parameters are required as input, as reported in Figure 1.

Figure 2 illustrates a schematic of the blade cascade quantities.

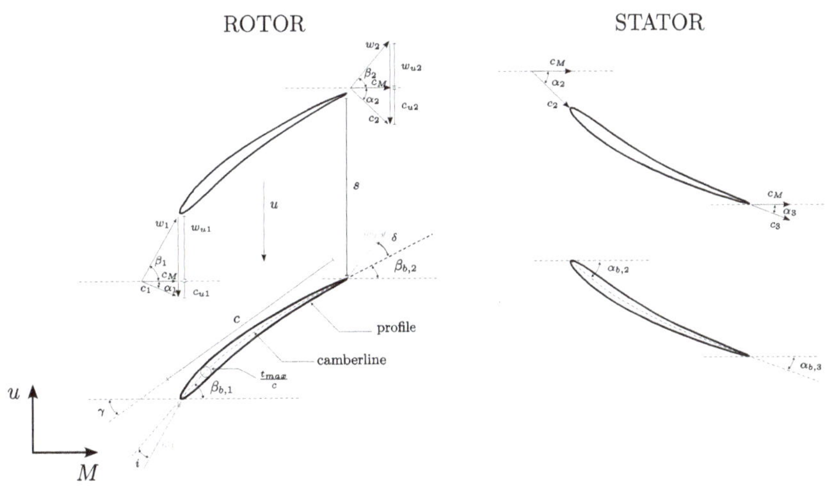

Figure 2. Blade geometrical and flow quantities for a sample axial-flow pump stage section.

The absolute and relative velocities are denoted by c and w, respectively. The absolute and relative flow angles are represented by α and β, with a subscript b used to distinguish them from blade metal angles. The differences between flow-related and geometrical parameters define the leading edge incidence (i) and trailing edge deviation (δ) angles.

$$i = \beta_1 - \beta_{b,1} \tag{1}$$

$$\delta = \beta_2 - \beta_{b,2} \tag{2}$$

Blade cascade flow behavior is characterized by velocity diagrams, showing velocity components in the meridional (M) and tangential (u) directions aligned with the machine axis and blade row speed, respectively. Key parameters include the profile maximum thickness ($t_{\max}$), normalized by the chord length (c); stagger angle (γ), calculated as the average between inlet and outlet metal angles; profile camber ($\theta = \beta_{b,1} - \beta_{b,2}$); tip clearance ($\delta_{\text{tip}}$); and blade solidity ($\sigma = c/s$), the ratio of chord to blade pitch. Blade pitch (s) is determined by the radial location and number of blades (N_b) as $2\pi r/N_b$.

A summary of the relevant blade cascades quantities adopted for ARES modeling is reported in Table 1.

Table 1. Summary of relevant blade cascades parameters.

Symbol	Description
w	relative flow velocity
c	absolute flow velocity
u	rotor velocity
c_M	meridional velocity
c_u	circumferential component of absolute velocity
α_1, α_2	inlet/outlet absolute flow angles
β_1, β_2	inlet/outlet relative flow angles
$\alpha_{b,1}, \alpha_{b,2}$	inlet/outlet stator blades metal angles
$\beta_{b,1}, \beta_{b,2}$	inlet/outlet rotor blades metal angles

2.2. Governing Equations

According to Serovy et al. [12], momentum conservation is expressed as follows:

$$g\frac{\partial h}{\partial r} = \frac{c_u^2}{r} - c_r\frac{\partial c_r}{\partial r} - c_M\frac{\partial c_r}{\partial M} \tag{3}$$

where g is the gravitational acceleration, h is the static head, and derivatives are computed along the radial, r, and axial, M, directions. Upon introducing the definition of the total head, $h_0 = h + (c_u^2 + c_M^2)/(2g)$, Equation (3) is then integrated between streamlines j and $j+1$. Neglecting the variation in the radial component, c_r, and including the definitions of both the energy exchange and the velocity diagram across a blade row, the finite difference solution of the problem may be formulated as follows:

$$Ac_{M,j+1}^2 + Bc_{M,j+1} + C = 0 \tag{4}$$

where the coefficients are defined below, depending on flow quantities at an axial station $i-1$ and on the head losses, $h_{0,loss}$:

$$A = 1 + \tan^2(\beta_{j+1})\left[1 + \frac{r_{j+1} - r_j}{r_{j+1}}\right] \tag{5}$$

$$B = -2u_{j+1}\tan(\beta_{j+1})\frac{r_{j+1} - r_j}{r_{j+1}} \tag{6}$$

$$\begin{aligned} C &= -c_{M,j}^2 - 2g(h_{0,j+1}^{(i-1)} - h_{0,j} - h_{0,loss,j+1}) \\ &\quad +2u_{j+1}^{(i-1)}c_{u,j+1}^{(i-1)} - u_{j+1}^2\frac{r_j}{r_{j+1}} \\ &\quad +c_{u,j}^2\left(\frac{r_{j+1}}{r_j} - 2\right) \end{aligned} \tag{7}$$

Equation (4) yields the meridional velocity at a certain streamline $j+1$. Thus, starting from a given condition at j_{start}, the spanwise distribution of c_M at a given axial station i can be computed. Starting from the inflow boundary conditions, the solution marches downstream row by row. Iterations are necessary, since the radial locations of the streamlines are recursively revised and so are the corresponding flow quantities.

Finally, to impose continuity, the axial velocity profile is verified to satisfy mass conservation. Here, the volume flow rate is computed using the quadrature:

$$Q_{comp} = 2\pi \sum_{j=1}^{Ns-1} \frac{(c_{M,j+1} + c_{M,j})}{2} (r_{j+1}^2 - r_j^2) \tag{8}$$

2.3. *Implementation*

ARES takes a formatted parameter file as input, which must contain fluid properties, complete hub-to-shroud geometrical characterization of the blades at all the computing stations, inflow boundary conditions, and solver convergence criteria regarding residuals threshold and maximum allowed iterations. Once the parameters are read, the code advances as charted in Figure 3. The steps are briefly summarized below:

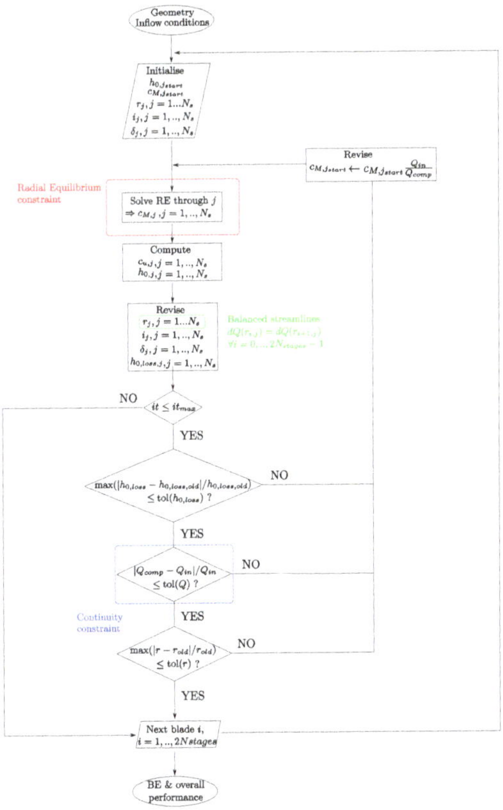

Figure 3. Flow chart of ARES meanline solver.

1. The computational domain is initialized at constant radial intervals, into which geometrical parameters and inflow boundary conditions are interpolated using a 3-point Lagrange technique.
2. Meridional velocity spanwise distribution is evaluated at the first computing station.

3. BE radial locations are updated along with the variables defined spanwise, including the empirical correlations.
4. Convergence is checked according to maximum iterations and residual thresholds. Three criteria requiring simultaneous achievement are selected: the continuity with the inflow flow rate and residuals on both the BE radius locations and the head losses distribution, the latter referred to the previous iteration.
5. Upon convergence, the code either moves to the next blade row or exits, thus printing the machine's global and local statistics.

2.4. Empirical Correlations

Several correlations are included to model flow effects otherwise neglected by the assumptions adopted to solve the steady inviscid governing equations. The overall head losses are given as follows:

$$h_{0,loss} = (\bar{\omega}_{pro} + \bar{\omega}_{sec} + \bar{\omega}_{tip} + \bar{\omega}_{EWBL})\frac{w_1^2}{2g} \tag{9}$$

where the phenomenon-specific coefficients denote, respectively, profile losses, $\bar{\omega}_{pro}$, secondary flows, $\bar{\omega}_{sec}$, tip leakage, $\bar{\omega}_{tip}$, and end-wall boundary layer (EWBL) effects, $\bar{\omega}_{EWBL}$.

Concerning off-design behavior, the model assumes a dependency on the difference between the actual and reference incidence, i^*. Thus, minimum loss profile losses and deviation are modified according to specific functions $f(i - i^*)$.

2.4.1. Minimum Loss Incidence

The parameter is modeled using Lieblein [10] measurements, from which the following correlation is derived:

$$i^* = K_{sh} K_{i,t} i_{0,10} + \tilde{n}\theta \tag{10}$$

Here, K_{sh} depends on the profile camberline shape, either Double Circular Arc (DCA) or NACA-65 equivalent; $K_{i,t}$ varies according to the profile thickness; $i_{0,10}$ is the reference incidence for a zero-camber, NACA 65-series, 10% thickness profile; and $\tilde{n}$ is the correlation slope. Coefficients definitions fitting the original data can be found in Aungier [20].

2.4.2. Deviation

The effective deviation, δ, is computed at any incidence starting from the minimum loss deviation, δ^*, with the latter defined according to Lieblein [10] as follows:

$$\delta^* = K_{sh} K_{\delta,t} \delta_{0,10} + \tilde{m}\theta \tag{11}$$

where K_{sh} is the parameter introduced above, $K_{\delta,t}$ depends on the profile thickness, $\delta_{0,10}$ is the reference deviation for a zero-camber, 65-series, 10% thickness aerofoil; and $\tilde{m}$ is the linear law slope. Coefficient correlations are proposed by Aungier [20]. The deviation is then computed by introducing additional effects as follows:

$$\delta = \delta_{2D,off} + \Delta_{3D}\delta + \Delta_{AVR}\delta \tag{12}$$

Here, $\delta_{2D,off}$ models the planar cascades deviation at off-reference incidence following the theory by Dong-run et al. [21,22]; $\Delta_{3D}\delta$ includes 3D flow dynamics based on Robbins et al. Johnson and Bullock [10]; and $\Delta_{AVR}\delta$ considers the impact of the Axial Velocity Ratio (AVR) across the cascades, which is modeled following Pollard and Gostelow [18].

2.4.3. Profile Losses

The profile losses at any operating condition rely on the minimum losses value, $\bar{\omega}^*_{pro}$, which is modeled using the correlation by Lieblein [10]:

$$\bar{\omega}^*_{pro} = 2\frac{\tilde{\theta}_2}{c}\frac{\sigma}{\cos\beta_2}\left(\frac{\cos\beta_1}{\cos\beta_2}\right)^2\left(\frac{2H_{TE}}{3H_{TE}-1}\right)\left[1-\frac{\tilde{\theta}_2}{c}\frac{\sigma H_{TE}}{\cos\beta_2}\right]^{-3} \tag{13}$$

where $\tilde{\theta}_2$ and H_{TE} denote, respectively, the wake momentum thickness and the wake form factor. The correlations to close Equation (13) were modified by Koch and Smith [19] and recently rearranged by Tournier and El-Genk [15]. Thus, the effective loss parameter, $\bar{\omega}_{pro}$, is modeled based on the incidence parameter, ξ, by Aungier [20], where the correlations to compute $f(\xi)$ are provided. Finally, the losses derive from the following:

$$\bar{\omega}_{pro} = \bar{\omega}^*_{pro} f(\xi) \tag{14}$$

2.4.4. Secondary Losses

Secondary flow losses are modeled using Herrig et al. [28] as expressed by Tournier and El-Genk [15]:

$$\bar{\omega}_{sec} = 0.018\sigma\frac{(\cos\beta_1)^2}{(\cos\beta_m)^3}C_L^2 \tag{15}$$

where β_m is the average relative flow angle, and C_L is the cascades lift coefficient.

2.4.5. Tip-Leakage Losses

In this case, the model adopted is the one proposed by Denton [29] as recently rearranged by Banjac et al. [16]. The theory directly relates the loss coefficient, $\bar{\omega}_{tip}$, to the overall entropy production energy exchange, $T\Delta S$, associated to the leakage mass flow. The correlation reads as follows:

$$\bar{\omega}_{tip} = \frac{2N_b T\Delta S}{w_1^2 \rho Q} \tag{16}$$

Here, a linear variation of the velocities across the tip profile is assumed. Thus, an integration is performed, including Lieblein [10] correlations, to obtain the entropy term.

2.4.6. End-Wall Boundary Layer Losses

External casing boundary layer losses are modeled following Howell [30], as proposed by Tournier and El-Genk [15]. The loss coefficient depends on the blade height, h_b, and it is defined as follows:

$$\bar{\omega}_{EWBL} = 0.0146\frac{c}{h_b}\left(\frac{\cos\beta_1}{\cos\beta_2}\right)^2 \tag{17}$$

Lumped models of the flow effects are linearly distributed along the blade radius. Table 2 summarizes the list of empirical correlations included in ARES.

Table 2. Summary of the empirical correlations implemented.

Parameter	Symbol	Reference
Ref. incidence	i^*	Lieblein [10]
Ref. deviation	δ^*	Lieblein [10], Aungier [20]
Deviation	δ	Dong-run et al. [21], Dong-run et al. [22], Robbins et al. [10], Pollard and Gostelow [18]
Profile losses	$\bar{\omega}_{pro}$	Lieblein [10], Koch and Smith [19], Aungier [20]
Secondary losses	$\bar{\omega}_{sec}$	Howell [30]
Tip leakage losses	$\bar{\omega}_{tip}$	Banjac et al. [16], Denton [29]
EWBL losses	$\bar{\omega}_{EWBL}$	Tournier and El-Genk [15]

3. Results and Discussion

Three test cases with available measurements from the literature are selected to asses the accuracy of the code. The corresponding primary parameters are reported in Table 3. It should be noted that, in the results discussion, the specific parameters follow the definitions of the corresponding reference papers. However, to facilitate the comparison among the case studies, the flow coefficient φ adopted in Table 3 is computed equivalently for any geometry and reads as follows:

$$\varphi = \frac{Q}{nD^3} \tag{18}$$

where n is the rotational regime, D is the rotor external diameter, and Q is the volumetric flow rate.

Before analyzing the code performance, the implementation was calibrated comparing empirical model combinations (Figure 4). Specifically, a sensitivity study was conducted to determine the minimum number of streamlines required to make the results independent of the channel discretisation. This investigation showed that incrementing N_s above 21 does not provide consistent variation of the predicted statistics (Figure 4a), and therefore, this quantity was adopted henceforth. Thus, all the computations reach convergence below 1×10^{-5} on all the residuals within 25 iterations. A typical residuals profile for near-design computations is reported in Figure 4b.

Table 3. Main parameters of the test cases adopted for the validation of ARES.

Parameter	R02 † [13]	HIREP † [25]	AxWJ-2 ‡ [31]
Blade number, N_b	16	7	6–8
Profile	DCA	DCA	NACA
Hub radius, r_{hub} [mm]	45.7	266.7	46.3–79.8
Tip radius, r_{tip} [mm]	114.3	533.3	152.4–127.7
Hub solidity, σ_{hub}	2.11	1.19	1.88–2.43
Tip solidity, σ_{tip}	0.84	0.56	1.72–0.97
Tip gap, δ_{tip} [mm]	0.5	3.3	0.5–0
Rotor speed, n [rpm]	∼3918.6	260	1400
Tip Reynolds number, Re_{1c}	2.15×10^6	5.5×10^6	5.36×10^6
Flow coefficient, $\varphi = Q/(nD^3)$	0.387	0.216	0.135

† Rotor only. ‡ Rotor and stator blades data.

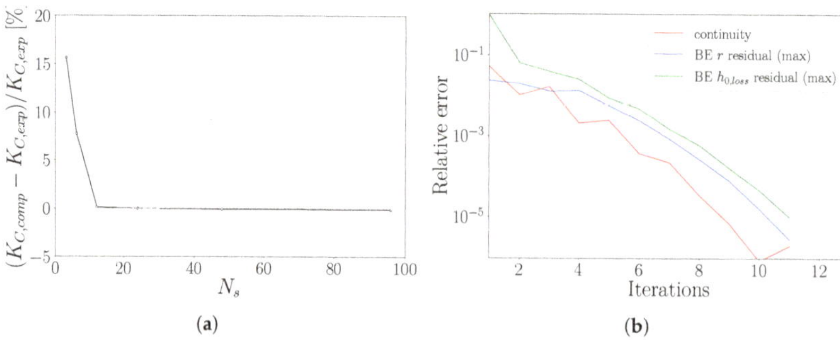

Figure 4. Sensitivity analysis of the streamlines number, N_s, (**a**) and typical convergence profile of a computation (**b**).

3.1. NASA Rotor 02

The NASA R02 is a straight-duct rotor intended for initial stages operations, following an inlet inducer. The geometry results from a modification of previous designs [32,33] and the blade element measurements were made available in Miller et al. [13]. The corresponding data are adopted as a reference for both the inflow boundary conditions and the outflow predictions comparison. Six different operating points (configurations 1 to 6) are selected within a range of flow coefficients, φ, ranging from 0.262 to 0.337. Here, the machine parameter is defined as $\varphi = Q/(A_{annulus}u_{tip})$, with $A_{annulus}$ denoting the rotor passage cross-sectional area. For this case study, the inflow quantities are provided as radial profiles instead of uniform distributions.

The resulting rotor map is reported in Figure 5, depicting the work coefficient, ψ, and the hydraulic efficiency, η, as functions of the flow coefficient. Here, the two parameters are respectively defined as follows:

$$\psi = \frac{gH}{u_{tip}^2} \tag{19}$$

$$\eta = \frac{H}{H_{id}} \tag{20}$$

where the head $H = \Delta p^0/(\rho g)$ is computed as the total pressure (p^0) jump across the rotor, u_{tip} is the tip rotor speed, and the ideal head rise is computed as $H_{id} = (c_{u,2}u_2 - c_{u,1}u_1)/g$.

The pressure rise capability trend is in line with the experimental measurements. However, the computed solution exhibits a stable overprediction, inducing a discrepancy with a $\Delta\psi = 0.13 \div 0.17$. However, with marginal improvements, the code portrays lower errors near the design operating conditions. This aspect is further emphasized by the efficiency curves comparison. In fact, from the experiments, the sudden drop of the hydraulic efficiency below the design point, combined with the smooth behavior of the work coefficient, suggests an abrupt increment of the measured torque. Differently, the code tends to smear out highly non-linear effects, recovering a smooth efficiency curve that constantly departs from the test data as the flow coefficient increases. As a result, the prediction error transitions from values below 2% to over 9%.

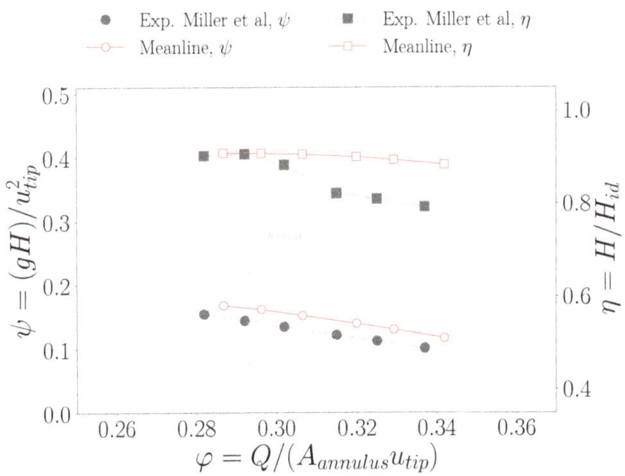

Figure 5. Work coefficient and hydraulic efficiency as functions of the flow rate coefficient for the NASA R02 rotor, comparing ARES solution and experimental values from Miller et al. [13].

To investigate local accuracy, the spanwise distributions of meridional and tangential velocities and outflow relative angle are adopted to compare between ARES computations and experiments at both off- (configuration 1, $\varphi = 0.337$) and near-design (configuration 5, $\varphi = 0.292$) conditions (Figure 6).

For configuration 1, the tangential velocity distribution exhibits a significant alignment with the test data, except for a marginal discrepancy in the span region between 60% and 90% (Figure 6a). Conversely, the meridional component is stably underpredicted across the entire blade radius, with the major separation observed at mid-channel and the accuracy loss mitigated near the extremes. The mutual behavior of the two velocity components can be analyzed through the relative flow angle distribution (Figure 6b). The predicted curve is generally in good agreement with measurements, depicting a maximum of 1° deviation above a span of 30%. At this location, such overprediction is induced by the underprediction of c_M, in contrast to the accurate estimate of c_u. On the other hand, above a span of 60%, the computed and measured curves tend to realign. This effect results from the underestimation of the tangential component, which is mitigated by a more accurate prediction of the meridional one.

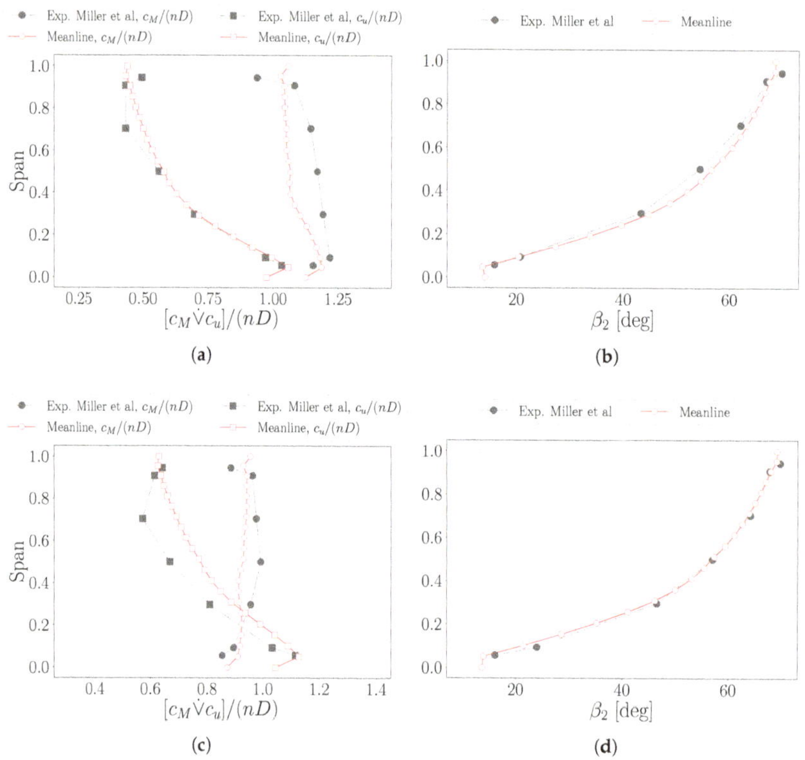

Figure 6. Spanwise distributions of meridional and tangential velocities (**left**) and outflow relative angle (**right**) for off-design, $\varphi = 0.337$, (**a**,**b**) and near-design, $\varphi = 0.292$, conditions (**c**,**d**). Comparison between present computations and test data from Miller et al. [13].

The most accurate efficiency prediction is achieved for flow configuration 5. Since the computed head rise is higher than in the experiments, the low deviation of the efficiency curve depends on the ideal head rise prediction, which is a quantity uniquely determined by velocity components. In fact, the tangential velocity exhibits overprediction across the entire span, with a discrepancy increasing from the hub to shroud, except for end-wall boundary layer effects reducing the computed and tested curves separation (Figure 6c). The resulting meridional velocity prediction evidently suggests an accuracy improvement, despite a stable underestimation. Similarly to the previous case, the mid-channel solution exhibits higher deviation from the test data. Conversely, the relative flow angle portrays a significant alignment with experimental measurements, with further reduction of the discrepancy, now located near span 10% (Figure 6d).

This analysis indicates that pressure and velocity predictions are not directly correlated. In fact, while flow directions were generally aligned with experiments, the pressure field was stably overpredicted.

3.2. HIREP

The HIgh REynolds number Pump (HIREP) facility was adopted to conduct measurements on the flow effects of an IGV installed upstream of a rotor [25,26]. Here, the test data between the two blades are retained as inflow boundary conditions, while computations are compared downstream of the rotor. This test case provides insights into the code accuracy, when operated under non-uniform, swirled inflow. An operating map is reconstructed, tracking the variation in the torque coefficient, K_C, as a function of the flow coefficient, φ (Figure 7), with the former defined as in the reference experiments [25]:

$$K_C = \frac{C}{\rho n^2 D^5} \tag{21}$$

where C is the rotor torque, and φ is defined as in Equation (18). Under near-design conditions, ARES exhibits significant accuracy. Conversely, the predicted maps depart from the experimental measurements, as off-design operations are considered. This behavior is more emphasized at higher values of φ, denoting a tendency to underestimate the rotor torque by up to $\Delta K_C = -28\%$. Differently, the discrepancy is reduced at lower values with a $\Delta K_C = +8\%$. The smooth monotonic trend of the curve is representative of the code stability over a wide range of operations.

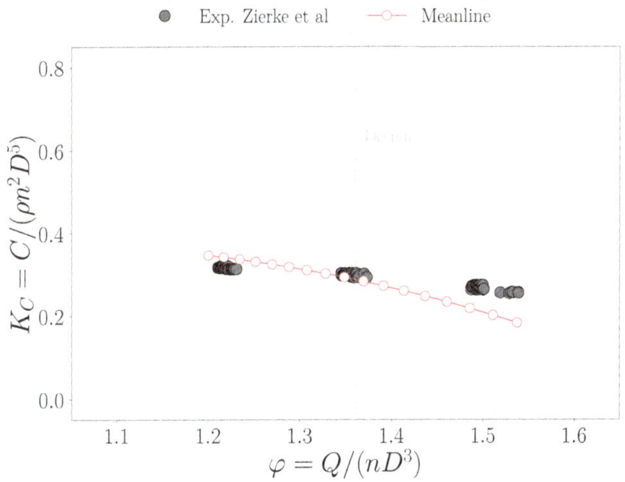

Figure 7. Torque coefficient as a function of the flow rate coefficient for the HIREP rotor, comparing ARES solution and experimental values from Zierke et al. [25].

As local statistics, spanwise distributions of normalized velocities and pressure coefficients are considered at design condition $\varphi = 1.36$ (Figure 8). From the analysis of the tangential velocity component (Figure 8a) ARES can be seen to overpredict the flow turning, leading to negative values of c_u. Differently, during experiments, this component was essentially suppressed downstream of the rotor. Thus, continuity forces the axial component to decrease, thereby producing the left-shift observable in the corresponding distribution. The total pressure curves underscore a tendency of the model to underestimate losses, depicting stably higher values than the test data (Figure 8b). However, the behaviour is not monotonic, with the greatest discrepancy occurring near midspan. Globally, the combination with the overprediction of the velocity distribution enables the model to recover a good agreement between the computed and measured static pressure fields. Anyhow, the curve is stably above experiments.

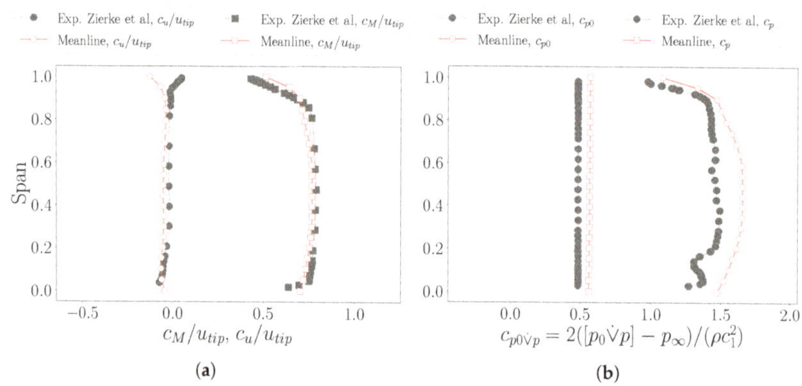

Figure 8. Spanwise distribution of the flow statistics, comparing ARES solution and experiments from Zierke et al. [25], for design conditions ($\varphi = 1.36$). Reported quantities include normalized tangential and axial velocities (**a**) and total and static pressure coefficients (**b**).

3.3. *AxWJ-2*

The test case represents a notional model for the validation of waterjet pump numerical models [27,31,34]. Here, the geometry is selected to evaluate the code accuracy for a rotor–stator configuration, featuring a shaped duct with a variable cross-sectional area. Computations are performed using uniform inflow boundary conditions derived from the processed mass flow rate.

In this case, the characteristic map reports the head rise coefficient, ψ, and the hydraulic efficiency, η, as functions of the flow coefficient, φ (Figure 9). While the latter follows Equation (18), the other two are defined below:

$$\psi = \frac{gH}{(nD)^2} \tag{22}$$

$$\eta = \frac{\rho Q g H}{C\omega} \tag{23}$$

where the angular velocity ω expresses the rotational regime in rad/s.

ARES confirms significantly accurate near-design conditions ($\varphi = 0.85$), with a 5% discrepancy in the efficiency estimate, while exaggerating the head losses at higher mass flow rates. Despite keeping stable over a wide range of operating points, at low mass flow rates, the code diverges before completing the span of the experimental map. Notably, before the last simulated point, the results depict an increasing agreement with the reference measurements. Specifically, around the best efficiency point ($\varphi = 0.77$), the efficiency is predicted with a 1.4% error.

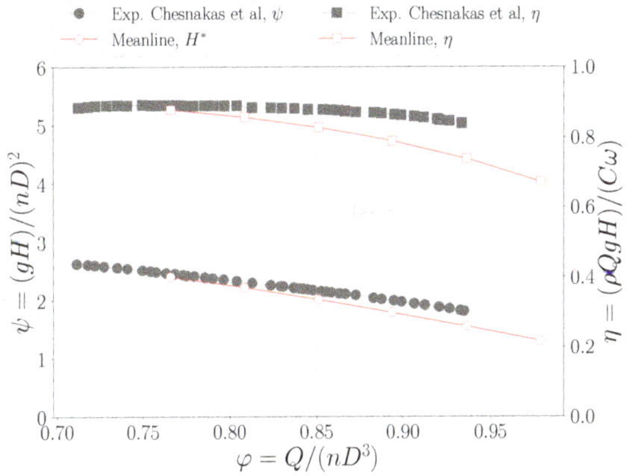

Figure 9. Overall head coefficient and efficiency as functions of the flow rate coefficient for the AxWJ-2 pump, comparing ARES solution and experimental values from Chesnakas et al. [27].

Thus, local Particle Image Velocimetry (PIV) measurements from Chesnakas et al. [27] are compared with computations at the station downstream of the rotor (Figure 10). In this case, the spanwise distribution of the axial velocity component (Figure 10a) significantly aligns with experiments, except for a velocity defect of the computed solution between the hub and 20% of the span. Then, the present method exhibits difficulty in predicting boundary layer dynamics near the shroud, where it tends to preserve the lower span evolution. Similar considerations hold for the tangential velocity component (Figure 10b). In fact, for a major portion near the midspan, the numerical solution and test data are almost superimposed, while the lower and upper regions exhibit higher discrepancies. Especially for the latter, the results suggest scarce modeling of end-wall dynamics inside rotating domains.

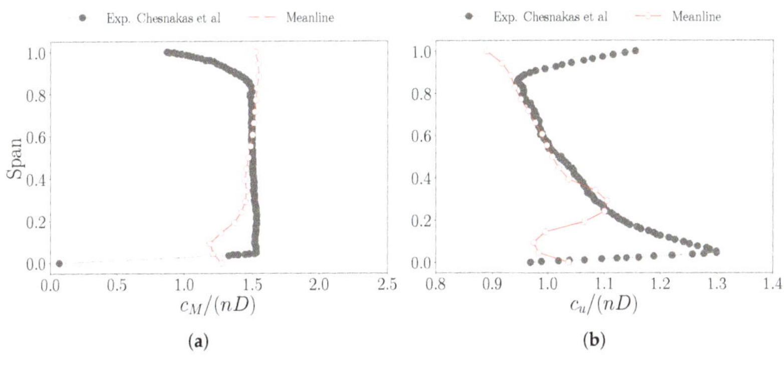

Figure 10. Spanwise distribution of the flow statistics, comparing ARES solution and experiments from Chesnakas et al. [27] downstream of the AxWJ-2 rotor, for design conditions ($\varphi = 0.85$). Reported quantities include normalized meridional (**a**) and tangential (**b**) velocities.

The same statistics are analyzed downstream of the stator blades (Figure 11). For this location, the disagreements are partially affected by the difference between the measurements station (at the nozzle exhaust) and the computations section (at the blade trailing edge), which differ due to the duct meridional shape variation. As a result, while the profile of the simulated axial velocity aligns with experiments (Figure 11a), the corresponding values are lower because the code does not model the additional downstream expansion.

Furthermore, the computed hub dynamics emphasizes boundary layer evolution, inducing exaggerate spanwise fluctuations in the lower span. Considering the tangential velocity component, the difference between the two solutions is large (Figure 11b). While the wake dynamics impacts the evolution near the duct axis, the experiments exhibit an almost rectified flow. Differently, the numerical solution recovers a stably positive solution. Although the magnitudes are comparable between the computations and test data, the distribution profiles depict substantially different flow behaviors.

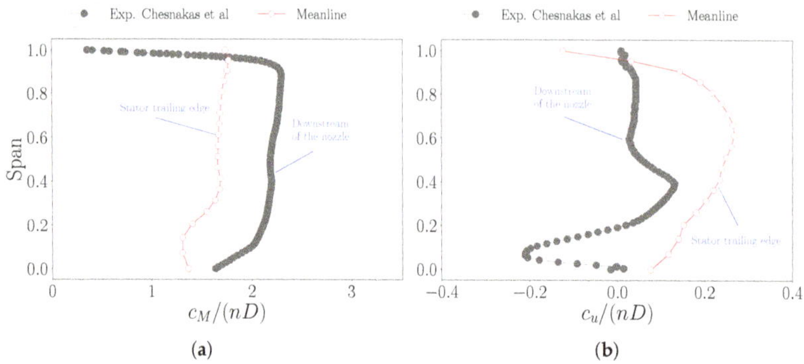

Figure 11. Spanwise distribution of the flow statistics, comparing ARES solution and experiments from Chesnakas et al. [27] downstream of the AxWJ-2 stator, for design conditions ($\varphi = 0.85$). Measurements refer to the nozzle exhaust, which further expands the flow from the station used for meanline predictions, at the blade trailing edge. Reported quantities include normalized meridional (**a**) and tangential (**b**) velocities.

The results for this case study align with the two previous investigations featuring a rotor-only configuration. The code demonstrates good accuracy in predicting the flow downstream of the rotating blades, whereas larger discrepancies arise when comparing the stator blade flow to experimental data.

4. Conclusions

ARES (Axial-flow pump Radial Equilibrium through Streamlines), a meanline code for designing axial-flow pumps for Outboard Dynamic-inlet Waterjets (ODW) applications, is introduced. The solver iteratively enforces flow momentum (aka Radial Equilibrium) and mass conservation laws to model the machine performance. In particular, several empirical correlations are adopted to model real-flow effects, such as off-design deviation, losses from profile shape, secondary flows, tip leakage, and the end-wall boundary layer (EWBL). A Blade Element Method (BEM) is used for the discretization process, addressing the meridional channel with planar cascades. This code is aimed at providing a fast and accurate tool for ODW axial-flow pump design. Specifically, the possibility to parameterize the geometry is a key ability that can be leveraged by optimization loops for the maximization of initial geometries performance.

The code's accuracy is evaluated using three test cases: the NASA Rotor 02, the HIgh REynolds number Pump (HIREP), and the Axial-flow WaterJet pump (AxWJ-2). Global and local statistics compare the numerical solution to available experiments. The code is found to be stable over a wide range of operating points, with accuracy decreasing as the mass flow rate exceeded the design value. At lower values, the efficiency is predicted within a 1% error compared to the test data, while at design conditions, the discrepancy exceeds 5%. Spanwise distributions of the velocities and pressure generally show good agreement with measurements at the rotor exit. However, further calibration is needed to improve loss predictions near the hub and shroud walls, especially downstream of the stator blades.

Thus, this study is an intermediate step toward developing a comprehensive tool for Outboard Dynamic-inlet Waterjets (ODW) design. After further calibration, the research will integrate the pump module into the nacelle design routine for the meanline prediction of coupled operations. Integrated into an optimization algorithm, the code will be able to be exploited for accurate high-performance designs.

Author Contributions: Conceptualization, F.A., F.D.V. and A.M.; methodology, F.A., F.D.V. and A.M.; software, F.A.; validation, F.A.; formal analysis, F.A.; investigation, F.A.; resources, F.A. and E.B.; data curation, F.A., F.D.V. and A.M.; writing—original draft preparation, F.A., F.D.V. and A.M.; writing—review and editing, F.D.V., A.M. and E.B.; visualization, F.A.; supervision, F.D.V., A.M. and E.B.; project administration, E.B.; funding acquisition, E.B. All authors have read and agreed to the published version of the manuscript.

Funding: This research was funded by the University of Padova through the project ECO-SPEED, which was co-funded by SEALENCE S.p.a., under the program UNI-IMPRESA 2022.

Data Availability Statement: The original contributions presented in this study are included in the article. Further inquiries can be directed to the corresponding author.

Conflicts of Interest: The authors declare no conflict of interest.

Abbreviations

The following abbreviations are used in this manuscript:

AVR	Axial Velocity Ratio
BE	Blade Element
BEM	Blade Element Method
CFD	Computational Fluid Dynamics
DCA	Double Circular Arc
EWBL	End-Wall Boundary Layer
HIREP	High Reynolds Number Pump
IGV	Inlet Guide Vane
ODW	Outboard Dynamic-inlet Waterjet
PIV	Particle Image Velocimetry
RE	Radial Equilibrium

References

1. Benini, E.; Gobbo, W. Propulsion Device with Outboard Waterjet for marine Vehicles. U.S. Patent 12,116,096, 15 October 2024.
2. Avanzi, F.; De Vanna, F.; Magrini, A.; Benini, E. Numerical characterisation of Outboard Dynamic-inlet Waterjets: Propulsive aspects of intake/pump integration. *Acta Mech.* **2024**. [CrossRef]
3. Park, W.G.; Yun, H.S.; Chun, H.H.; Kim, M.C. Numerical flow simulation of flush type intake duct of waterjet. *Ocean Eng.* **2005**, *32*, 2107–2120. [CrossRef]
4. Jung, K.H.; Kim, K.C.; Yoon, S.Y.; Kwon, S.H.; Chun, H.H.; Kim, M.C. Investigation of turbulent flows in a waterjet intake duct using stereoscopic PIV measurements. *J. Mar. Sci. Technol.* **2006**, *11*, 270–278. [CrossRef]
5. Suryanarayana, C.; Satyanarayana, B.; Ramji, K.; Saiju, A. Experimental evaluation of pumpjet propulsor for an axisymmetric body in wind tunnel. *Int. J. Nav. Archit. Ocean Eng.* **2010**, *2*, 24–33. [CrossRef]
6. Suryanarayana, C.; Satyanarayana, B.; Ramji, K.; Nageswara, M.R. Cavitation studies on axi-symmetric underwater body with pumpjet propulsor in cavitation tunnel. *Int. J. Nav. Archit. Ocean Eng.* **2010**, *2*, 185–194. [CrossRef]
7. Shi, S.; Tang, W.; Huang, X.; Dong, X.; Hua, H. Experimental and numerical investigations on the flow-induced vibration and acoustic radiation of a pump-jet propulsor model in a water tunnel. *Ocean Eng.* **2022**, *258*, 111736. [CrossRef]
8. Jiang, J.; Ding, J.; Lyu, N.; Luo, H.; Li, L. Control volume determination for submerged waterjet system in self-propulsion. *Ocean Eng.* **2022**, *265*, 112594. [CrossRef]

9. Menegozzo, L.; Benini, E. Meanline Calculation of Surge Margin Loss due to Inlet Flow Distortion. In Proceedings of the Turbo Expo: Power for Land, Sea, and Air, Virtual, 21–25 September 2020; Volume 2A. [CrossRef]
10. Johnson, I.A.; Bullock, R.O. *Aerodynamic Design of Axial-Flow Compressors*; Technical Report NASA-SP-36; NASA: Washington, DC, USA, 1965. Available online: https://ntrs.nasa.gov/api/citations/19650013744/downloads/19650013744.pdf (accessed on 3 March 2025).
11. Henderson, R.E.; McMahon, J.F.; Wislicenus, G.F. *A Method for the Design of Pumpjets*; Technical Report NOw 63-0209-C-7; Pennsylvania State Univ State College Ordnance Research Lab: State College, PA, USA, 1964. Available online: https://apps.dtic.mil/sti/pdfs/AD0439631.pdf (accessed on 3 March 2025).
12. Serovy, G.; Kavanagh, P.; Okiishi, T.; Miller, M. *Prediction of Overall and Blade-Element Performance for Axial-Flow Pump Configurations*; Technical Report NASA CR-2301; NASA: Washington, DC, USA, 1973. Available online: https://ntrs.nasa.gov/api/citations/19730023457/downloads/19730023457.pdf (accessed on 3 March 2025).
13. Miller, M.J.; Okiishi, T.H.; Serovy, G.K.; Sandercock, D.M.; Britsch, W.R. *Summary of Design and Blade-Element Performance Data for 12 Axial-Flow Pump Rotor Configurations*; Technical Report NASA-TN-D-7074; NASA: Washington, DC, USA, 1973. Available online: https://ntrs.nasa.gov/api/citations/19730006255/downloads/19730006255.pdf (accessed on 3 March 2025).
14. Furuya, O.; Chiang, W.L. *A New Pumpjet Design Theory*; Technical Report N00014-85-C-0050; David Taylor Research Center: Bethesda, MD, USA, 1988. Available online: https://apps.dtic.mil/sti/pdfs/ADA201353.pdf (accessed on 3 March 2025).
15. Tournier, J.; El-Genk, M.S. Axial flow, multi-stage turbine and compressor models. *Energy Convers. Manag.* **2010**, *51*, 16–29. [CrossRef]
16. Banjac, M.; Petrovic, M.V.; Wiedermann, A. Secondary Flows, Endwall Effects, and Stall Detection in Axial Compressor Design. *J. Turbomach.* **2015**, *137*, 051004. [CrossRef]
17. Banjac, M.; Petrovic, M.V. Development of Method and Computer Program for Multistage Axial Compressor Design: Part I—Mean Line Design and Example Cases. In Proceedings of the Turbo Expo: Power for Land, Sea, and Air, Oslo, Norway, 11–15 June 2018; Volume 2C. [CrossRef]
18. Pollard, D.; Gostelow, J.P. Some Experiments at Low Speed on Compressor Cascades. *J. Eng. Power.* **1967**, *89*, 427–436. [CrossRef]
19. Koch, C.C.; Smith, L.H., Jr. Loss Sources and Magnitudes in Axial-Flow Compressors. *J. Eng. Power.* **1976**, *98*, 411–424. [CrossRef]
20. Aungier, R.H. *Axial-Flow Compressors: A Strategy for Aerodynamic Design and Analysis*; ASME Press: New York, NY, USA, 2003.
21. Wu, D.; Teng, J.; Qiang, X.; Feng, J. An Implicit Off-Design Deviation Angle Correlation of Axial Flow Compressor Blade Elements. In Proceedings of the Turbo Expo: Power for Land, Sea, and Air, Charlotte, NC, USA, 26–30 June 2017; Volume 2B, p. V02BT41A005. [CrossRef]
22. Wu, D.; Teng, J.; Qiang, X.; Zheng, T. Development of an off-design deviation angle prediction model for full blade span in axial flow compressors. *Proc. Inst. Mech. Eng. Part G J. Aerosp. Eng.* **2019**, *233*, 3170–3183. [CrossRef]
23. Guo, J.; Zhang, Y.; Chen, Z.; Feng, Y. CFD-based multi-objective optimization of a waterjet-propelled trimaran. *Ocean Eng.* **2020**, *195*, 106755. [CrossRef]
24. Avanzi, F.; Baù, A.; De Vanna, F.; Benini, E. Numerical Assessment of a Two-Phase Model for Propulsive Pump Performance Prediction. *Energies* **2023**, *16*, 6592. [CrossRef]
25. Zierke, W.C.; Straka, W.A.; Taylor, P.D. *The High Reynolds Number Flow Through an Axial-Flow Pump*; Technical Report TR 93-12; Applied Research Laboratory: Ferguson Township, PA, USA, 1993. Available online: https://apps.dtic.mil/sti/pdfs/ADA273844.pdf (accessed on 3 March 2025).
26. Zierke, W.C.; Straka, W.A. Flow visualization and the three-dimensional flow in an axial-flow pump. *J. Fluids Eng.* **1996**, *12*, 250–259. [CrossRef]
27. Chesnakas, C.J.; Donnelly, M.J.; Pfitsch, D.W.; Becnel, A.J.; Schroeder, S.D. *Performance Evaluation of the ONR Axial Waterjet 2 (AxWJ-2)*; Technical Report NSWCCD-50-TR-2009/089; Naval Surface Warfare Center Carderock Division: Bethesda, MD, USA, 2009. Available online: https://apps.dtic.mil/sti/pdfs/ADA516369.pdf (accessed on 3 March 2025).
28. Herrig, L.J.; Emery, J.C.; Erwin, J.R. *Systematic Two-Dimensional Cascade Tests of NACA 65-Series Compressor Blades at Low Speeds*; Technical Report NACA-TN-3916; NASA: Washington, DC, USA, 1957. Available online: https://ntrs.nasa.gov/api/citations/19930084843/downloads/19930084843.pdf (accessed on 3 March 2025).
29. Denton, J.D. The 1993 IGTI Scholar Lecture: Loss Mechanisms in Turbomachines. *J. Turbomach.* **1993**, *115*, 621–656. [CrossRef]
30. Howell, A.R. Fluid Dynamics of Axial Compressors. *Proc. Inst. Mech. Eng.* **1945**, *153*, 441–452. [CrossRef]
31. Michael, T.J.; Schroeder, S.D.; Becnel, A.J. *Design of the ONR AxWJ-2 Axial Flow Water Jet Pump*; Technical Report NSWCCD-50-TR-2008/066; Naval Surface Warfare Center Carderock Division: Bethesda, MD, USA, 2008. Available online: https://apps.dtic.mil/sti/pdfs/ADA489739.pdf (accessed on 3 March 2025).
32. Crouse, J.E.; Soltis, R.F.; Montgomery, J.C. *Investigation of the Performance of an Axial-Flow-Pump Stage Designed by the Blade-Element Theory: Blade-Element Data*; Technical Report NASA-TN-D-1109; NASA: Washington, DC, USA, 1961. Available online: https://ntrs.nasa.gov/api/citations/19980228119/downloads/19980228119.pdf (accessed on 3 March 2025).

33. Miller, M.J.; Crouse, J.E.; Sandercock, D.M. Summary of Experimental Investigation of Three Axial-Flow Pump Rotors Tested in Water. *J. Eng. Power.* **1967**, *89*, 589–599. [CrossRef]
34. Marquardt, M.W. *Summary of Two Independent Performance Measurements of the ONR Axial Waterjet 2 (AxWJ-2)*; Technical Report NSWCCD-50-TR-2011/016; Naval Surface Warfare Center Carderock Division: Bethesda, MD, USA, 2011. Available online: https://apps.dtic.mil/sti/tr/pdf/ADA540499.pdf (accessed on 3 March 2025).

fluids

MDPI

Article

Predictive Analysis of Structural Damage in Submerged Structures: A Case Study Approach Using Machine Learning

Alexandre Brás dos Santos [1,*], Hugo Mesquita Vasconcelos [1], Tiago M. R. M. Domingues [1], Pedro J. S. C. P. Sousa [1,2], Susana Dias [1], Rogério F. F. Lopes [1,2], Marco L. P. Parente [2], Mário Tomé [3], Adélio M. S. Cavadas [3] and Pedro M. G. P. Moreira [1]

1 INEGI, Institute of Science and Innovation in Mechanical Engineering and Industrial Engineering, Campus da FEUP, R. Dr. Roberto Frias 400, 4200-465 Porto, Portugal
2 Department of Mechanical Engineering, Faculdade de Engenharia, Universidade do Porto, R. Dr. Roberto Frias s/n, 4200-465 Porto, Portugal
3 PROMETHEUS, School of Technology and Management (ESTG), Polytechnic Institute of Viana do Castelo, Avenida do Atlântico n° 644, 4900-348 Viana do Castelo, Portugal
* Correspondence: absantos@inegi.up.pt

Abstract: This study focuses on the development of a machine learning (ML) model to elaborate on predictions of structural damage in submerged structures due to ocean states and subsequently compares it to a real-life case of a 6-month experiment with a benthic lander bearing a multitude of sensors. The ML model uses wave parameters such as height, period and direction as input layers, which describe the ocean conditions, and strains in selected points of the lander structure as output layers. To streamline the dataset generation, a simplified approach was adopted, integrating analytical formulations based on Morison equations and numerical simulations through the Finite Element Method (FEM) of the designed lander. Subsequent validation involved Fluid–Structure Interaction (FSI) simulations, using a 2D Computational Fluid Dynamics (CFD)-based numerical wave tank of the entire ocean depth to access velocity profiles, and a restricted 3D CFD model incorporating the lander structure. A case study was conducted to empirically validate the simulated ML model, with the design and deployment of a benthic lander at 30 m depth. The lander was monitored using electrical and optical strain gauges. The strains measured during the testing period will provide empirical validation and may be used for extensive training of a more reliable model.

Keywords: fluid–structure interaction; benthic lander; structural health monitoring; machine learning models

1. Introduction

Offshore structures play an important role in exploring and producing valuable energy resources from marine environments. The reliability and safety of these structures are of utmost importance and rely on the accurate prediction and assessment of potential structural damage due to ocean conditions. To address these challenges, Structural Health Monitoring (SHM) has emerged as a growing priority, combining sensor networks, advanced engineering materials, and computational knowledge systems, enabling engineers to evaluate the condition and performance of structures over time or estimate their remaining in-service life [1].

However, deploying sensor systems in submerged environments presents significant challenges related to durability, accessibility, and data reliability. The harsh underwater conditions can lead to issues such as corrosion, biofouling, and physical damage to sensors, compromising their longevity and performance. Additionally, the underwater environment poses obstacles to reliable data transmission, further complicating effective monitoring [2].

To overcome these physical limitations, virtual sensors have emerged as a transformative approach. These computational models leverage real-time data from a limited number of physical sensors, numerical simulation and machine learning techniques, to infer structural responses and conditions in areas where direct measurements are impractical or impossible. These tools enable comprehensive monitoring of critical regions of a structure while reducing the dependence on extensive physical sensor networks [3,4].

To support SHM efforts, empirical and numerical models have been used in coastal engineering, including sea level rise, hydrodynamics, and wave propagation [5]. These models incorporate complex, time-dependent interactions like wave-current and wave-structure dynamics, making them essential for generating data to develop SHM models. However, their computational intensity and inherent simplifications often limit their application for real-time monitoring. Recent advancements in machine learning (ML) are opening new developments in SHM applications to overcome these limitations, enabling more effective predictive maintenance and real-time damage detection. Although the amount of available data is increasing, the lack of sufficient and high-quality data is still a challenge, impacting the effectiveness and reliability of SHM systems. For instance, a recent study has applied deep learning techniques, such as a BRANN-based deterministic model for fragility assessment of deepwater drilling risers under storm conditions [6].

The SHM process consists of four key stages: load events sensing and determination of whether damage was imparted to the structure, identification of the precise loading location, evaluation of the extent and severity of the damage and, finally, prediction of the remaining life of the structure, taking the evaluated damage into account. This methodology is particularly useful for high-profile mechanical systems such as marine structures where performance is critical and on-site damage detection is challenging or even impossible [7].

To address these challenges and bridge the research gap in SHM for submerged structures and real-time monitoring, this study presents a methodology for predicting structural damage caused by ocean waves and currents. By combining wave parameters collected from an oceanic buoy with structural responses, the approach employs an ML model to deliver timely and reliable damage assessments. Acting as a virtual sensor, this model offers a new tool to improve real-time SHM in submerged environments.

This research is part of the Marewind Project, which aims to advance innovative solutions for offshore structures. Within this framework, and to address the challenges previously mentioned, a benthic lander was designed and manufactured to achieve three distinct objectives. The first two are related to testing materials: testing of antifouling coatings and testing new formulations for a concrete ballast system. The last goal, and the one that this paper is focused on, is the development of ML algorithms for predicting structural damage in submerged structures due to the inherent effects of ocean waves and currents. The data to train the Machine Learning models were initially generated using analytical and numerical methods. This process utilised historical data from a metocean buoy, provided by the Hydrographic Institute of Portugal, to model various sea states, characterised by wave parameters such as wave height, wave period and wave direction.

The subsequent sections detail the design of the benthic lander, the experimental apparatus and the methods developed to generate the data for training the machine learning models. A flowchart of the methodology is shown in Figure 1.

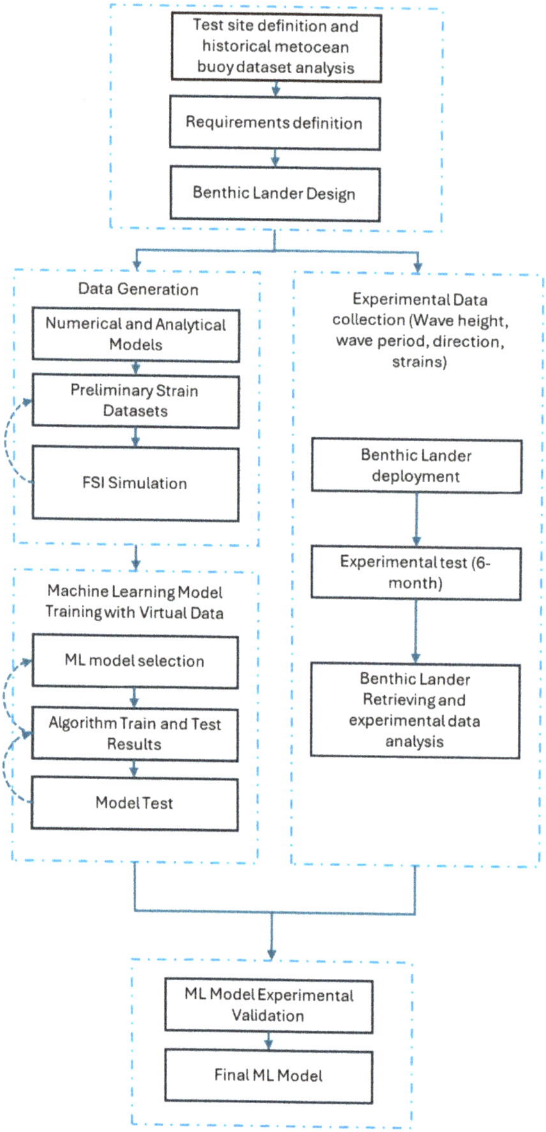

Figure 1. Schematic representation of the workflow.

2. Benthic Lander Design and Experimental Apparatus

The case study's test site is located near the Portuguese coastline, near Sines. A dataset with historical data from an oceanic buoy that is operating near the test site was acquired. This dataset comprises data about the ocean state, such as wave height and wave period, containing data from January 2017 to February 2024, with approximately 82,000 sea state measurements. This oceanic buoy is part of the Portugal Hydrographic Institute oceanic buoys network operating all over the Portuguese coastline. The historical dataset was analysed using wave theories such as the Airy Wave theory and the Stokes Wave theory [8]. The analysis supported the definition of the most critical scenario the lander could encounter, and generation of data for the machine learning models.

The selection of which wave theory to use was made according to the Le Méhauté diagram [9], which indicates the region of validity of different wave theories for various water depths and wave parameters.

The lander was deployed in the location illustrated in Figure 2. Simultaneously, a spotter buoy was anchored to the seabed at the same location. The spotter buoy measured the ocean states during the 6-month experiment, providing more accurate data regarding ocean conditions, and thus improving the reliability of the ML model.

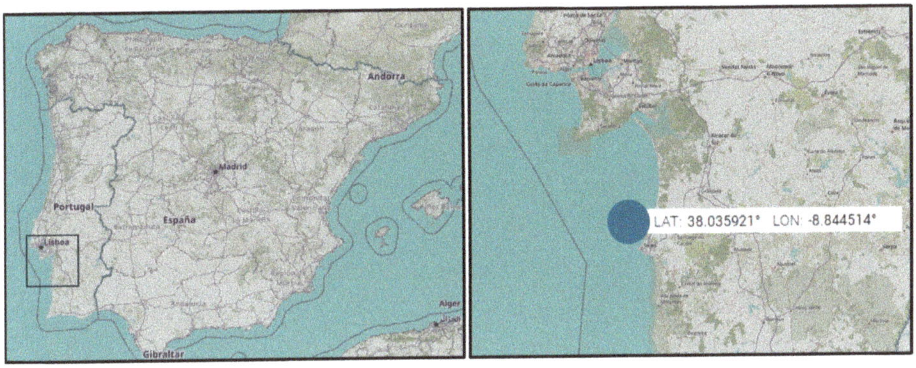

Figure 2. Lander deployment location. Adapted with permission from Ref. [10]. 2024, OpenStreetMap.

The lander is a welded structure of structural steel S275JR (EN 1.0044) treated with hot-dip galvanisation with a total mass of approximately 500 kg. The structure is tripod-shaped and was designed to be deployed into the benthic zone of the seabed at a 30 m depth. This design allows the assembly of steel plates on each leg, which increases the total weight of the structure, and the assembly of the electronic equipment in a central container. The lower centre of gravity and the high mass ensure the structure remains upright in challenging underwater conditions, including extreme currents and waves, allowing excellent stability on the ocean floor. The structure predominantly consists of circular tube profiles, except for the plates where the strain gauges are located.

The lander is monitored using two independent systems: a primary system comprising optical strain gauges, Fiber Bragg Grating (FBG) sensors and a secondary system of conventional electrical strain gauges. These sensors were mounted on four specific plates at the centre of the structure. Figure 3 shows the assembly process of the strain gauge sensors as well as the final structure.

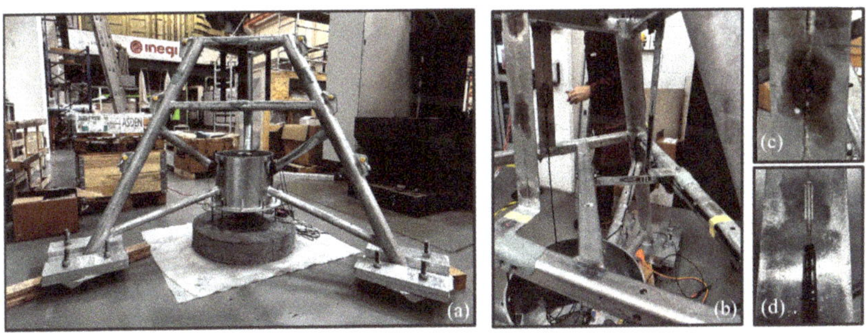

Figure 3. Final structure. (**a**) Overall lander structure; (**b**) specific plates mounted on the structure where strain gauges are mounted; (**c**) primary system—optical strain gauges' assembly; (**d**) secondary system—electrical strain gauges' assembly.

Two critical scenarios were identified to guarantee the success of the experiment. The first scenario addressed the stages of deploying the structure in the ocean and retrieving it. In this phase, the primary considerations were the total weight of the structure along with any mounted equipment, which together pose the most significant challenge for the deployment operation. The second scenario focused on the impact of ocean currents and waves on the structure, due to the pressures exerted by different sea states. For each scenario, a Finite Element Model (FEM) of the lander was developed.

While the load calculation was straightforward in the first scenario, the second scenario required a more detailed analysis with the calculation of the loads for each individual component. Figure 4 illustrates the adopted sub-division.

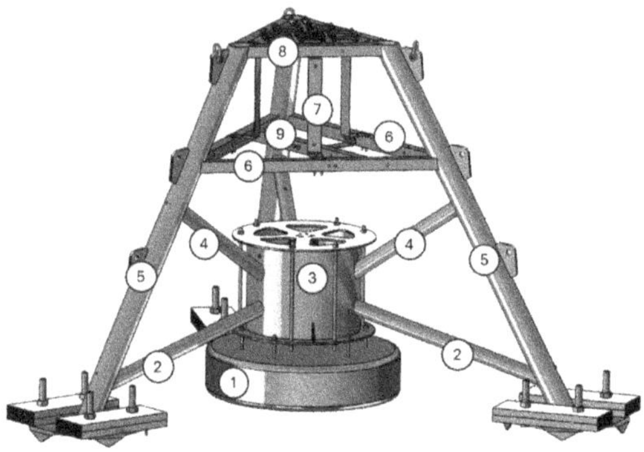

Figure 4. Division of lander structure into subsections for load calculation.

As a widely used method for wave loading prediction, Morison's Equation was employed to calculate the loads suffered by each component of the lander. It comprises two main components: the inertial force (F_I), and the drag force (F_D) [11].

$$F = F_D + F_I, \tag{1}$$

$$F = 0.5\rho C_D |v| v A_p + C_M \rho V_{vol} \dot{v}, \tag{2}$$

where ρ is the density, v is the fluid velocity, A_p is the projected area of the structure perpendicular to the flow and V_{vol} is the displaced volume.

The values of the drag (C_D) and inertia (C_M) coefficients are typically determined experimentally, being a function of the Reynolds number (Re), the Keulegan–Carpenter number (KC) and the roughness (Δr). The recommended practical coefficient values for the coefficients were assumed considering each part's geometry and its Reynolds number [12].

The critical ocean condition identified on the historical dataset is outlined in Table 1. However, to ensure the structural integrity of the entire experimental test, a safe approach was adopted, and a 5 m/s velocity magnitude was considered for the numerical simulation.

Table 1. Wave parameters for the selected ocean state.

Wave Period	Wave Height	Wavelength	Depth
9.39 s	6.73 m	105 m	30 m

Results and Discussion

Figure 5 shows the results obtained in the numerical simulation, validating the structure design.

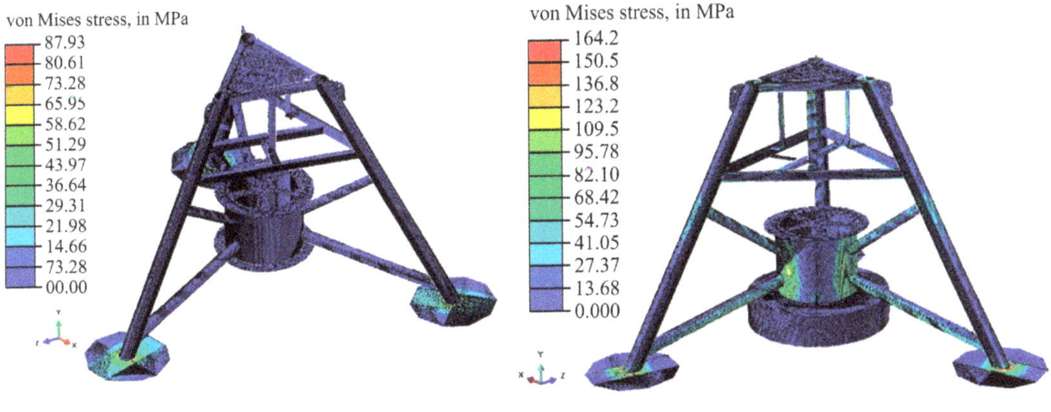

(a) (b)

Figure 5. FEA results; (**a**) first scenario: structure hanging—max von Mises stress 88 MPa located on the leg base of the lander; (**b**) second scenario: structure deployment—max von Mises stress of 164 MPa located on the round tube at the centre of the lander.

The analysis showed that the highest calculated von Mises stress is located at the base of the structure and the tube in the centre, for the first and second scenarios, respectively. These results were compared with the material properties (yield stress of 275 MPa) to ensure that the lander can withstand the demands in each scenario.

3. FSI Methodology

Physical modelling of a structure subjected to hydrodynamic loads is usually performed via a two-stage procedure: a CFD simulation followed by an FEM analysis.

The approach for Fluid–Structure Interaction (FSI) simulations combines the Finite Volume Method (FVM) with the volume of fluid (VOF) method to create the CFD models. For these analyses, the commercial software Ansys Fluent was used. In the CFD models, pressure fields on the lander structure were calculated and then applied to a Finite Element Method (FEM) model to assess structural responses to hydrodynamic wave loads. For the FEM analysis, the commercial software Ansys Mechanical was used.

Considering the extension of the problem, the main challenge in designing the CFD numerical model was to create an accurate simulation with reduced computational cost. To meet these goals, the CFD model was split into two parts. This method lowered the computing effort needed, avoiding an entire ocean-depth 3D CFD model of the test site.

The first part of the approach focused on a fully developed 2D domain, with a depth corresponding to the full scale of the test site's bathymetry and a length sufficient to characterise the wavy flow while minimising undesirable numerical effects. For this purpose, a numerical wave tank was developed. The primary objective was to generate velocity profiles at each time step that accurately reflected the impact of surface waves at the bottom of the domain.

The velocity profiles, generated in a 2D numerical wave tank, were applied as inlet boundary conditions to a 3D model that incorporates the lander structure. This approach significantly reduced computational cost, as the 3D domain could be limited in size and considered fully submerged, eliminating the need for the computationally intensive VOF multiphase model. The 3D CFD model, restricted to this configuration, produces pressure fields that are subsequently used in the structural analysis of the lander within a one-way FSI framework.

3.1. Two-Dimensional Numerical Wave Tank: Velocity Profile Generation

The configuration of the numerical wave tank model is organised into three primary stages: (1) the geometry setup, where the physical dimensions of the model domain are defined; (2) the mesh generation, in which the computational grid is created and refined near the still water level (SWL) to capture the free surface dynamics accurately; and (3) the wave and physics setup, which involves specifying the type of analysis, domain properties, wave generation and other parameters related to air–water interactions. The surface tension at the air–water interface is considered negligible. The air temperature is set to 25 °C, corresponding to a density of 1.185 kg/m^3. The fluid temperature is also set to 25 °C, with a density of 1030 kg/m^3 to represent saltwater conditions.

3.1.1. Governing Equations

The solver in ANSYS Fluent 2023 is based on the finite volume method (ANSYS Inc., Canonsburg, PA, USA, 2023). This method divides the computational domain into a series of control volumes and discretises the governing equations, enabling iterative solutions over each control volume. Through this process, approximate values of the flow variables are computed at discrete points across the domain. The governing equations are the continuity equation for constant density and the Navier—Stokes equations for an incompressible, constant viscosity fluid, given by [13]:

$$|\nabla| \cdot \vec{u} = 0, \tag{3}$$

$$\frac{\partial \cdot \vec{u}}{\partial t} = -\vec{u} \cdot I\vec{u} + \frac{1}{\rho Re}\nabla \cdot \left[(\mu + Re\mu_t)\left(\nabla\vec{u} + \left(\nabla\vec{u}\right)^T\right)\right] - \frac{\nabla P}{\rho} + \frac{\vec{a}}{F_r}, \tag{4}$$

where $\vec{u}, P, \rho, \mu, \mu_t$ $\vec{a}$ and F_r are the non-dimensional velocity, pressure, density, dynamic viscosity and dynamic turbulent/eddy viscosity fields, gravitational acceleration and Froude number, respectively.

The volume of fluid (VOF) method is employed to track the free surface position. This method calculates the volume fraction of each fluid, present within each control volume. The volume fraction is determined using the following equation:

$$\frac{\partial \alpha}{\partial t} + \nabla\left(\alpha\vec{V}\right) = 0, \tag{5}$$

where $\vec{V}$ represents the velocity vector, and α denotes the water volume fraction. The mixture density is calculated based on the volume fraction using the following equation:

$$\rho = \alpha\rho_\omega + (1-\alpha)\rho_a, \tag{6}$$

where ρ_ω represents the water density and ρ_a the air density.

3.1.2. Geometry and Boundary Conditions

The geometry domain and location of boundary conditions used in the 2D simulations can be seen in Figure 6, where it is important to highlight the presence of refinement zones near the water–air interface and inflation near the wall (bottom of the sea) to control parameters related to the turbulence model (κ-ω SST) and keep y^+ close to unity.

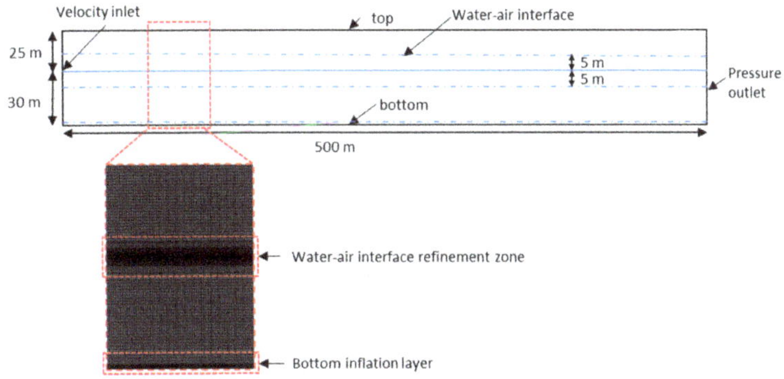

Figure 6. Numerical wave tank geometry and mesh refinement zones.

An inlet velocity method was applied to the model. A pressure outlet boundary condition was implemented at the far field boundary for wave absorption. Similar to the approaches used by Kim et al. [14], a damping zone featuring a numerical beach was established near the far field boundary to dissipate wave energy and minimise reflections within the simulation domain.

ANSYS Fluent code offers a static boundary method in the open channel flow and open channel wave boundary condition volume of fluid sub-models. The generation of incoming waves was achieved by defining a velocity inlet boundary condition coupled with an open channel wave boundary condition. This boundary method enabled the generation of the regular waves.

3.1.3. Simulation Parameters

Simulations were carried out using a transient approach. Within each time step, a minimum residual of 10^{-4} was imposed. Temporal discretisation is crucial for the quality of the solution in computational simulations. Large time step intervals can lead to wave damping, while small intervals significantly extend the computational time required to reach a solution. Therefore, it is essential to select the time step interval carefully.

Multiple studies have proposed recommended values for the optimal time step interval. Marques Machado et al. [15] concluded that the optimal time step is reached by dividing the wave period by 50 intervals (T/50). This conclusion aligns with the recommendation from another work [16] that suggests a maximum interval of T/40. However, other researchers [17,18] have suggested using even smaller time step values, recommending intervals equal to or less than T/100. A time step interval of T/100 was chosen for all the sea states.

A summary of the numerical parameters used in the setup for the 2D numerical wave tank can be found in Table 2.

Table 2. Simulation setup parameters for the numerical wave tank.

Parameter	Value
Solver	Pressure-based
Time Formulation	Transient
Turbulence model	$\kappa - \omega$ SST
Scheme	PISO
Multiphase model	VOF
Time step (s)	T/100
Number of time steps	10 Wave Period
Max. iterations per time step	100

3.1.4. Computational Mesh

A multizone method was used to create a structured mesh, in which the domain is divided into quadrilateral elements. A central zone with a finer mesh was established to improve the resolution of steep gradients at the water–air interface (refer to Figure 6). This refinement was accomplished using the edge sizing feature, allowing for increased divisions within the targeted area.

The computational mesh was optimised taking into consideration both the highest wavelength and wave height found in the metocean dataset. Various criteria have been proposed to determine the appropriate mesh size for simulations. Among these, Havn [18] suggests several criteria, which are summarized in Table 3.

Table 3. Grid size parameters for mesh refinement in the numerical wave tank.

Parameter	Value
Cells per wavelength	$\frac{\lambda}{\Delta x} > 100$
Cells per wave amplitude	$\frac{a}{\Delta z} > 10$
Aspect Ratio	$\frac{\Delta x}{\Delta z} < 10$

Three different meshes were created to evaluate the influence of mesh dimensions on the study's results, and they are described in Table 4. These meshes follow the recommendations suggested by Haven [18]. Apart from differences in their dimensions, all the constructed grids share qualitative features similar to those shown in Figure 6.

Table 4. Grid parameters for different mesh configurations in the numerical wave tank.

Mesh ID	Number of Divisions	Δx [m]	$\frac{\lambda}{\Delta x}$	Number of Division Interface z	Δz [m]	$\frac{a}{\Delta z}$
1	525	0.952	131.3	25	0.4	8.41
2	800	0.625	200.08	31	0.322	10.45
3	1200	0.417	299.89	50	0.2	16.825

Number of Divisions represents the total number of grid cells along the x-direction. Δx [m] is the grid spacing in the x-direction, and λ/Δx indicates the ratio of wavelength to grid spacing, reflecting the resolution for capturing wave propagation. Number of Division Interface z refers to the number of grid cells along the z-direction at the water–air interface. Δz [m] is the grid spacing in the z-direction, and a/Δz denotes the ratio of wave amplitude to grid spacing, which determines the accuracy of wave height representation.

The numerical results are validated by comparing the surface elevation of the waves, generated within the numerical wave tank, against the theoretical profiles derived from Stokes' second-order wave theory. The analysis focuses on specific locations within the tank, namely at x = λ and x = 2λ, where λ represents the wavelength.

The results from three different mesh resolutions are compared in Figures 7 and 8, revealing that the surface elevations across all meshes are similar. The discrepancies observed between the numerically simulated and theoretically predicted surface elevations can be attributed to some factors inherent to the numerical modelling of wave dynamics. One notable factor is the interaction of waves with the tank boundaries, particularly given the significant ratio of the wavelength to the tank length of the wave simulated.

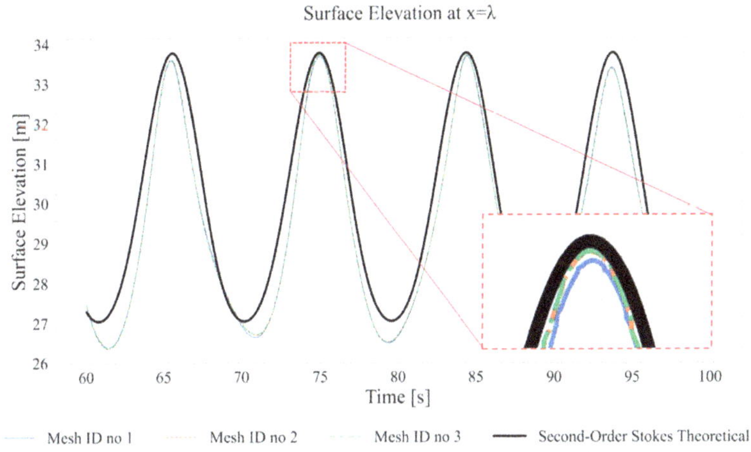

Figure 7. Free surface elevation comparison between numerical and theoretical wave at x = λ.

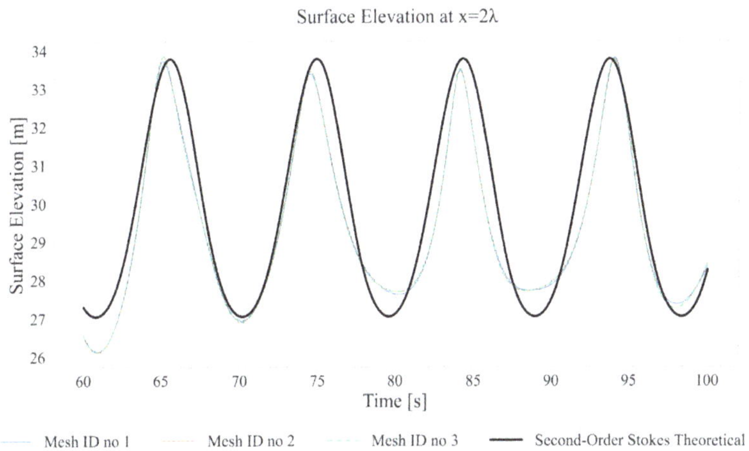

Figure 8. Free surface elevation comparison between numerical and theoretical wave at x = 2 λ.

3.2. Three-Dimensional Hydrodynamic Model: Pressure Distribution and FSI

The velocity profiles obtained in the previous numerical wave tank are used to set the velocity inlet for the 3D restricted domain. This restricted domain reduced the computational cost. The 3D domain has a limited dimension and is totally submerged, without the need for the high-demanding VOF multiphase model.

3.2.1. Computational Domain and Boundary Conditions

A computational domain with the shape of a cube that encloses the lander structure was created. The dimensions of the domain are $5 \times 5 \times 5$ m (in width $\times$ depth $\times$ height). An inlet boundary condition was applied to two faces and a pressure outlet was applied to the other two faces of the domain. A symmetry boundary condition was applied on top, and a wall boundary condition was applied to the bottom. The boundary conditions and the spatial arrangement of the lander structure within the 3D computational domain are represented in Figure 9.

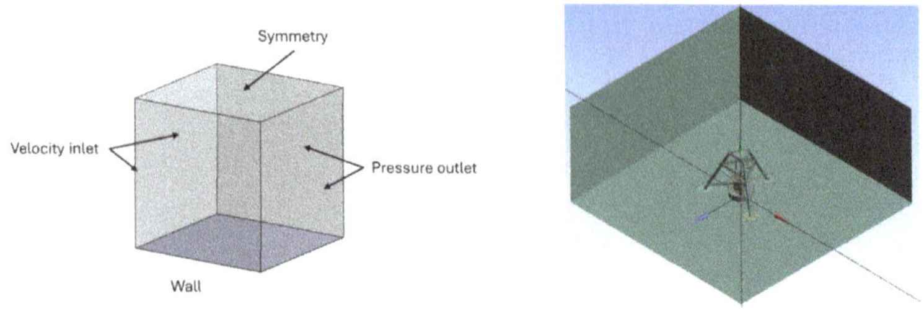

(a) (b)

Figure 9. Three-dimensional computational domain layout. (**a**) Applied boundary conditions, (**b**) lander structure placed at the centre of the domain.

3.2.2. Grid Generation and Grid Independence Study

Grids for the CFD domain were created using the Poly-Hexcore Mosaic meshing technology developed by ANSYS. Mosaic meshing in Ansys Fluent improves CFD simulations by combining various mesh types to optimise both mesh quality and computational efficiency. It integrates poly-hexcore meshes in the bulk region with high-quality layered poly-prism meshes near boundaries, thereby reducing cell count and increasing solver speed [19–21].

In Figure 10, the implementation of Mosaic technology in the 3D mesh generation of the domain near the lander structure is shown, demonstrating the use of octree hexahedral elements within the bulk region (a), isotropic poly-prisms in the boundary layer (b) and mosaic polyhedral elements facilitating the transition between these zones (c).

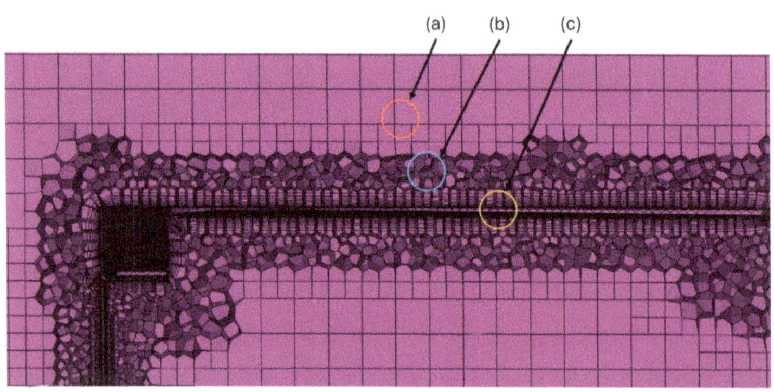

Figure 10. Mesh refinement in the vicinity of the interfaces. (**a**) Octree hexahedral elements used for bulk meshing; (**b**) isotropic poly-prism elements applied near wall boundaries for accurate boundary layer resolution; (**c**) mosaic polyhedral elements employed for efficient and smooth transition between regions, optimising both accuracy and computational cost in the simulation.

The complexity of the geometry posed a challenge to the generation of a fully structured mesh. Local size controls were applied within bodies of influence to achieve a finer mesh around the lander structure. Figure 11 demonstrates how the body of influence enables a seamless transition from the coarser mesh in the outer domain to the finer mesh near the lander structure.

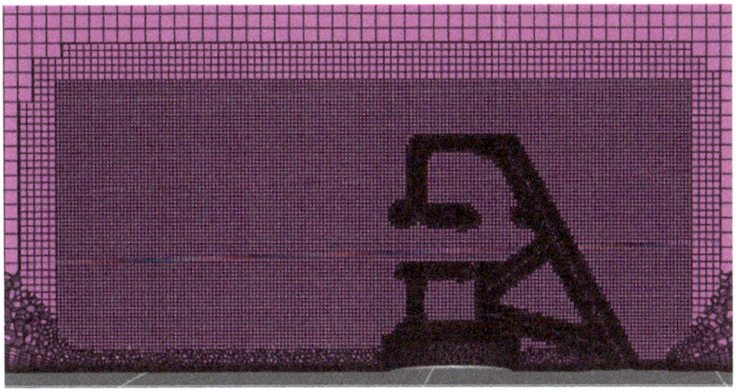

Figure 11. Application of a body of influence for mesh refinement.

The resultant pressure and velocity values from the CFD simulations serve as input parameters of the structural study, which implies that the CFD results are required to be accurate. With this in mind, a study regarding grid convergence was performed. The parameter used for the analysis was the drag force on the direction of the flow.

Five meshes with different degrees of refinement were created. Table 5 shows the drag force versus the number of elements for each created different mesh. The variation parameter regarding the most refined mesh $\varepsilon_i = \frac{\overline{P}_{coarser} - \overline{P}_{finer}}{\overline{P}_{finest}}$ was calculated and is shown in Table 5. The drag force was measured after 12 s of simulation.

Table 5. Different meshes created and results of mesh independence study.

Mesh ID	Number of Cells	Drag Force [N]	ε_i (%)
1	2,104,970	112.4	−9.36
2	4,436,569	125.2	−8.05
3	5,207,808	136.2	−0.29
4	5,511,176	136.6	−0.07
5	8,424,436	136.7	-

A relative difference in drag force of 0.29% between mesh number 3 and mesh number 4 and 0.07% between mesh number 4 and mesh number 5 was detected. Therefore, mesh number 4, comprising 5,511,176 cells, was chosen, supporting the flow dynamics simulation around the lander without the excessive computational demand of mesh number 5.

Mesh number 4 incorporates a body of influence with a target mesh size of 17 mm. Local sizing at the lander surface was adjusted to a maximum face size of 4 mm, and a boundary layer with 10 layers and a growth rate of 1.2 was added to capture detailed flow characteristics. A maximum skewness of 0.85 and an average skewness of 0.03 were obtained, as well as a minimum orthogonal quality of 0.15 and an average of 0.96.

3.2.3. Simulation Parameters

The turbulence model was the $\kappa - \omega$ method, with enhanced wall function. A mesh of about five million elements was used in all simulations. The time step used in the 3D model had to match the one selected in the numerical wave tank simulation for the corresponding wave under analysis. A summary of the numerical parameters used in the setup for the 3D simulations can be found in Table 6.

Table 6. Simulation setup parameters for the 3D domain simulations.

Parameter	Value
Solver	Pressure-based
Time Formulation	Transient
Turbulence model	$\kappa - \omega$ SST
Scheme	SIMPLE
Time Step (s)	T/100
Number of Time Steps	10 Wave Period
Max. Iterations per Time Step	100

The hydrodynamic analysis conducted via CFD produced pressure fields as the primary outputs. These pressure fields were subsequently applied to the structural model in a one-way FSI framework, serving as boundary conditions for the FEM simulation to evaluate the structural response to hydrodynamic loads. Figure 12 illustrates the pressure field imported from the CFD analysis, which is applied to the structural model for further analysis.

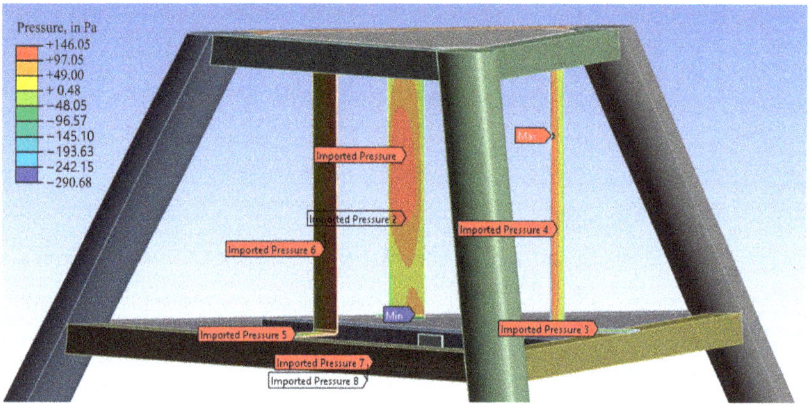

Figure 12. Imported pressure from the CFD analysis.

4. Preliminary Dataset Generation

To address the limitations posed by time-consuming FSI simulations, an alternative approach and a preliminary dataset of strains were created to generate enough data to develop and select the ML models. One major issue is ensuring data quality, as poor-quality data can lead to inaccurate or misleading model predictions [22].

The adopted approach uses a Finite Element Model of the structure to calculate the strains. Similar to the design validation, the loads to apply on the FEM model were calculated with the Morison equations. Figure 13 displays the locations where strain measurements were taken, highlighting the four dedicated plates and the specific points selected for strain calculations.

The approach computes the load for each sea state in the historical wave dataset using wave theories and the Morison equation. Water particle velocities were calculated at a height of one meter, corresponding to the strain gauge installation height. A steady current velocity of 0.5 m/s was included to account for ocean currents. The loads on each plate were computed using the Morison equation, with drag and added mass coefficients obtained from DNV-RP-C205 (Appendices E and D, respectively), which are suitable for thin plates perpendicular to the flow.

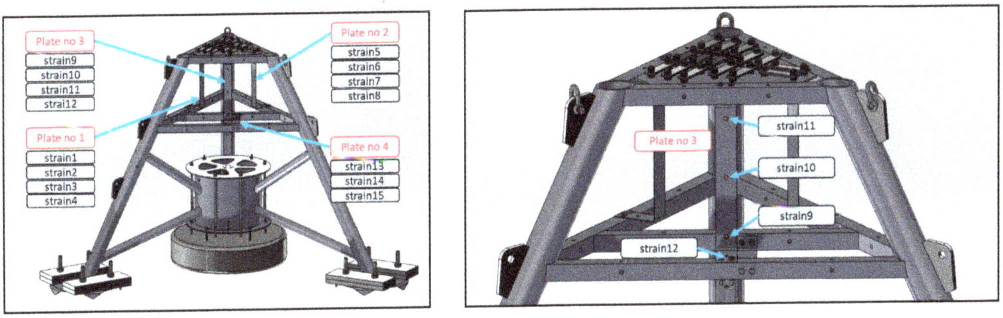

(a) (b)

Figure 13. Strain gauges' position on the lander structure; (**a**) general strain gauge position overview and plate identification; (**b**) detailed view of the strain gauges' positions for plate No 4.

Strain computations were then performed using a Finite Element Model (FEM) of the structure. Figure 14 illustrates the application of the loads on the dedicated plates, as well as the discretisation of the plates into finite elements for the structural analysis.

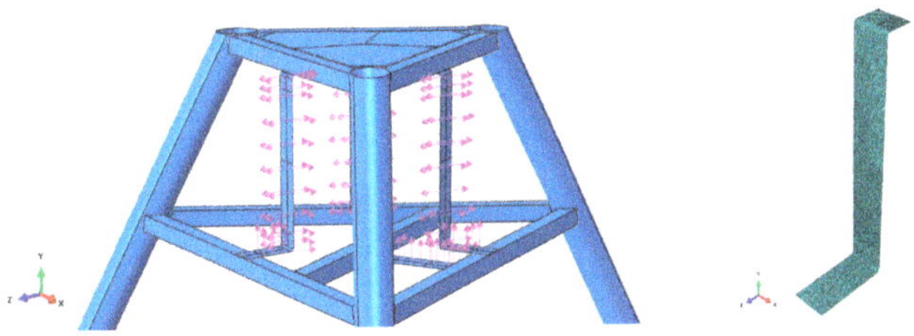

(a) (b)

Figure 14. Lander structure load and mesh details; (**a**) representation of the loads applied to the FEM model; (**b**) a general overview of the mesh used in the FEM simulations of a dedicated plate.

The resulting dataset includes strain values calculated at 15 points across four plates, corresponding to the locations where strain gauges were installed on the lander structure in the experimental study. Figure 15 shows the strains calculated at location 2 on plate 1 as a representative example.

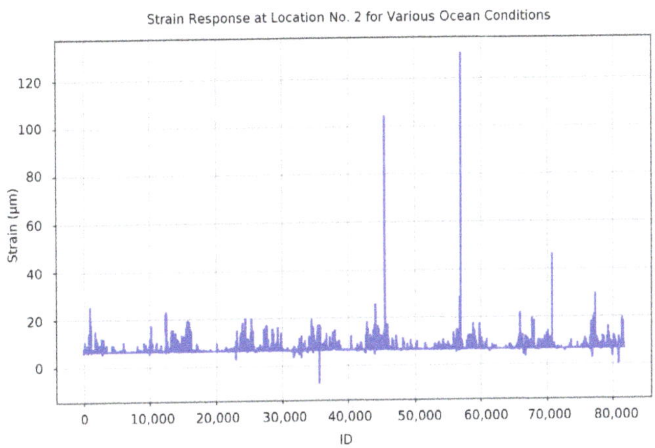

Figure 15. Strain response for all ocean conditions at location no. 2 on plate no. 1.

Strains were calculated for cases involving only the current effect and for the selected sea state, as shown in Tables 7 and 8. These results were compared with those derived from preliminary datasets using the FEM results. While the magnitudes differed, a correlation in strain behaviour was observed across both approaches. Specifically, both methods consistently identified high- and low-strain areas across the plates. Moreover, agreement was found in detecting compressive and tensile states along the vertical plates. However, this correlation did not extend to the horizontal plate, where the analytical and FEM methods failed to capture the expected strain behaviour.

Table 7. Strain comparison resulting from current effect.

Plate Number	Location ID	Preliminary Dataset Results	FSI Results
Plate 1	STRAIN 1	−4.1188	−4.9787
	STRAIN 2	6.1024	7.1612
	STRAIN 3	−5.7090	−7.2144
	STRAIN 4	1.5425	2.1410
Plate 2	STRAIN 5	2.0614	3.1670
	STRAIN 6	−3.0528	−4.6700
	STRAIN 7	2.8567	4.7764
	STRAIN 8	−0.7763	−1.2090
Plate 3	STRAIN 9	2.0603	2.7780
	STRAIN 10	−3.0485	−4.1100
	STRAIN 11	2.8517	4.2820
	STRAIN 12	−0.7830	−1.1960
Plate 4	STRAIN 13	−0.0156	0.1867
	STRAIN 14	−0.0084	−0.5300
	STRAIN 15	−0.0027	0.2605

Table 8. Strain comparison resulting from the selected ocean state.

Plate Number	Location ID	Preliminary Dataset Results	FSI Results
Plate 1	STRAIN 1	−13.6271	−52.478
	STRAIN 2	20.1919	75.074
	STRAIN 3	−18.8938	−77.058
	STRAIN 4	5.1253	24.583
Plate 2	STRAIN 5	25.3005	71.284
	STRAIN 6	−37.5133	−103.94
	STRAIN 7	35.0922	103.57
	STRAIN 8	−9.4765	−28.712
Plate 3	STRAIN 9	−11.6953	−47.745
	STRAIN 10	17.3130	70.118
	STRAIN 11	−16.1931	−71.221
	STRAIN 12	4.4529	17.861
Plate 4	STRAIN 13	0.0983	1.821
	STRAIN 14	0.0195	−1.020
	STRAIN 15	0.0314	1.403

5. AI Predictive Models

The machine learning algorithms developed for the prediction of forces to which the lander is subjected to are presented in this chapter.

In this study, various machine learning algorithms were evaluated to identify the most effective models for predicting structural damage based on wave parameters and strain data. Among the models tested, particular emphasis was placed on the Artificial

Neural Network Multilayer Perceptron (ANN MLP) and random forest algorithms due to their demonstrated proficiency in handling the complex and nonlinear data patterns of the strains and wave dataset. The ANN MLP is renowned for its ability to model intricate relationships through multiple layers of interconnected neurons, while the random forest algorithm excels in ensemble learning by constructing a multitude of decision trees for robust predictions [23,24].

The methodologies employed for data preprocessing, model training, validation and evaluation are detailed in the subsequent sections, providing a comprehensive overview of our analytical approach.

5.1. Dataset Description

The dataset consists of 81,724 samples. Each sample is composed of three inputs, namely Wave Height (WS_30) [m], Wave Period (THS_30) [s] and Direction (WAVE_THTP) [°]; and fifteen outputs corresponding to the fifteen different strains calculated along the lander. This way, an 81,724 × 18 matrix has been imported for machine learning development.

A Pearson correlation analysis was conducted, resulting in the correlation matrix shown in Figure 16. This matrix provides insight into the linear relationship between pairs of variables. Each value represents the degree of correlation between two variables, ranging from −1 to 1. A value of −1 indicates a perfect inverse correlation, 1 indicates a perfect direct correlation, and 0 signifies no linear relationship. Along the diagonal, where each variable is compared to itself, the correlation is always 1.

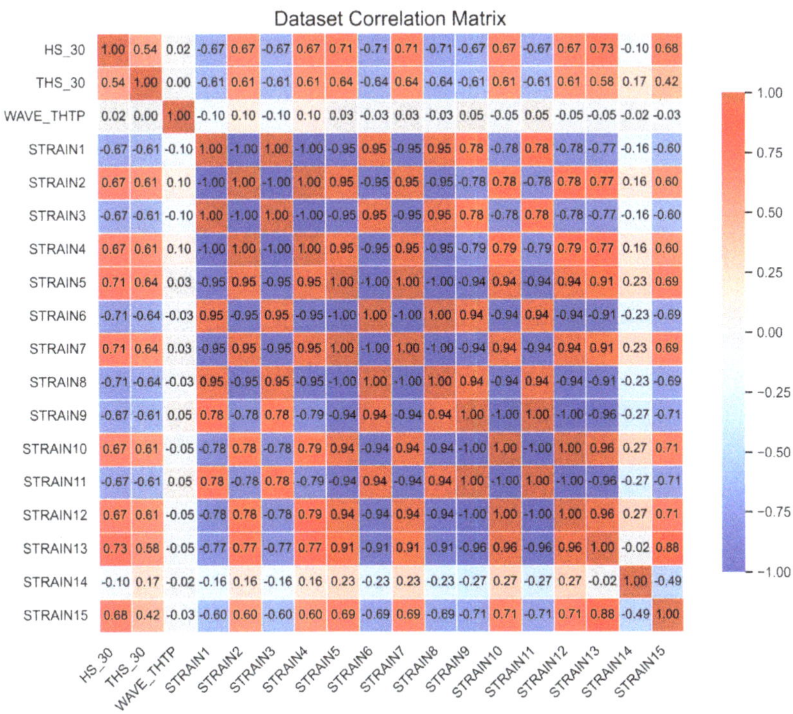

Figure 16. Pearson's correlation matrix.

Variables with an inverse correlation are marked in blue, while those with a direct correlation are shown in red. The analysis reveals that 9 out of 15 outputs exhibit full relation with other variables. For instance, strain 1 is inversely correlated with strain 2 and strain 4 but directly correlated with strain 3. Similarly, strain 5 and strain 9 are inversely correlated with strain 6 and strain 8, and strain 10 and strain 12, respectively, while being directly correlated with strain 7 and strain 11. These correlations have a physical basis, as illustrated in Figure 16. Each correlation group corresponds to one of the three plates depicted in Figure 13.

To analyse data distribution, the mean, standard deviation, minimum and maximum values, number of zeros and 25, 50, and 75 percentiles of all the variables have been computed. The data, shown in Table 9, indicate that for all variables, most of the measurements are close to the mean value of that variable while standard deviation and min/max values are considerably far away from that mean. That suggests that the sample behaves as a normal distribution, which has the majority of the samples next to the mean value and then the frequency of the data is reduced as it goes further from this value. This is coherent as the pattern of the inputs is replicated in the outputs. It can also be observed that the number of zeros in the outputs is null

Table 9. Statistical analysis of the strain and wave dataset.

Validation Set	Wave Height	Wave Period	Wave Direction	Strain 1	Strain 2
Mean	1.6695	6.7941	299.3296	−4.6397	6.8742
Std	0.8126	1.8251	22.9865	1.1155	1.6532
Min	0.2600	2.8000	1.0000	−87.9362	−7.6519
25%	1.0600	5.4000	293.0000	−4.6710	6.1025
50%	1.5200	6.6000	304.0000	−4.1736	6.1835
75%	2.1000	8.0000	311.0000	−4.1190	6.9199
Max	14.1500	18.6800	359.0000	5.1527	130.3429

5.2. Data Pre-Treatment

The dataset was divided into training (70%), validation (15%) and test (15%) sets using a binning process to ensure that each subset accurately represented the overall data distribution. This stratified division is essential for maintaining the statistical properties across all subsets, thereby improving the models' ability to generalise to new, unseen data. A visual representation of this dataset division is provided in Figure 17, while the statistical parameters for the training, testing and validation datasets are detailed in Tables 10–12, respectively.

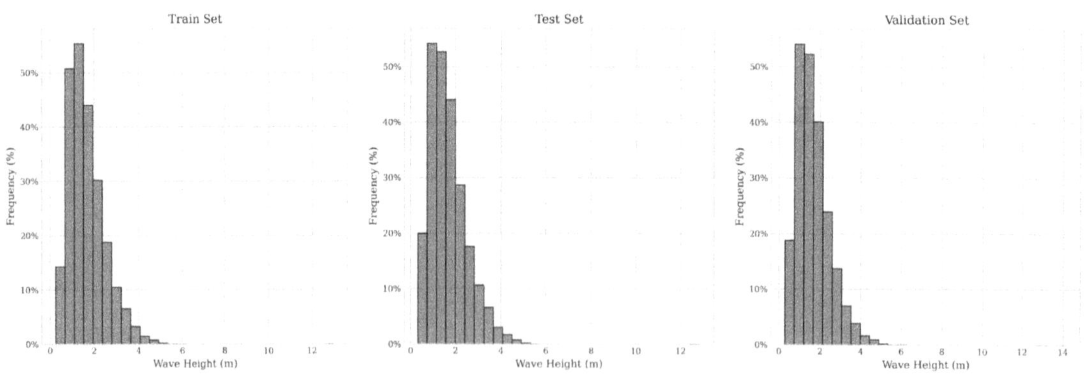

Figure 17. Histograms of the data splits for machine learning model development. (**a**) Training dataset, (**b**) testing dataset and (**c**) validation dataset.

Table 10. Statistical parameters for the training dataset.

Train Set	Wave Height (m)	Wave Period (s)	Wave Direction (°)
Mean	1.669	6.794	299.384
Std	0.811	1.823	28.862
Min	0.260	2.800	1.000
25%	1.060	5.400	293.000
50%	1.520	6.600	304.000
75%	2.100	8.000	311.000
Max	12.970	18.680	359.000
Count		57,206	

Table 11. Statistical parameters for the testing dataset.

Test Set	Wave Height (m)	Wave Period (s)	Wave Direction (°)
Mean	1.669	6.798	299.228
Std	0.817	1.833	22.149
Min	0.290	3.000	1.000
25%	1.060	5.400	293.000
50%	1.520	6.600	302.000
75%	2.100	8.000	311.000
Max	14.150	18.680	359.000
Count		12,259	

Table 12. Statistical parameters for the validation dataset.

Validation Set	Wave Height (m)	Wave Period (s)	Wave Direction (°)
Mean	1.669	6.798	299.227
Std	0.817	1.833	22.149
Min	0.290	3.000	1.000
25%	1.060	5.400	293.000
50%	1.520	6.600	302.000
75%	2.100	8.000	311.000
Max	14.150	13.000	359.000
Count		12,259	

Prior to executing the predictive algorithms, data normalisation was performed to reduce disparities in scale among the features. Normalisation is crucial because it adjusts the scales of input features, preventing variables with larger magnitudes from unduly influencing the model outcomes. This is especially important for algorithms like ANNs which are sensitive to the scale of input data due to the optimisation processes they employ [25].

Standard scaling was applied to the training dataset wave height and period, and the obtained scaling mechanism was then used on the test and validation sets. Standard scaling transforms features to have a mean of zero and a standard deviation of one. This transformation is beneficial for algorithms that assume data are normally distributed, helping improve model performance and stability [26]. The wave direction was split into sine and cosine components, ensuring their values range from -1 to 1, which helps maintain directional continuity and improves model input consistency.

5.3. Models

5.3.1. Random Forest Regressor

The Random forest algorithm (RFR) is an ensemble learning method that constructs multiple decision trees during training and outputs their aggregated prediction for regression tasks. This model enhances predictive accuracy and controls overfitting by leveraging the diversity of individual trees.

To optimise the performance of the RFR model, four hyperparameters were tuned:

- Number of Estimators: This parameter determines the number of decision trees in the forest. Increasing the number of estimators generally improves model performance by reducing variance but also increases computational cost.
- Maximum Tree Depth: Controlling the depth of the individual trees helps prevent overfitting by limiting the complexity of the model. Shallower trees reduce the risk of capturing noise in the data.
- Maximum Features: This parameter sets the number of features to consider when looking for the best split at each node. Introducing randomness in feature selection enhances model generalisation by reducing correlation among trees.
- Minimum Number of Samples Required to Split an Internal Node: This hyperparameter specifies the minimum number of samples needed to split an internal node. Increasing this value can prevent the model from capturing noise, thereby reducing overfitting and decreasing computational effort.

A grid search combined with cross-validation was employed to identify the optimal hyperparameter values.

The hyperparameter tuning process identified the optimal values, which are presented in Table 13.

Table 13. Hyperparameters' best-fit value for random forest regressor algorithm.

Hyperparameter	Best-Fit Value
Number of estimators	1000
Minimum samples split	2
Maximum features	1.0

The performance of the RFR model was evaluated on the test dataset. The following evaluation metrics were used [27]:

- Mean Squared Error (MSE): measures the average squared difference between the predicted and actual strain values.

$$\mathrm{MSE} = \frac{1}{n}\sum_{i=1}^{n}(y_i - \hat{y}_i)^2 \tag{7}$$

- Root Mean Squared Error (RMSE): provides the error magnitude in the same units as the output variable.

$$\mathrm{RMSE} = \sqrt{\frac{1}{n}\sum_{i=1}^{n}(y_i - \hat{y}_i)^2} \tag{8}$$

- Coefficient of Determination (R^2): indicates the proportion of variance in the strain measurements explained by the model.

$$R^2 = 1 - \frac{\mathrm{SSR}}{\mathrm{SST}} \tag{9}$$

where SSR and SST are the sum of squares regression and the sum of squares total, respectively, and are given by:

$$SSR = \sum_{i=1}^{n} (\hat{y}_i - \overline{y})^2 \quad (10)$$

$$SST = \sum_{i=1}^{n} (y_i - \overline{y})^2 \quad (11)$$

where y_i is the actual value (true value) of the dependent variable for the i-th observation, $\hat{y}_i$ is the predicted value of the dependent variable for the i-th observation, generated by the model, n is the total number of observations (data points) in the dataset and $\overline{y}$ is the mean of the actual values

Figure 18 compares the predicted strain values with the actual strains from the test set for the strain gauge position number 2.

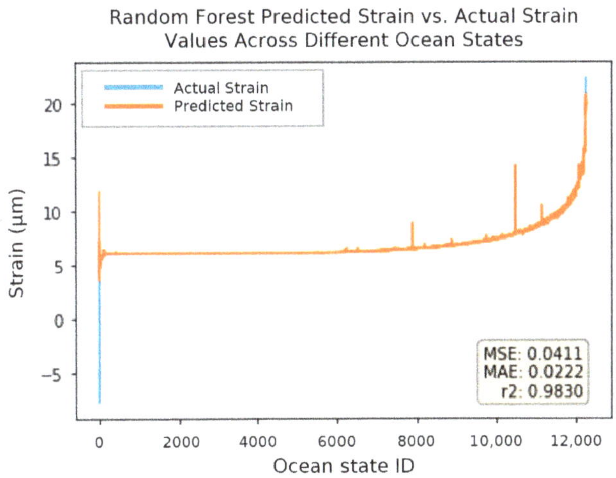

Figure 18. Comparison of the random forest regressor's predicted and actual strain values across various ocean states for strain gauge position number 2.

The model achieved an average R^2 score of 0.96 across all 15 strain outputs, demonstrating high predictive accuracy. The MSE and RMSE values were acceptably low, indicating that the model effectively captured the nonlinear relationships between the oceanographic parameters and the structural strains. The performance metrics of the decision tree regressor algorithm, with averaged results across all strain variables, are provided in Table 14.

Table 14. Performance metrics for decision tree regressor algorithm. Averaged results considering all strain variables.

Metric	Value
Global MSE	0.012
Global R^2	0.96
Computational Time	5.4 min

5.3.2. Artificial Neural Network: Multilayer Perceptron

Various architectures are typically developed for specific data types and tasks in artificial neural networks. Convolutional Neural Networks (CNNs) handle spatial data, making them ideal for image and video analyses. Recurrent Neural Networks (RNNs) excel in processing sequential data, such as time series or natural language. However, when it comes to generic regression problems, Multilayer Perceptrons (MLPs) often emerge as a superior choice. MLPs, with their layered structure and fully connected nodes, are good at modelling complex nonlinear relationships without the specialised data preprocessing required by CNNs or RNNs. This simplicity, coupled with their capability to approximate virtually any continuous function, makes MLPs particularly valuable for straightforward regression tasks where interpretability and computational efficiency are key [6]. An ANN is accurate if its predictions are equal to or extremely close to the actual outputs associated with a specific event. Predictions are evaluated using loss functions, and the algorithm optimises network weights and biases to minimise the loss function.

In this work, Mean Squared Error (MSE) was the most suited loss function for the problem. Mean Squared Error (MSE) is particularly effective as a loss function for regression because it emphasises larger errors more than smaller ones, which helps in reducing variance and improving the model's generalisation on unseen data. Additionally, minimising MSE during training directly correlates with maximising the R^2 score, as R^2 essentially measures the proportion of variance in the dependent variable that is predictable from the independent variables, and a lower MSE indicates a higher proportion of explained variance, thereby increasing the R^2 value.

To optimise the network's weights in such a way that the loss function is minimised, backpropagation is employed, a method that ensures the network learns effectively from its errors. As previously explained, a forward pass process happens when training MLPs where input data are passed through the network, layer by layer, until they reach the output layer where the final predictions are made. Then, the error is quantified through the loss function, providing a starting point for adjustments to improve network accuracy.

MLPs are flexible neural networks with several parameters that can be adjusted. It is essential to identify the optimal parameters for a specific dataset using the validation dataset. Optuna was employed to optimise the hyperparameters of the neural network model. The optimisation process considered variables such as the number of layers, units per layer, activation functions, dropout rates, optimiser types, learning rates, epochs and batch sizes. By dynamically adjusting these hyperparameters, Optuna determined the optimal values that effectively minimised the validation loss on the validation dataset. This automated approach facilitated efficient exploration of the hyperparameter space, enhancing model performance without the need for manual tuning [28].

As detailed in Table 15, the best-performing MLP model included deeper layers, a change in activation function and a different optimizer, resulting in significantly improved performance metrics, including a lower RMSE and a better R^2 score.

Table 15. Hyperparameters and results from the GridSearch optimisation and the best-performing MLP model.

	Best MLP
N° Hidden Layers	2
First Hidden Layer	831
Second Hidden Layer	978
Epochs	227,204
Activation Function	Leaky ReLU
Optimisation Algorithm	Adam
Learning Rate	0.008480
Batch size	645
Global RMSE	0.00601
Global R^2	0.99957

Figure 19 compares the predicted strain values with the actual strains from the test set.

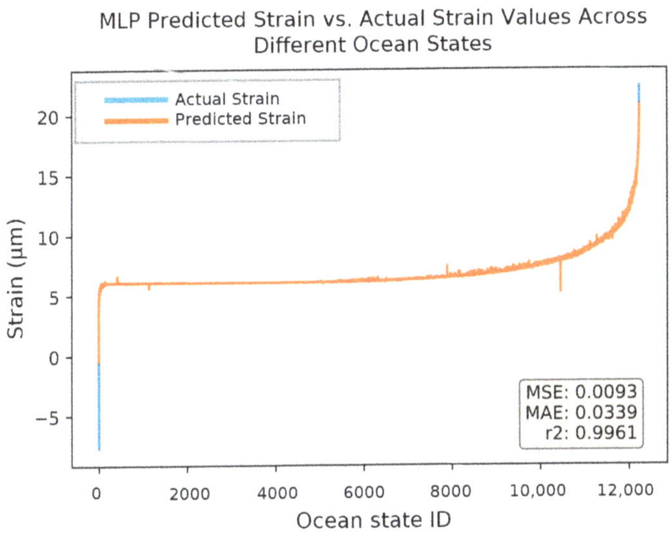

Figure 19. Comparison of predicted and actual strain values from the MLP model across various ocean states for strain gauge position number 2.

6. Discussion

This study aimed to develop a machine learning model for predicting structural damage in submerged structures due to ocean waves and currents. By combining numerical simulations and machine learning (ML) techniques, the effectiveness of these approaches for Structural Health Monitoring (SHM) applications was assessed. The goal was to generate reliable models that could predict critical structural responses, such as stress and strain, under marine conditions and the process of generating reliable data for training purposes.

When comparing strain predictions between current-only scenarios and specific sea states, the correlation between the analytical and FEM method and the FSI simulations demonstrates general consistency in the high- and low-strain regions. This agreement is crucial, as it indicates that the analytical and FEM method, despite their simplifications, capture key trends in the structural response. However, the discrepancies observed on the horizontal plates highlight limitations in the analytical approach, mainly due to assumptions made about load distribution.

The two proposed machine learning approaches presented very good results and are detailed in Table 16. This table provides a direct comparison of the models, illustrating their performance in terms of accuracy, computational efficiency and complexity.

The performance metrics of the evaluated machine learning models highlight trade-offs between accuracy, computational efficiency and the ability to handle different data patterns. Random forest regressors can capture complex patterns, offering high accuracy with a low computation time. Finally, the MLP delivered the highest accuracy but required significant computational resources (2 days) and an even higher development time. These results underscore the importance of selecting models that align with the complexity and patterns of the available data, ensuring that the chosen algorithms can capture the underlying relationships while maintaining computational feasibility in AI development for SHM.

Table 16. Performance comparison of machine learning algorithms studied.

Algorithm	R^2	RMSE	Training Computational Time	Hyperparameters
Random Forest	0.960	0.012	5.4 min	Number of estimators, maximum depth, minimum samples split, maximum features
MLP	0.995	0.011	2 days	Loss functions, optimization algorithms, learning rate, number of layers and neurons, among others

The reliability of machine learning models is fundamentally dependent on the quality and diversity of the data used during training. While empirical methods, such as real-scale experiments, provide the most realistic and accurate data, their high cost and logistical challenges often limit their use in extensive studies. In this study, numerical methods, including CFD and FEM simulations, were used to generate large volumes of training data efficiently. These methods offer a viable alternative, allowing the modelling of complex systems that reduce reliance on physical experiments. However, the importance of empirical validation cannot be overstated.

Future work will focus on the analysis of the experimental data collected from the benthic lander, which will be essential for validating the machine learning models originally trained with numerical data. This validation is essential to confirm the models' ability to predict structural responses under real-world marine conditions. Subsequently, a new machine learning model will be developed and trained using the experimental data in order to enhance the model's predictive accuracy. In parallel, further work is necessary to optimise the computational domain for future Fluid–Structure Interaction (FSI) simulations, balancing accuracy with computational efficiency.

While the framework proposed in this study was designed for a specific structure within the Marewind project, improving the generalizability of SHM models remains an important challenge. Future research is needed to develop methodologies to expand the applicability of these models to diverse structures and operational conditions, ensuring their robustness and adaptability.

Author Contributions: A.B.d.S. and H.M.V.: Conceptualization, methodology, validation, investigation, writing—original draft preparation. R.F.F.L., M.L.P.P., M.T. and A.M.S.C.: Conceptualization. S.D., P.J.S.C.P.S. and T.M.R.M.D.: Conceptualization, writing—review and editing, supervision. P.M.G.P.M.: Project administration. All authors have read and agreed to the published version of the manuscript.

Funding: This work was developed in the scope of the project "MAREWIND—Materials solutions for cost Reduction and Extended service life on WIND offshore facilities", which has received funding from the European Union's Horizon 2020 research and innovation programme under grant agreement N° 952960. The author Rogério F. F. Lopes gratefully acknowledges the "Fundação para a Ciência e a Tecnologia" (FCT) for the funding of the PhD scholarship 2022.13216.BD.

Data Availability Statement: Restrictions apply to the availability of these data. Data were obtained from Hydrographic Institute of Portugal and are available upon request directly to the Hydrographic Institute of Portugal.

Acknowledgments: The authors would like to thank the Hydrographic Institute of Portugal for their invaluable support in the experimental work, including the deployment and retrieval of the benthic lander, as well as their contribution of resources, which greatly enhanced the success of the experimental work.

Conflicts of Interest: The authors declare no conflicts of interest.

References

1. Pengfei, L.; Dong, X. 1—AE health monitoring technique for composite wind turbine blade: A state-of-art review. In *Acoustic Emission Signal Analysis and Damage Mode Identification of Composite Wind Turbine Blades*; Liu, P., Ed.; Elsevier: Amsterdam, The Netherlands, 2023; pp. 1–37. Available online: https://www.sciencedirect.com/science/article/pii/B9780323886529000042 (accessed on 29 May 2024).
2. Silva-Campillo, A.; Pérez-Arribas, F.; Suárez-Bermejo, J.C. Health-Monitoring Systems for Marine Structures: A Review. *Sensors* **2023**, *23*, 2099. [CrossRef] [PubMed]
3. Jiang, K.; Han, Q.; Du, X.; Ni, P. Structural dynamic response reconstruction and virtual sensing using a sequence to sequence modeling with attention mechanism. *Autom. Constr.* **2021**, *131*, 103895. [CrossRef]
4. Dasari, K.; Dogra, A. Designing a Virtual Twin for Structural Health Monitoring by Integrating BIM and Digital Twin Framework. In *Low Carbon Materials and Technologies for a Sustainable and Resilient Infrastructure*; Pancharathi, R.K., K.Y. Leung, C., Chandra Kishen, J.M., Eds.; Springer Nature: Singapore, 2024; pp. 263–278.
5. Van Rijn, L.C. Unified View of Sediment Transport by Currents and Waves. I: Initiation of Motion, Bed Roughness, and Bed-Load Transport. *J. Hydraul. Eng.* **2007**, *133*, 649–667. [CrossRef]
6. Zhang, N.; Chang, Y.; Shi, J.; Chen, G.; Zhang, S.; Cai, B. Fragility assessment approach of deepwater drilling risers subject to harsh environments using Bayesian regularization artificial neural network. *Ocean Eng.* **2021**, *225*, 108793. [CrossRef]
7. Giurgiutiu, V. 1—Introduction. In *Structural Health Monitoring*; Giurgiutiu, V., Ed.; Academic Press: Burlington, NJ, USA, 2008; pp. 1–11. Available online: https://www.sciencedirect.com/science/article/pii/B9780120887606500027 (accessed on 29 May 2024).
8. Karadeniz, H.; Saka, M.P.; Togan, V. Water Wave Theories and Wave Loads. In *Stochastic Analysis of Offshore Steel Structures: An Analytical Appraisal*; Karadeniz, H., Ed.; Springer: London, UK, 2013; pp. 177–252. [CrossRef]
9. Mehaute, B.L. *An Introduction to Hydrodynamics and Water Waves*; Springer Science & Business Media: Berlin/Heidelberg, Germany, 2013.
10. OpenStreetMap Contributors. OpenStreetMap 2024. Available online: https://www.openstreetmap.org (accessed on 2 December 2024).
11. Chakrabarti, S.K. *Hydrodynamics of Offshore Structures*, 3rd ed.; revised reprint; Computational Mechanics: Plainfield, IL, USA, 1987.
12. Clauss, G.; Lehmann, E.; Östergaard, C. *Offshore Structures*; Springer: London, UK, 1992. Available online: http://link.springer.com/10.1007/978-1-4471-3193-9 (accessed on 11 June 2024).
13. Galdi, G.P. An Introduction to the Navier-Stokes Initial-Boundary Value Problem. In *Fundamental Directions in Mathematical Fluid Mechanics*; Galdi, G.P., Heywood, J.G., Rannacher, R., Eds.; Birkhäuser: Basel, Switzerland, 2000; pp. 1–70. [CrossRef]
14. Kim, S.-Y.; Kim, K.-M.; Park, J.-C.; Jeon, G.-M.; Chun, H.-H. Numerical simulation of wave and current interaction with a fixed offshore substructure. *Int. J. Nav. Archit. Ocean Eng.* **2016**, *8*, 188–197. [CrossRef]
15. Marques Machado, F.M.; Gameiro Lopes, A.M.; Ferreira, A.D. Numerical simulation of regular waves: Optimization of a numerical wave tank. *Ocean Eng.* **2018**, *170*, 89–99. [CrossRef]
16. Ning, D.Z.; Teng, B. Numerical simulation of fully nonlinear irregular wave tank in three dimension. *Int. J. Numer. Methods Fluids* **2007**, *53*, 1847–1862. [CrossRef]
17. Silva, M.C.; Vitola, M.; Pinto, W.T.; Levi, C. Numerical simulation of monochromatic wave generated in laboratory: Validation of a cfd code. In Proceedings of the 23 Congresso Nacional de Transport Aquaviario Construcao naval Offshore, Rio de Janeiro, Brazil, 25–29 October 2010; pp. 1–12.
18. Havn, J. Wave Loads on Underwater Protection Covers. 2011. Available online: https://ntnuopen.ntnu.no/ntnu-xmlui/handle/11250/238072 (accessed on 17 April 2024).
19. Zore, K.; Sasanapuri, B.; Parkhi, G.; Varghese, A. Ansys mosaic poly-hexcore mesh for high-lift aircraft configuration. In Proceedings of the 21st AeSI Annual CFD Symposium, Bangalore, India, 8–9 August 2019.
20. Ansys Mosaic Technology Combines Disparate Meshes. Available online: https://www.ansys.com/resource-center/white-paper/ansys-fluent-mosaic-technology (accessed on 18 April 2024).
21. Arocena, V.M.; Danao, L.A.M. Improving the Modeling of Pressure Pulsation and Cavitation Prediction in a Double-Volute Double-Suction Pump Using Mosaic Meshing Technology. *Processes* **2023**, *11*, 660. [CrossRef]

22. Aldoseri, A.; Al-Khalifa, K.N.; Hamouda, A.M. Re-Thinking Data Strategy and Integration for Artificial Intelligence: Concepts, Opportunities, and Challenges. *Appl. Sci.* **2023**, *13*, 7082. [CrossRef]
23. Goodfellow, I. *Deep Learning*; MIT Press: Cambridge, MA, USA, 2016.
24. Breiman, L. Random forests. *Mach. Learn.* **2001**, *45*, 5–32. [CrossRef]
25. Cabello-Solorzano, K.; Ortigosa de Araujo, I.; Peña, M.; Correia, L.; Tallón-Ballesteros, A.J. The Impact of Data Normalization on the Accuracy of Machine Learning Algorithms: A Comparative Analysis. In Proceedings of the 18th International Conference on Soft Computing Models in Industrial and Environmental Applications (SOCO 2023), Salamanca, Spain, 5–7 September 2023; García Bringas, P., Pérez García, H., Martínez de Pisón, F.J., Martínez Álvarez, F., Troncoso Lora, A., Herrero, Á., Calvo Rolle, J.L., Quintián, H., Corchado, E., Eds.; Springer Nature: Cham, Switzerland, 2023; pp. 344–353.
26. sklearn.preprocessing.StandardScaler. Scikit-Learn. Available online: https://scikit-learn.org/stable/modules/generated/sklearn.preprocessing.StandardScaler.html (accessed on 19 April 2024).
27. 3.4. Metrics and scoring: Quantifying the Quality of Predictions. Scikit-Learn. Available online: https://scikit-learn.org/stable/modules/model_evaluation.html (accessed on 28 November 2024).
28. Optuna—A Hyperparameter Optimization Framework. Optuna. Available online: https://optuna.org/ (accessed on 21 October 2024).

fluids

MDPI

Review

Literature Review on Single and Twin-Screw Extruders Design for Polymerization Using CFD Simulation

Elham Delvar [1,2], Inês Oliveira [1,2], Margarida S. C. A. Brito [1,2,*], Cláudia G. Silva [1,2], Arantzazu Santamaria-Echart [3], Maria-Filomena Barreiro [3] and Ricardo J. Santos [1,2,*]

1 Associate Laboratory in Chemical Engineering (ALiCE), Faculty of Engineering, University of Porto, 4200-465 Porto, Portugal; up202111358@edu.fe.up.pt (E.D.); up201906077@edu.fe.up.pt (I.O.); cgsilva@fe.up.pt (C.G.S.)
2 Laboratory of Separation and Reaction Engineering—Laboratory of Catalysis and Materials (LSRE-LCM), Faculty of Engineering, University of Porto, 4200-465 Porto, Portugal
3 CIMO, LA SusTEC, Instituto Politécnico de Bragança, Campus de Santa Apolónia, 5300-253 Bragança, Portugal; asantamaria@ipb.pt (A.S.-E.); barreiro@ipb.pt (M.-F.B.)
* Correspondence: mbrito@fe.up.pt (M.S.C.A.B.); rsantos@fe.up.pt (R.J.S.)

Abstract: This work presents a comprehensive review of the evolution in modeling reactive extrusion (REx), tracing developments from early analytical models to advanced computational fluid dynamics (CFD) simulations. Additionally, it highlights the key challenges and future directions in this field. Analytical models to describe the velocity profiles were proposed in the 1950s, involving certain geometrical simplifications. However, numerical models of melt polymeric flow in extruders have proven to be crucial for optimizing screw design and predicting process characteristics. The state-of-the-art CFD models for single and twin-screw extruders design address the impact of geometry (type of mixing elements and geometrical simplifications of CFD geometries), pressure and temperature gradients, and quantification of mixing. Despite the extensive work conducted, modeling reactive extrusion using CFD remains challenging due to the intricate interplay of mixing, heat transfer, chemical reactions, and non-Newtonian fluid behavior under high shear and temperature gradients. These challenges are further intensified by the presence of multiphase flows and the complexity of extruder geometries. Future advancements should enhance simulation accuracy, incorporate multiphase flow models, and utilize real-time sensor data for adaptive modeling approaches.

Keywords: reactive extrusion process; CFD simulation; polymerization; extruder designing

1. Introduction

Reactive extrusion is a fundamental process in polymer manufacturing where the material is forced to flow through a flow restriction, a die; a polymerization reaction takes place during the extrusion process. This process involves the polymerization, modification, or functionalization of polymers within the extruder. Reactive extrusion is widely used to modify the polymer molecular structure through crosslinking and grafting, to add reinforcements (fibers or fillers) to polymers during extrusion, or to promote the reaction between monomers to produce the final polymer. This technique is used to develop advanced materials, including high-performance composites, biodegradable plastics, and specialty polymers with enhanced chemical or mechanical properties.

The extruder consists of a hollow cylindrical barrel, where a screw is inserted and where the material is continuously dragged to the outlet through a die. Thus, the screw is the most important component of an extruder and is divided into three main geometric

 https://doi.org/10.3390/fluids10010009

zones: feeding, compressing, and mixing zones. These screw elements in the extruders facilitate fluid conveyance such as feeding, mixing, and compartmentalization of reactants into different zones [1,2]. The reactants must be homogenized, and the reactions must have enough time to complete before the process is complete. At the barrel's inlet, the reactants are fed into the extruder. In a standard extruder, the reactants are injected as a fluid or fed as solids in a hopper and at a specific stoichiometric ratio [3,4].

Over the years, Computational Fluid Dynamics (CFD) has emerged as a powerful tool for predicting and understanding the flow of polymer melting in extruders. It allows researchers to optimize their performance under different operating conditions without the need for expensive experimentation, saving material and energy costs. CFD simulations in REx facilitate the analysis of the complex interactions between fluid flow, heat transfer, and chemical reactions. The fluid flow affects the efficiency of mixing between reactants, and thus, CFD simulations enable the visualization of velocity fields, shear rates, and mixing patterns inside the extruder. These studies enable to optimize the screw design and geometry to enhance mixing and to ensure the uniform gradients of temperature and mass fraction of reactants. Since REx involves exothermic or endothermic reactions, the analysis of temperature gradients through CFD is crucial. CFD can predict the local overheating or underheating that can cause degradation and enable the identification of the best strategies to maintain a uniform temperature. Finally, CFD simulations enable the prediction of potential issues that arise during scale-up related to mixing or incomplete reaction and adjust process parameters accordingly.

REx usually involves the flow of polymer melting and the conveying of solids since pre-polymers or monomers in the solid state are fed to the extruder. Wilczyński, Nastaj [5] introduced a novel continuous concept for global modeling coupling, Discrete Element Modeling (DEM) for solid conveying and CFD modeling for molten polymer flow. This new approach solves the conservation equations as a continuum for the entire system without the need to segment the process into different zones (solid conveying, melting, and melt conveying). This enables a more precise prediction of extrusion behavior. Lewandowski and Wilczyński [6] reviewed the modeling approaches for the specific case of twin-screw extruders. Their work presents a comprehensive global model encompassing solid transport, polymer plasticating, and the flow of molten polymer. The authors highlight that advancements in these models are largely driven by the integration of DEM/CFD modeling techniques.

This paper presents a comprehensive literature review tracing the evolution from analytical expressions to CFD models predicting flow behavior in single and twin-screw extruders. The review is structured as follows: (i) an overview of extruder types, including single and twin-screw extruders; (ii) analytical models proposed in the literature to describe the flow in extruders using geometrical simplifications; (iii) a detailed review of CFD simulations, focusing on extruder geometries, mixing quantification, pressure and temperature gradients, and extruder design. Finally, the paper discusses the current challenges in implementing CFD models for REx and outlines future directions in this field.

1.1. Single-Screw Extruder

Single-screw extruders are characterized by just one screw located in the barrel. This equipment is easily manufactured, has a straightforward design, and is mechanically reliable. It uses frictional stress to promote the transport of solid materials, such as polymer pellets and particle fillers, and viscous stress to transport melt materials [7].

The single-screw extruder is widely used for simple applications, such as melting, plasticizing, and discharging melt for the manufacture of films, pipes, and profiles.

The schematic representation of the single-screw extruder is depicted in Figure 1. The major section of the extrusion system involves a screw which is rotating inside a barrel.

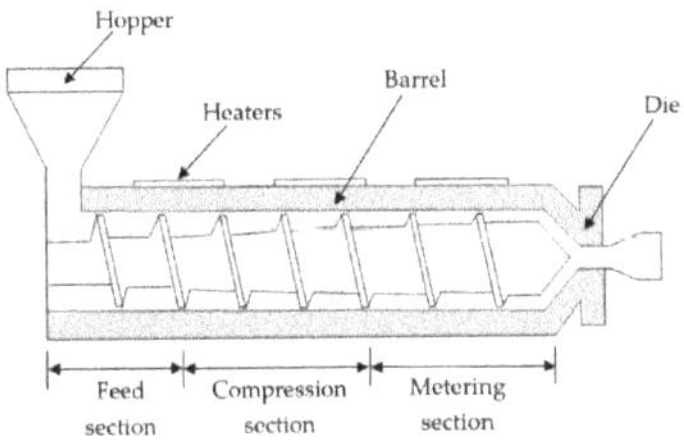

Figure 1. Schematic of components in Single-screw extruder.

1.2. *TwinScrew Extruder*

Twin-screw extruders are characterized by two screws located in the barrel. Twin-screw extruders are mainly used for processing heat-sensitive polymers, such as PVC, and in specialized applications like compounding, polymer filling, and reinforcement.

These devices can be categorized based on the screw rotation: a corotating extruder, when the screws turn in the same direction, counter-rotating extruder, when they rotate in opposite directions. Figure 2 shows a cross-section of two types of screws in the twin-screw extruders.

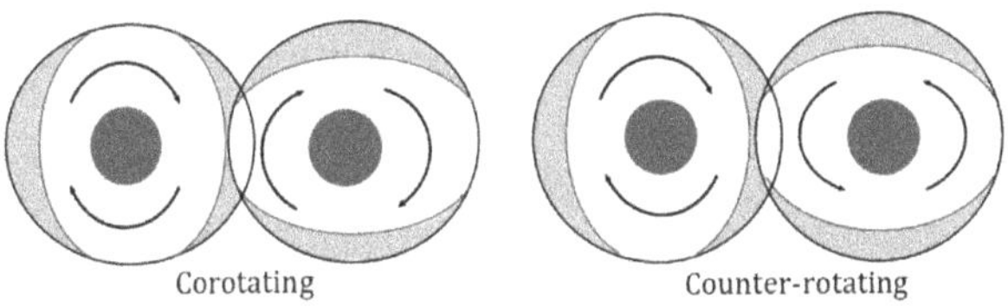

Figure 2. Cross section of corotating and counter-rotating extruder.

Furthermore, the distance between the screw shafts is also an important feature of screws. Non-intermeshing consists of tangential or separated screws, and intermeshing includes partial or full intermeshing, such as self-wiping screws. The screw configurations in co-rotating and counter-rotating extruders are shown in Figure 3.

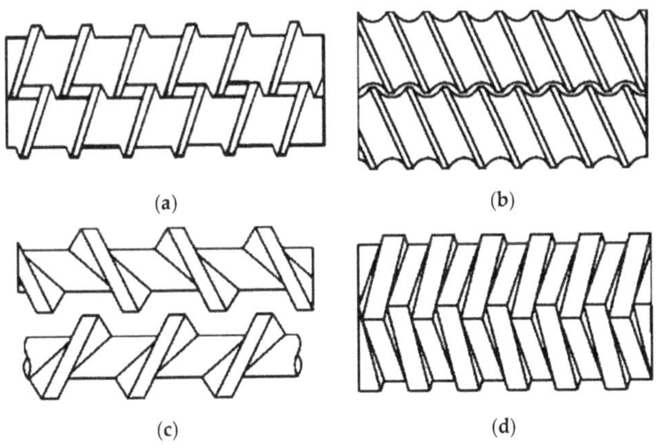

Figure 3. Intermeshing configuration for (**a**) partially intermeshing corotating, (**b**) closely intermeshing corotating, (**c**) tangential or non-intermeshing counter-rotating, and (**d**) closely intermeshing counter-rotating. Reprinted with permission from [8] (Permission reference 5899281196245).

Therefore, six types of twin-screw extruders can be used: tangential corotating, partially intermeshing corotating, closely intermeshing corotating, and the same types for counter-rotating extruders, respectively.

The applications of intermeshing corotating extruders depend on their speed. High-speed corotating extruders are used to add additives (colorants, fillers, reinforcements, stabilizers, etc.), devolatilization to remove solvents and reactive extrusion. Low-speed intermeshing corotating extruders are used to produce profiles and pipes. Non-intermeshing corotating extruders are not used in practice due to their poor mixing efficiency. Intermeshing counter-rotating extruders are used to compound PVC and other resin systems. Non-intermeshing counter-rotating extruders are efficient material conveyance and mixing, and therefore, they are used for devolatilization and chemical reactions.

1.3. Single-Screw Versus Twin-Screw Extruder

There are some differences between the processes occurring in single-screw and twin-screw extruders. The flow mechanism in twin-screw extruders differs significantly from those occurring in single-screw extruders. The complex flow patterns in twin-screw extruders enhance mixing and devolatilizing capabilities, heat transfer, and polymer melting efficiency. Moreover, the mixing mechanisms in twin-screw extruders are more efficient in intermeshing counter-rotating extruders and almost absent in non-intermeshing screws. Therefore, the twin-screw extruders are used in more complicated applications as homogenizers, dispersants of pigments and additives, reactive composition, concentration, evaporation, alloying, and polymerization [9].

On the other hand, in single-screw extrusion, the drag flow is predominant, and the transport efficiency is poor. This leads to poor mixing efficiency and a slow melting process.

Single and twin-screw extruders have different feeding processes. There are two types of feeding processes: gravity-fed (flood-feeding extrusion) or controlled by metering (starve-fed extrusion). Flood-feeding is typically used in single-screw extrusion where the feed hopper is filled, and the extruder screw rotation speed determines the feed rate. Starve feeding is typically used in twin-screw extrusion, where a feeder is used to meter the material into the system at a predetermined rate where rotation speed and throughput are independent of each other.

The implementation of twin-screw extruders for REx is more well established in the industry. Extensive control of residence time distribution and mixing and increased heat and mass transfer capability are the main reasons for this matter [10].

2. Analytical Models for Extruders

Both single and twin-screw extruders have complex geometries, making it difficult to analyze the flow in these devices. Analytical models have been proposed since the 1950s, introducing geometric conceptualizations that allowed for the derivation of analytical flow profiles. These models have been used as geometrical simplifications even for CFD models; therefore, it is crucial to present them here for better comprehension of the literature review on CFD for screw extruders.

2.1. Single-Screw Extruders

A base case for geometrical conceptualization is shown in Figure 4, which is the geometry of a typical metering section of a single-screw extruder. The main design parameters are the channel width, W, the flight width, e, the barrel diameter, D, the metering depth, H, the flight clearance, δ, and the helix angle, θ.

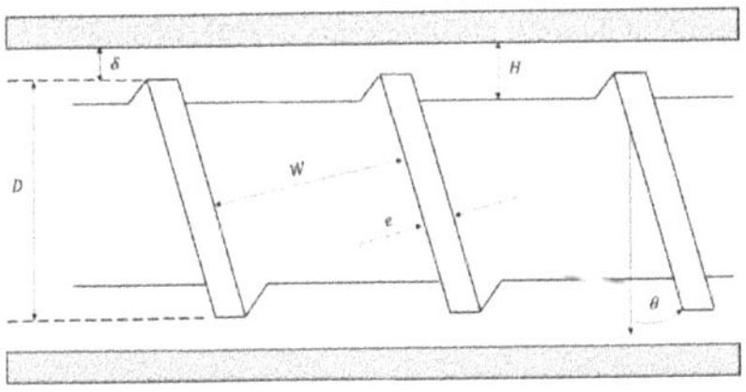

Figure 4. Geometry of a typical metering section of a single-screw extruder.

The melt flows within single-screw extruders were pioneered by Rowell and Finlayson [11] through an examination of flow dynamics in a "screw pump" for viscous oils. This analysis is based on drag and pressure flow concepts for an isothermal Newtonian fluid. Later, Maillefer [12] and Carley, Mallouk [13] applied this analysis to the extrusion of polymers, where the flow was analyzed by simplifying the screw geometry to a rectangular channel, which is known as the flat plate approximation. So, the screw should be visualized as uncoiled and laid flat, as it is represented in Figure 5, as well as the respective coordinate system.

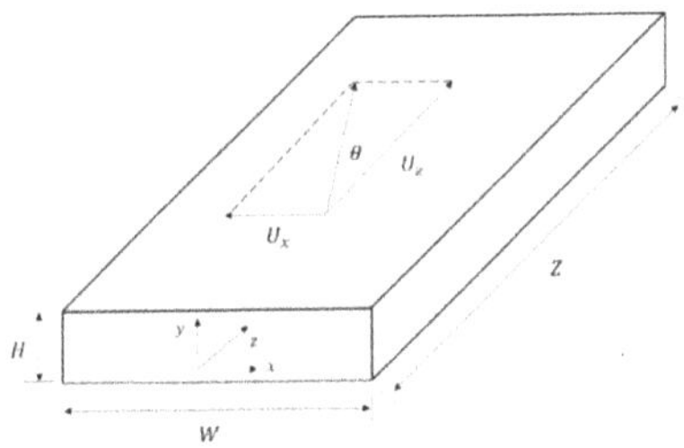

Figure 5. Unwound screw channel with the respective coordinate system.

If the channel's depth is significantly smaller than the outer diameter of the barrel, then the channel's curvature can be neglected. In this model, the side walls represent the leading and trailing surfaces of the screw flight, the upper surface corresponds to the screw root, and the bottom wall represents the inner surface of the barrel. The screw root moves across the channel at the screw's helix angle with a velocity $U = \pi DN$, where N is the screw rotational speed in rpm. The velocity U has two components of the screw's speed, U_x and U_z [14].

The formulation of the flows is based on several key assumptions: the flow is laminar and isothermal, the fluid is Newtonian and incompressible, and there is no slip at the walls [15]. The equations of motion, expressed in rectangular coordinates, assume that inertial terms are negligible in the steady state. Additionally, the equations are simplified using the lubrication approximation, meaning the velocities in the z-direction are fully developed and depend solely on the x and y coordinates. There is no y-component in the velocity terms, as its value is negligible for screws with a small channel depth-to-width ratio H/W. Therefore, the equations are given by:

$$0 = -\frac{\partial p}{\partial z} + \mu\left(\frac{\partial^2 u_z}{\partial x^2} + \frac{\partial^2 u_z}{\partial y^2}\right) \quad (1)$$

$$0 = -\frac{\partial p}{\partial x} + \mu\left(\frac{\partial^2 u_x}{\partial x^2} + \frac{\partial^2 u_x}{\partial y^2}\right) \quad (2)$$

The pressure gradient is constant and can be replaced by $\Delta p/Z$, where Δp is the pressure drop along the extruder and Z is the helical length. The dimension of the unwound screw, Z, is related to the axial length of the metering zone of the extruder, L, through the following:

$$Z = \frac{L}{\sin(\theta)}$$

In Equation (1), the boundary conditions include $u_z = 0$ at the screw root and $u_z = -U\cos(\theta)$ on the barrel surface. Carley, Mallouk [13] solved the resultant flow between two parallel plates in relative motion, while omitting the effect of the screw flights. Equation (3) represents the integrated solution:

$$u_z = U_z\frac{y}{H} - \frac{1}{2\mu}y(H-y)\frac{\Delta p}{Z} \quad (3)$$

where $U_z = U\cos(\theta)$ and μ is the molecular viscosity. The velocity profile in Equation (3) involves two independent solutions. The first term, $U_z\frac{y}{H}$, also known as the drag flow, does not depend on the viscosity of the fluid and represents the maximum pumping capacity of the extruder. In the drag flow, the velocity changes linearly throughout the depth of the channel. The latter term, $\frac{1}{2\mu}y(H-y)\frac{\Delta p}{Z}$, commonly referred to as the pressure flow, depends on the viscosity and results in a parabolic distribution. The pressure flow only exists if the flow is restricted at the outlet through the presence of a die. Combining the drag and pressure flows yields the net velocity at each point. These velocity profiles are shown in Figure 6.

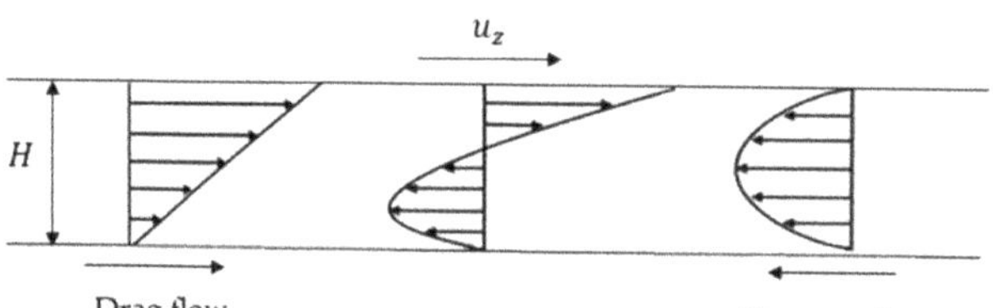

Figure 6. Velocity profiles in the parallel plane.

The axial component of the drag flow conveys the fluid forward along the channel while the pressure flow builds up in the reverse direction due to the resistance offered by the die at the outlet.

In addition to the discussed velocity profiles, there is also the transverse flow, analyzed, for the first time, by Mohr, Saxton [14] considering a Newtonian fluid. For Equation (2), the boundary conditions are $u_x = 0$ on the screw root and $u_x = -U\sin(\theta)$ on the barrel surface. Also, there can be no net flow of material in the x direction due to the sidewalls, i.e., the effect of the flight clearance, is omitted in this analysis. Therefore, $\int_0^H u_x dy = 0$, and the solution, which is known as the transverse flow, is represented as follows:

$$u_x = U_x\frac{y}{H}\left(2 - \frac{3y}{H}\right) \quad (4)$$

where $U_x = -U\sin(\theta)$. The respective velocity profile is shown in Figure 7.

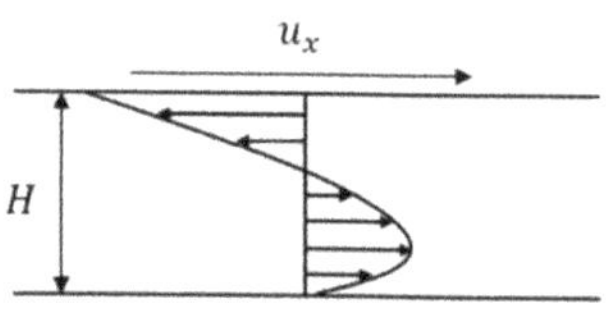

Figure 7. Velocity profile in the transverse plane.

The transverse flow moves the fluid toward the leading flight, after which the fluid is pushed downward to return to the bottom wall, creating a circulating motion that is very important for the mixing process within the extruder. As a result, the fluid element follows a helical path with an oval cross-section [14].

Concerning the discharge in the single-screw extruder, the outlet flowrate can be determined by integrating the product of u_z with the area of the screw channel from the top to the bottom, which results in $\int_0^H u_z W dy$ [13]. The outlet flow rate is given as follows:

$$Q = \frac{1}{2}\pi DHWNcos(\theta) - \frac{WH^3}{12Z}\frac{\Delta p}{\mu} \tag{5}$$

Equation (5) results from the difference between the flow rate for drag flow, $Q_D = \frac{1}{2}\pi DHWN\cos(\theta)$, and for the pressure flow, $Q_P = -\frac{WH^3}{12Z}\frac{\Delta p}{\mu}$. If there is no restriction, the extruder is in a state of "free discharge", or "open discharge", the pressure flow is zero and the flow rate at the outlet is given by $Q = Q_D$. If the equipment is closed off completely, there is no discharge, $Q = 0$, and the pressure build-up is maximum. This means that the drag flow is exactly offset by the pressure flow. As for the transverse component of the velocity, since it moves across the channel, it does not contribute to the output [16].

Furthermore, a leakage flow, Q_L, occurs in the clearance between the flight and the barrel due to the pressure difference created by the material being pushed forward by the screw. Therefore, the total flowrate of the extruder should be expressed as the difference between the flowrate for drag flow and the flowrates for pressure and leakage flows, $Q = Q_D - Q_P - Q_L$. However, since Q_L is usually much smaller than the other flow rates, it is often neglected when discharged by the extruder.

The power required by the extruder is also an important variable to consider. The power can be determined by the force's F work required to rotate the screw at a velocity U, resulting in $\dot{W} = FU$. The power is given as follows [17]

$$\dot{W} = \left(\frac{4\mu {U_x}^2}{H} + \frac{\mu {U_z}^2}{H} + \frac{U_z H}{2}\frac{\Delta p}{Z}\right)LW \tag{6}$$

Therefore, the basic performance variables that characterize the extruder are the outlet flowrate Q, the pressure buildup Δp and the power $\dot{W}$.

2.2. Twin-Screw Extruders

Studies regarding flow and geometry within the melt-conveying section of twin-screw extruders are scarce in the literature. The complexity of the flow also makes theoretical models for twin-screw extrusion less advanced compared to those for single-screw processes. The geometry and flows of the melt-conveying section of the two most common types of twin-screw extruders are analyzed here for self-wiping corotating (SWCOR) and closely intermeshing counter-rotating (CICTR).

2.2.1. Self-Wiping Corotating Twin-Screw Extruders (SWCORs)

Erdmenger [18] was one of the first to analyze its flow in the corotating extrusion field, concluding that material travels in the extruder according to the flow pattern disclosed in Figure 8. Also, it was demonstrated the close distance between screws enables them to clean each other surfaces of crust or layers of material caused by the pushing of material over to its mate. This improves the mixing efficiency and reduces the residues. Hence, the term "self-wiping" is used for this type of equipment.

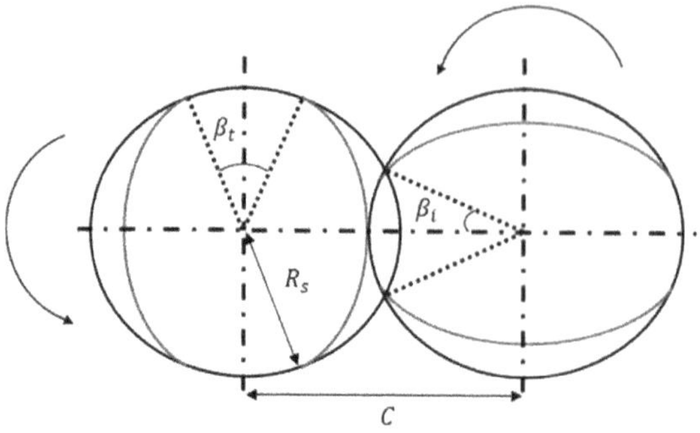

Figure 8. Cross section of a self-wiping intermeshing twin-screw extruder.

To gain further knowledge about melt-conveying flows in SWCORs, it is first necessary to describe their respective geometry. Booy [19] derived the cross-section geometry of a fully wiped corotating twin-screw extruder, which is represented in Figure 8, along with the centerline distance between two screws, C, screw radius, R_s, half the angle of intermeshing between two screws, β_i, and the tip angle of each screw, β_t.

Moreover, the cross-section of SWCORs presents a distinctive shape that is influenced by its diameter, centreline distance, and number of parallel channels or tips, n, due to the condition that one screw must wipe its counterpart, regardless of its position. This unique shape is determined by a kinematic analysis which provides equations that link design parameters and serve as the foundation for determining and selecting the screw dimensions. According to Figure 8, the centreline distance, C, screw radius, R_s, and intermesh angle, β_i, are related as follows:

$$C = 2R_s \cos \beta_i \tag{7}$$

Furthermore, β_i is related to β_t and n as follows:

$$\beta_i = \frac{\pi}{2n} - \frac{\beta_t}{2} \tag{8}$$

For the flow analysis, the channel cross-section geometry in a direction normal to the flights, which is shown schematically in Figure 9, must be considered. In Figure 9, it is represented the circumferential angle, φ, the channel's depth, H, and tip angle, β_t.

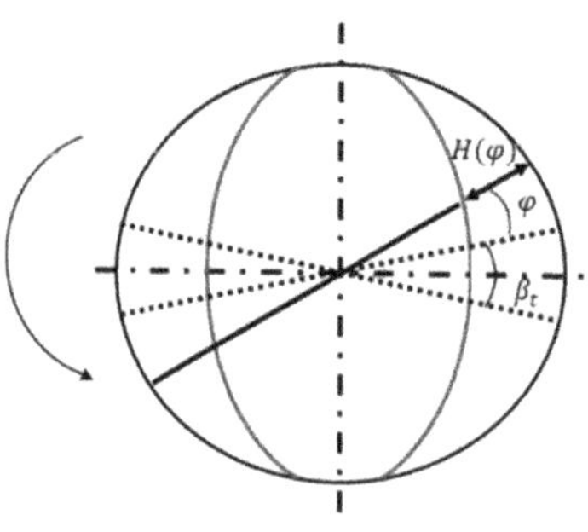

Figure 9. Channel's depth as function of the circumferential angle.

The channel's depth, H, is related to the circumferential angle, φ, through the following expression:

$$H(\varphi) = R_s(1 + \cos \varphi) - \sqrt{C^2 - {R_s}^2 \sin^2 \varphi} \tag{9}$$

The cross-channel depth profile can be plotted along the x axis, as it was represented for single-screw extruders. Figure 10 is a screw cross section in a plane perpendicular to the axis, representing both the x axis and channel's depth, $H(x)$.

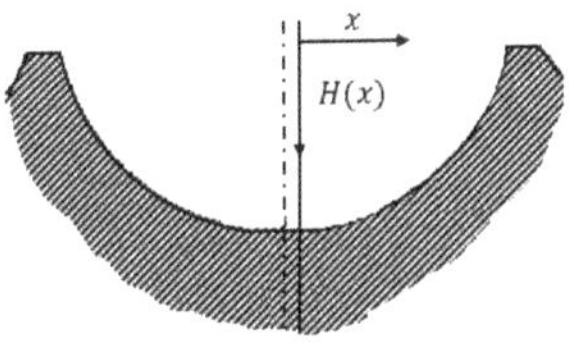

Figure 10. Screw cross-section plane perpendicular to the axis.

Denson and Hwang Jr. [20] modified Equation (9) to account for the helix angle and the flights' width, through the relation, $x = \frac{L\varphi\cos\theta}{2\pi} + \frac{e}{2}$. So, the expression for $H(x)$ is as follows:

$$H(x) = 2R_s - C \text{ for } 0 \leq x \leq \frac{e}{2} \tag{10}$$

$$H(x) = R_s\left[1 + \cos\left(\frac{2\pi\left(x - \frac{e}{2}\right)}{L\cos\theta}\right)\right] - \sqrt{C^2 - {R_s}^2\sin^2\left(\frac{2\pi\left(x - \frac{e}{2}\right)}{L\cos\theta}\right)} \text{ for } \frac{e}{2} \leq x \leq \frac{W}{2} \tag{11}$$

$$H(x) = 0 \text{ for } \frac{w}{2} \leq x \leq \frac{W + e}{2} \tag{12}$$

The open cross-sectional area between the barrel and the screw, A_p, proposed by Booy [19], is given by subtracting the cross-sectional area of the barrel, A_b, by the cross-sectional area of the two screws, $2A_s$, resulting in the following:

$$A_p = A_b - 2A_s \tag{13}$$

where,

$$A_b = 2(\pi - \beta_i){R_s}^2 + R_s C\sin\beta_i \tag{14}$$

$$A_s = n\left[\beta_i C^2 - CR_s\sin\beta_i\right] + \frac{1}{2}n\beta_t\left(C^2 + 2{R_s}^2 - 2CR_s\right) \tag{15}$$

$$A_p = 4{R_s}^2\left[\left(n - \frac{1}{2}\right)\beta_i + \left(n + \frac{1}{2}\right)\sin\beta_i\cos\beta_i - \pi\cos^2\beta_i + (\pi - 2n\beta_i)\cos\beta_i\right] \tag{16}$$

For the flows within screw elements in SWCORs, Denson and Hwang Jr. [20] examined an isothermal flow of incompressible Newtonian liquids for fully filled channels while neglecting the intermeshing region, which is valid when the flight tip angle, β_t, is small. Booy [21] and Szydlowski and White [22] extended this research by including the effects of the intermeshing region on the velocity field. White and Szydlowski [23] considered composite models of the flow rate and pressure profiles in modular twin-screw extruders involving both screw and kneading disks, which are mixing elements. Tayeb, Vergnes [24] developed a model based on solving thermal balance and Stokes equations in the melt-pumping section for an extruder with a deep screw channel.

This paper will discuss only the analysis, neglecting the intermeshing region [20]. As a simplification, similarly to the flat plate approximation of single-screw extruders, both the screw and barrel can be uncoiled. In this model, the barrel is treated as a flat plate moving with a velocity U, at an angle θ, relative to the stationary screw. However, for twin-screw extruders, the length of one unwound channel is the length of one turn of the screw. So, the model is a set of unwound channels alternating between each screw, as represented in Figure 11. Moreover, the equations of motion and their respective boundary conditions are the same as for single-screw extruders, and therefore, the volumetric flow rate for each channel is the same as Equation (5). Consequently, to obtain the total flow rate, one must add the flow rates of all independent channels, given that there are $(2n-1)$ parallel channels [21].

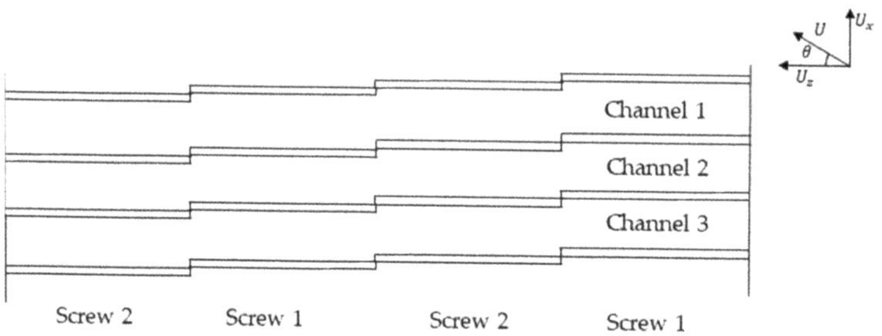

Figure 11. Unwound channels of screw elements.

Nevertheless, in most extrusion processes, screws are only partly filled with material, which is primarily related to the feeding scheme of the extruder since it is usually performed with metered feeding. Baird and Collias [25] demonstrated expressions for the calculation of velocity profiles and outlet flow rates depending on how much liquid is present within the extruder. The degree of fill of the channel is defined as the ratio of the filled channel cross-sectional area, A_f, to the total cross-sectional area, A, $f=\frac{A_f}{A}$. To obtain A, Equation (11) must be integrated across the width of the channel, which is given by

$$A=4R_s^2\left[\frac{1}{2}\beta_i-\left(\beta_i+\frac{1}{2}\beta_t\right)\cos^2\beta_i+\left(\frac{1}{2}\sin\beta_i+\frac{1}{2}\beta_t\right)\cos\beta_i\right]\sin\theta \tag{17}$$

The velocity profiles in the z and x directions are as follows:

$$u_z=\frac{yU_z}{H} \tag{18}$$

$$u_x=\frac{3U_xy^2}{H^2}+\frac{4U_xy}{H}+U_x \tag{19}$$

To obtain the volumetric flow rate, Equation (18) should be integrated over the cross section that is filled with fluid, which results in the following:

$$Q=(2n-1)fA\pi R_sN\cos\theta \tag{20}$$

Note, from Equation (20), the flow rate is dependent on the degree of fill and screw parameters. On the other hand, the velocity profiles are independent of f, which implies that regardless of whether the screw channels are fully filled or partly filled with liquid, the axial velocity remains the same. This suggests that flow dynamics within the screw are primarily influenced by rotational speed and screw design and not by the volume of liquid present.

2.2.2. Closely Intermeshing Counter-Rotating Extruders (CICTR)

Kiesskalt [26] and Schenkel [27] denominated for the first time the intermeshing counter-rotating extruders as positive displacement pumps, with their flow rate determined by the screw geometry and screw speed. Kaplan and Tadmor [28] developed a theoretical model for flow in non-intermeshing twin-screw extruders under the assumption that this equipment can result in a "three parallel plates" model, which leads to expressions for flow rate in terms of drag and pressure flow. Other works have analyzed the flow's mechanisms while considering the intermeshing level in the extruder [22,29].

The material transport in counter-rotating twin-screw extruders is made through a positive displacement mechanism, which is unique to this type of extrusion. Positive displacement depends on the degree of screw intermeshing, and it is most pronounced in CICTR. The flow takes place in a confined space known as the "C-shaped chamber", also known as the "C-chamber", where the maximum flowrate is as follows:

$$Q_{max} = 2nNV \tag{21}$$

where V is the volume of the C-chamber [25]. However, counter-rotating extruders may present leakage in certain regions of the screw and flights. Doboczky [30,31] and Janssen [32,33] analyzed the pumping characteristics of counter-rotating twin-screw extruders. Different types of flow in counter-rotating extruder are shown in Figure 12. Four key zones where leakage flow can occur were identified: flight flow (Q_f) through the gap between the flight and the barrel, calendaring flow (Q_c) between the bottom of the channel of one screw and the flight of the other screw, tetrahedron flow (Q_t) through the gap that goes from one screw to the other between the flanks of the flights of the two screws and side flow (Q_s) through the gap between the flanks of the screws perpendicular to the plane through the screw axis [32]. Moreover, Equation (20) can be corrected as follows:

$$Q = 2nNV - 2nQ_s - 2Q_f - Q_t \tag{22}$$

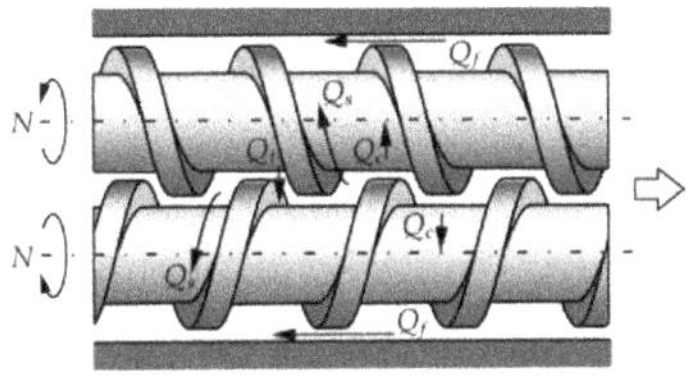

Figure 12. Types of flow in counter-rotating extruder, Q_f is leakage flow (screw flight and barrel), Q_c is calendaring flow (screw flight and opposite screw), Q_t is tetrahedron region, and Q_s is leakage flow (between the flanks of the screw). Reprinted permission from [34] (Permission reference 5898741218096).

3. CFD Simulation of Extruders

3.1. Single-Screw Extruders

Studies about single-screw extruders' computer modeling date back to the 1960s decade when Klein and Tadmor [35] assembled a mathematical model for the simulation of a plasticizing single-screw extruder, which was validated by experimental data. Additionally, Acur and Vlachopoulos [36] presented one of the first computer models for single-screw extruders, aiming at simulating the extrusion process of polymer while incorporating several components such as solids flow in the feed hopper and modified models for solids conveying and melting zone. Later on, Amellal, Lafleur [37] developed a model for a steady-state computer simulation of a single-screw extruder, which included the three zones of the extruder: solid-conveying, plasticizing, and metering section, which was crucial for the prediction of the extruder performance and to optimize its design and operating conditions. Then, Macgregor, Vlachopoulos [38] used a simulation methodology for conventional and barrier screws in order to predict the flow rate, pressure, and temperature profiles along the channel of the extruder.

These early works on extruder modeling paved the way for modern CFD models, which represent more sophistically complex three-dimensional flows and are going to be presented in the following sections.

3.1.1. Geometry

The geometry of single-screw extruders plays a major role in determining how material flows in the extruder and governs how they are transported, melted, and mixed. So, key geometric parameters, such as pitch, flight depth, and channel width and length, highly influence flow behavior and should be customized to meet the needs of the desired final product.

A literature review of the geometry used in the simulation of single-screw extruder is summarized in Table 1. As shown in Table 1, several geometries have been considered for single-screw extruders for CFD models. The flat plate model and helical geometries are geometrical simplifications, while simulations using z pin-type extruder, Dulmage-type screw, Saxton mixer, Faceted mixer, and Block-head mixer consider the actual geometry of the extruder.

Table 1. Summarized geometry of single-screw extruder used in literature.

Model	CFD Method	Geometry of CFD Model	Boundary Conditions	Research Focus
Flat plate model	FEM *	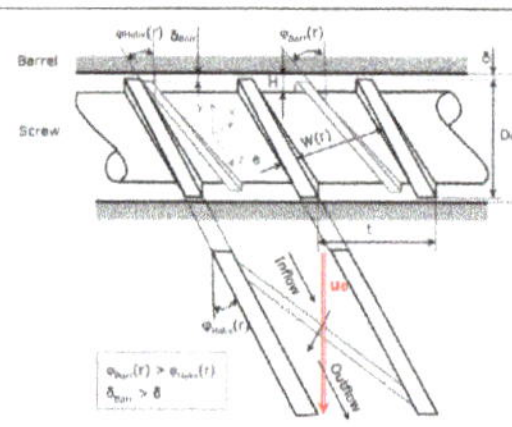 Reprinted with permission from Aigner, Köpplmayr [39]—Permission reference 5899410796913.	- Constriction of inlet mass flow rate. - Zero static pressure at the outlet. - Constant velocity at barrel surface.	Development of an approach to describe the flow behavior in barrier screws through CFD simulations in order to calculate the mass and pressure flow of different screw geometries.
	FDM **	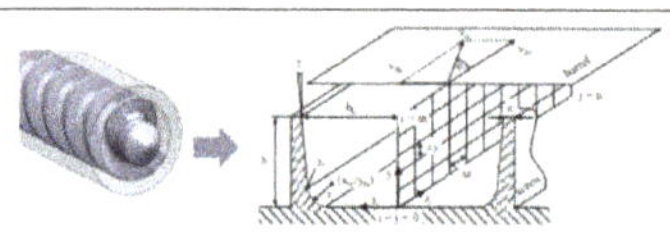 Reprinted with permission from Stueker and Schoeppner [40]—Permission reference 5899400520239.	- Stationary screw (bottom plate) ($u = 0$). - Moving barrel (top plate) with velocities of $U_x = -U\sin\theta$ and $u_z = U\cos\theta$. - Constant pressure gradient between inlet and outlet.	Analysis of the flow behavior of non-Newtonian polymer melts through the incorporation of real cross-sectional geometries of the screw channel.
	FEM	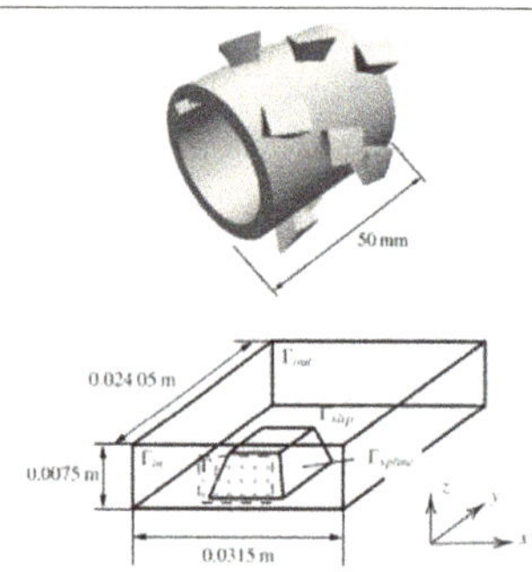 Reprinted with permission from Hube, Behr [41]—Permission references 5912470139350 and 5914101501361.	- Velocity of $U = \pi DN$ at the top plate (barrel surface). - Linear velocity profile, $u = \frac{U}{H}z$ at inflow. - Slip boundary conditions on both sides of the channel. - No-slip conditions on the bottom plate and mixing element. - Stress-free boundary conditions at the outflow. - Dirichet boundary conditions for the temperature on the top and bottom plates.	Introduction of a novel approach of shape optimization for the design of individual mixing elements with the help of the unwound mixing section to reduce computational complexity.

Table 1. *Cont.*

Model	CFD Method	Geometry of CFD Model	Boundary Conditions	Research Focus
Helical geometry and flat plate model	FVM ***	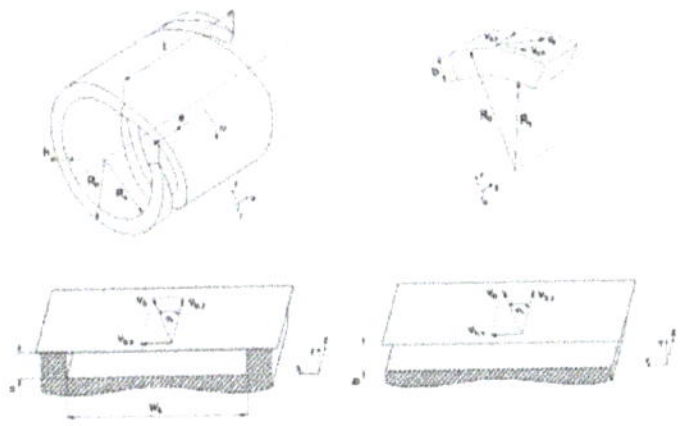 Reprinted with permission from Marschik and Roland [42]—Permission reference 5912460254971.	- Periodic boundary conditions between the inlet and outlet ($u_{in} = u_{out}$). - Velocities of the barrel of $U_x = U\sin\theta$ and $U_z = U\cos\theta$ for both models. - No-slip condition on the walls.	Comparison between flat plate model and curved screw channel through the prediction of the pumping capability of the extruder for a power-law fluid.
	FEM	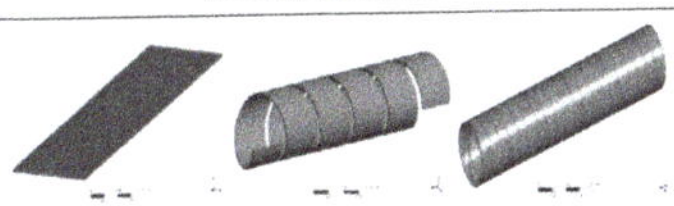 Reprinted from Ershov and Trufanova [43] (CC BY license).	- Moving screw with constant velocity defined at walls. - Constant temperature for body, screw, and input material.	Investigation of the effect of the shape of the channel and its effect on the flow mechanisms.
Helical geometry	FEM	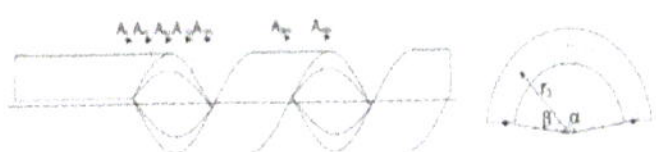 Reprinted from Kadyirov, Gataullin [44]. (CC BY license).	- Developed velocity profile at the inlet in the coaxial channel. - Fixed velocity of $U = -\pi DN$ at the tube wall. - No-slip conditions at the channel walls and surfaces of the screw shaft and flight.	Investigation of non-Newtonian solutions, specifically their effect on the flow structure and pressure drop.
	FEM	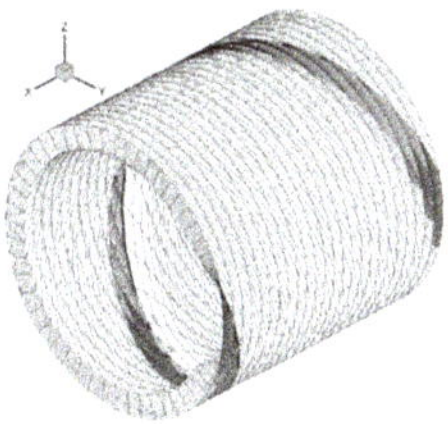 Reprinted with permission from Lim, Hwang [45]—Permission reference 5912441001952.	- Periodic boundary conditions between the inlet and outlet ($u_{in} = u_{out}$). - Stationary screw ($U = 0$). - Velocities of $u_x = 0,\ u_y = -U\sin\theta$ and $u_z = -U\cos\theta$ at the barrel wall. - Constant pressure gradient in the axial direction. - No-slip condition at the walls.	Introduction of a novel modeling technique to reduce simulation time and extruder's length by using a partial periodic unit of the geometry in the metering section.
	FEM	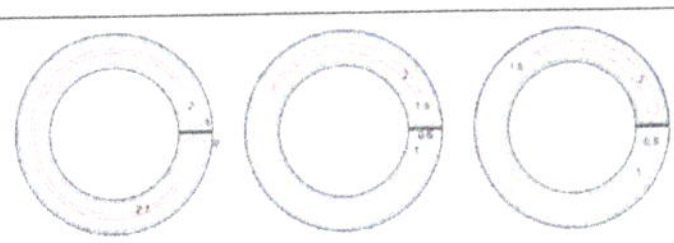 Reprinted from Vachagina, Kadyirov [46] (CC BY license).	- Developed velocity profile at channel inlet. - No-slip condition at the walls.	Modeling of a helical coordinate system to reduce the 3D problem to 2D for the study of melt flows through the helical geometry.
	FVM	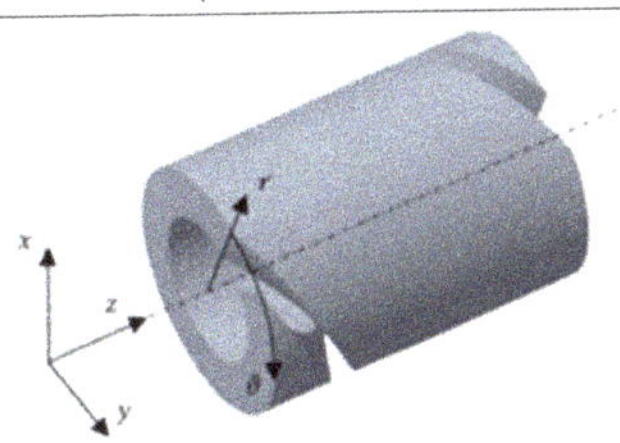 Reprinted from Herzog, Roland [47] (CC BY license).	- Periodic boundary conditions between the inlet and outlet ($u_{in} = u_{out}$). - Constant velocity at the barrel and zero velocity at the screw core. - Constant pressure gradient between inlet and outlet.	Introduction of a modeling method by including the 3D curved shape and the rotational motion of the screw channel boundaries.

Table 1. *Cont.*

Model	CFD Method	Geometry of CFD Model	Boundary Conditions	Research Focus
	FVM	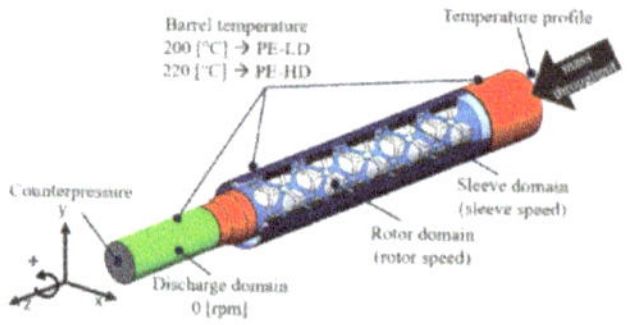 Reprinted from Janßen and Schiffers [48] (CC BY license).	- Mass throughput and temperature profile at the inlet. - Counterpressure at outlet. - Sleeve and rotor speed at sleeve and rotor domain, respectively. - Stationary discharge domain. - Constant barrel temperature.	Investigation of the influence of different materials on the performance of mixing sleeves and provide a new technique to predict the rotation speed of free-rotating mixers.
Pin-type extruder	FEM	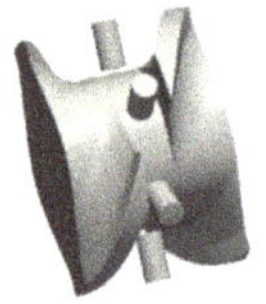 Reprinted with permission from Schöppner, Schadomsky [49]—Permission reference 5912460610087.	- Moving screw with speed adjusted to each simulation. - Constant barrel temperature. - Effects of wall sticking were not included.	Simulation of mixing behavior of rubber extruders through the investigation of new pin designs and geometries for process optimization.
Dulmage-type screw	FVM	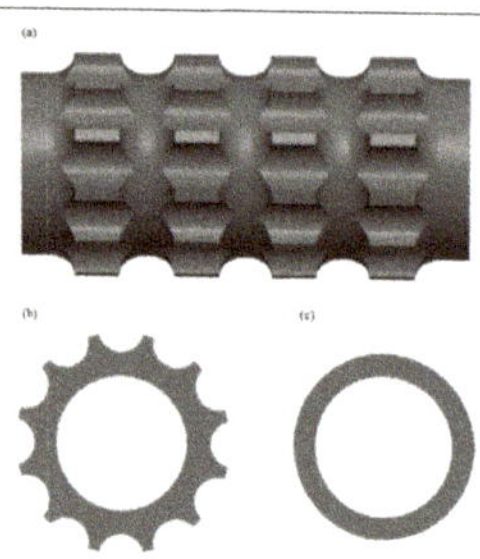 Reprinted with permission from Kimura, Nakayama [50]—Permission reference 5912460924947.	- Uniform axial velocity below the value for the flow rate at the inlet (from 10 kg/h and 30 kg/h). - Pressure of 0 Pa at outlet. - Zero velocity gradient in the normal direction. - Constant barrel surface temperature.	Investigation of mixing in a Dulmage-type screw to determine how it influences the operation conditions.
Saxton mixer	FVM	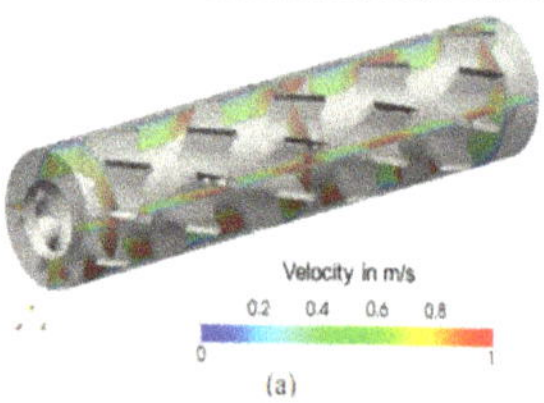 Reprinted with permission from Erb, Celik [51]—Permission reference 5912461232889.	- Moving screw with constant rotational speed.	Description of dispersive mixing through the calculation of the length stretch using an Eulerian frame and a Lagragian approach to apply in a 3D Saxton mixer.
Faceted mixer	*Not specified*	 Reprinted with permission from Frank, Hollenhorst [52]—Permission reference 5912461412482.	- *Boundary conditions were not explicitly detailed.*	Design and optimization of the mixing geometry of a faceted mixer through CFD simulation for an examination of pressure throughput behavior for several designs.

Table 1. *Cont.*

Model	CFD Method	Geometry of CFD Model	Boundary Conditions	Research Focus
Pin barrel extruder	FEM	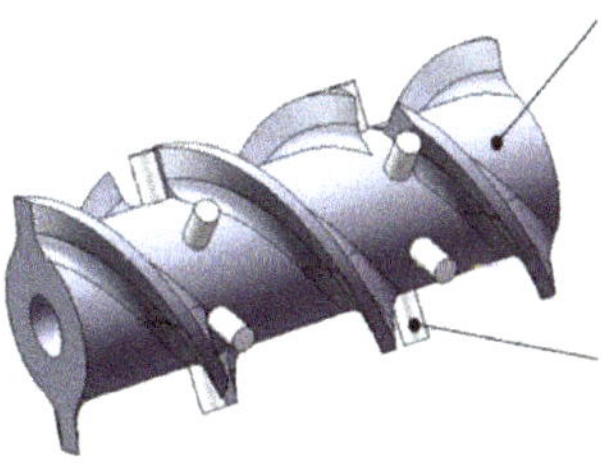 Reprinted from Wang, Pan [53] (CC BY license).	- Moving the screw with the rotational speed of N. - Calculated volume flow rate at the inlet, derived from the extrusion of output and material density. - Free flow at the outlet. - No-slip condition at the walls.	Proposal of a simple approximate mathematical model through the validation with simulation results for the quantitative assessment of elongational flow within extruders.
Block-head mixer	FEM	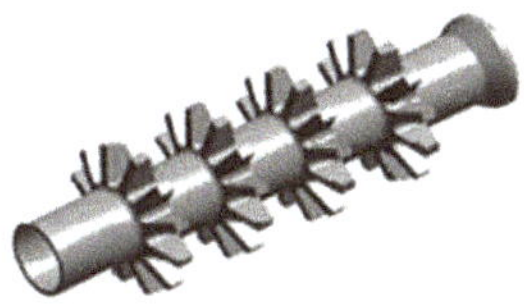 Reprinted with permission from Marschik, Osswald [54]—Permission reference 5912470463105.	- Mass flowrate with a specific screw speed N. - Constant normal force with tangential velocities at the inlet. - Normal and tangential forces at the outlet. - No-slip condition at the walls.	Investigation of the pumping and mixing ability of a block-head mixer through the analysis of several geometrical parameters.

* FEM—Finite Element Method. ** FDM—Finite Difference Method. *** FVM—Finite Volume Method.

3.1.2. Quantification of Mixing

The mixing quality of the polymer in the melt-conveying section of single-screw extruders has significant effects on the properties of the final products. Over the years, several researchers have evaluated the mixing ability of single-screw extruders through CFD simulations to optimize mixing performance [51,55–58].

Two mixing concepts are distinguished in polymerization reactions: distributive and dispersive. Distributive mixing is related to the uniform distribution of fluids, and thus, the main mechanisms are stretching and folding. The efficient distributive mixing avoids the localized concentration gradients, ensuring a uniform reaction rate and consistent polymer properties. On the other hand, dispersive mixing is related to the reduction of particle/droplet size. The main mechanisms are related to the strong forces that promote the break of cohesive bonds. Distributive mixing is crucial for polymerization processes where fillers or reinforcement agents are introduced.

Several factors affect the "goodness" of mixing within the extruder. Frank and Schöppner [59] studied the balance between pressure consumption and mixing quality through CFD simulations. The goal was to develop a mixer combining a good relationship between these two parameters while using a faceted mixing section as a case study. Additionally, the effect of non-Newtonian polymers on flow patterns can influence the mixing performance on single-screw extruders and was recently investigated through the Lattice Boltzmann model, providing a reference frame to apply in the chemical industry [60].

CFD methods can be employed to evaluate the distributive and dispersive mixing quality, which is essential for achieving uniform melt and material homogenization [61–63]. Recently, Schulz, Vorjohann [64] optimized the distributive mixing in single-screw extruders by comparing two computational approaches for its prediction: the mean distance of streamlines and passive scalar integration, showing that the latter offers more advantages, particularly in its integration into the CFD environment and its better performance in longer mixing sections.

Kubik, Zatloukal [65] used as metrics for dispersive mixing quantification of the average stress criterion. This parameter is determined by the integral of stress (shear or elongational) along the chosen pathline over the residence time of the pathline. The analysis was performed using fluids with different rheological behaviors: the Carreau–Yasuda model and the White–Metzner model.

Residence time distribution (RTD) is a key measure for fluid non-homogeneities in extrusion. Moritzer, Wittke [66] developed a CFD model that predicts shear stresses and average residence times of materials within the extruder, enabling the accurate simulation of the internal conditions of the extruder and the prediction of extrusion performance. In addition, the RTD for the 3D circulatory flow expressed in terms of physical and geometrical parameters, such as flow rate, helix angle, aspect ratios, and many others, was presented by Kim [67], which is also very useful for understanding and predicting the RTD systematically through numerical analysis. Olofsson, Roland [68] introduced a CFD model designed for partially filled single-screw extruders for starve-feeding, which is a common industrial case. This work enables the characterization of the extruder fill length and RTD, paving the way for more accurate modeling and control of starve-feeding extrusion systems.

Additionally, Mateboer and Hummel [69] assessed distributive mixing using tracer particles and Shannon entropy calculations in CFD simulations, demonstrating that these methods effectively distinguish mixing quality across different extrusion simulations and identify the extruder that provides the most efficient mixing. Recently, Hopmann and Schön [70] addressed the importance of flow visualization of mixing processes by proposing an experimental setup while validating the findings through CFD simulations. This work made a significant contribution to the field of mixing examination of single-screw extruders, offering a novel setup capable of high spatial resolution, allowing for better distributive and dispersive mixing analysis.

CFD results of single-screw extruders combining chemical reactions and flow modeling are scarce in the literature. Nevertheless, the mixing characteristics of steady non-isothermal reactive flows in the metering section were investigated through the influence of some key parameters, concluding that while screw speed had minimal impact on flow efficiency, both the pressure-to-drag flow and channel aspect ratios significantly influenced the mixing [71]. Later, Roy and Lawal [72] performed an analysis using the flat plate model, investigating how several parameters such as power-law index, flowrate, and homogeneous and heterogeneous reaction parameters affect concentration profiles, axial conversion, and Sherwood number.

3.1.3. Modelling of Pressure and Temperature Gradients

Pressure and temperature profile studies provide insights for understanding extrusion processes. Moreover, some studies integrate the temperature and pressure distribution profiles since they directly influence how the material flows, the melting behavior, and the overall quality of the final product. CFD has been implemented to study the reference frame dependence of melt temperature rise in single-screw extruders, mainly whether it differs when the barrel rotates, and the screw is fixed and vice-versa [73]. This study incorporates a CFD model that enables the simulation of the 3D flow and temperature rise, concluding that the kinematics and the melt temperature rise are equal for either a rotating screw or a rotating barrel.

Other studies have included both pressure and temperature to provide a more complete knowledge of the flow's mechanisms [74,75]. For instance, Roland, Marschik [76] evaluated different data-based modeling approaches to predict the 3D non-linear throughput-pressure relationship of metering channels in single-screw extruders.

Abdel-Ghany and Ebeid [77] studied the screw performance at different operating temperatures and pressures by comparing two screw geometries: a traditional and a rapid compression one. Results showed that while pressure gradient has little effect on the screw's displacement, increasing temperature leads to an increase in displacement in both types of screws. Recently, Olofsson, Roland [68] also incorporated temperature and pressure gradients through the simulation of the flow and thermal behavior of a starve-feeding single-screw extruder, which helps to predict temperature distributions in both screw and barrel, providing a better understanding of how the material heats up and cools down during the process. It also addresses the pressure distribution by accounting for the partially filled nature of the extruder, predicting how the pressure builds up and varies along the length of the screw.

3.1.4. Validation of Extruder Throughput

The validation of expression (Equation (5)) that predicts the flow rate of a single-screw extruder is here described from a flat plate approximation. The validation was performed using a 3D CFD model using ANSYS/Fluent to solve the Navier–Stokes Equations, namely the continuity equation,

$$\nabla \cdot \mathbf{u} = 0 \tag{23}$$

and the momentum equation,

$$\rho\left(\frac{\partial \mathbf{u}}{\partial t} + \mathbf{u} \cdot \nabla \mathbf{u}\right) = -\nabla p + \mu \nabla^2 \mathbf{u} \tag{24}$$

where $\mathbf{u}$ is the velocity vector, ρ and μ are density and viscosity, respectively, p is the pressure, and t is the flow time. The flow is isothermal in this work, so the energy equation was not solved. This set of equations neglects the body forces and considers an incompressible fluid, i.e., the density is constant.

Figure 13 shows the 3D CFD model of helical geometry and its respective coordinate system, where the dimensions used for the simulations were based on a single-flighted standard extruder (Fisher Scientific) with a diameter of D = 19 mm, a channel's width of W = 11.55 mm, a length of Z = 209.85 mm, a helix angle of θ = 14.11° and a flight clearance of δ = 0.5 mm [78].

Figure 13. Spiral model of the single-screw extruder.

The flow was initially studied at a steady state. The method used for the pressure-velocity coupling was the coupled method. For the spatial discretization of gradient, pressure, and momentum, the Least Squares Cell Based, Second Order, and Second Order Upwind were used, respectively.

Regarding the initial solution for the steady state simulation, all flow variables were initialized with a value of zero, and 10^{-6} was set as a convergence criterion. The simulated fluid has a density of ρ = 1000 kg m^{-3} and a viscosity of μ = 0.1 Pa s.

The boundary conditions are the velocities imposed on the walls, which consist of one screw rotating within a barrel. The screw is moving at a rotational speed of N = 9.55 rpm. The tangential velocity at each position of the extruder wall was $U = \omega R_{screw}$, with the components U_x and U_z. Pressure conditions were defined at the inlet and at the outlet.

The validation of the CFD model was conducted by comparing the resulting outlet flowrate from the simulation to the one described earlier (Equation (5)); for the cases where the presence of a die is not simulated, $\phi = 0$, and is simulated, $\phi = -1/3$ and $\phi = -2/3$. Table 2. shows the theoretical results, along with the ones provided by CFD simulations, for the flow rate.

Table 2. Outlet flowrates for the different values of ϕ.

ϕ	Equation (5)/m^3 s^{-1}	CFD Simulation/m^3 s^{-1}
0	7.80×10^{-8}	7.99×10^{-8}
−1/3	5.20×10^{-8}	5.68×10^{-8}
−2/3	2.60×10^{-8}	3.61×10^{-8}

The results are, overall, in agreement with the ones proposed by the literature (Equation (5)). As it was expected, the outlet flow rate decreases as the pressure imposed at this surface increases, which results from the die configuration. However, deviations from the computed values of Equation (5) are more noticeable for higher pressure drops, which may be due to the conceptualization approaches of the model, resulting in inaccuracies in the results.

3.2. *Twin-Screw Extruders*

The complex design of twin-screw extruders, including intermeshing or non-intermeshing screws, screw profile, and barrel configuration, significantly influences flow dynamics, mixing efficiency, heat transfer, and pressure distribution. CFD models enable the optimization of screw design and process parameters, providing insights into performance enhancements for various polymer processing applications, from compounding to reactive extrusion.

3.2.1. Geometry

CFD simulations have been implemented to simulate the conveying zone (melting and mixing zone) and extrusion zone. A 3D modeling of the conveying zone is essential to capture detailed flow characteristics. Due to its complexity, simulations require fewer simplifications to the geometry. Table 3. shows a summarized research review on the geometry of twin-screw extruders.

Table 3. Summarized geometry of twin-screw extruder used in literature.

Rotation Screw Direction	CFD Method	Geometry of CFD Model	Boundary Conditions	Research Focus
Corotating screws	FEM	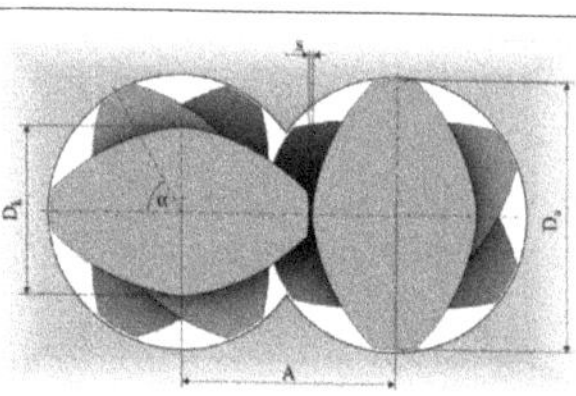 Reprinted with permission from Stritzinger, Roland [79] (Permission reference 5912480209160).	- Stationary barrel walls; - Zero velocity components at the barrel surface; - Velocity at the screw surface: $U = 2\pi Nr$; - Periodic boundary conditions between inlet and outlet.	Optimizing the extrusion process using a 3D model of a corotating intermeshing extruder with a kneading block for melt conveying to improve efficiency and power reduction.

Table 3. *Cont.*

Rotation Screw Direction	CFD Method	Geometry of CFD Model	Boundary Conditions	Research Focus
	FEM	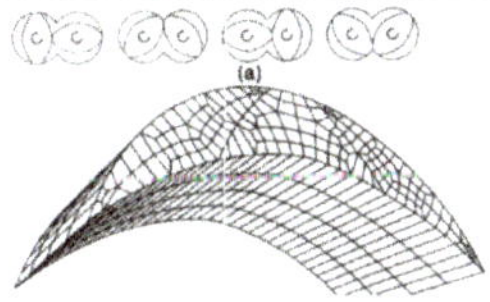 Reprinted with permission from Van Der Wal, Goffart [80] (Permission reference 5912490345777).	- Axial velocity is zero, $U_{screw} = 0$; - U_{barrel} is tangential according to the rotational speed; - No slip conditions at the barrel and screw surfaces; - Fully filled flow domain.	A 3D CFD model for a corotating extruder considers a simplification of the kneading block and studies the effect of parameters on mixing, such as shear rate and flow profile, by geometric simplification.
	SPH ****	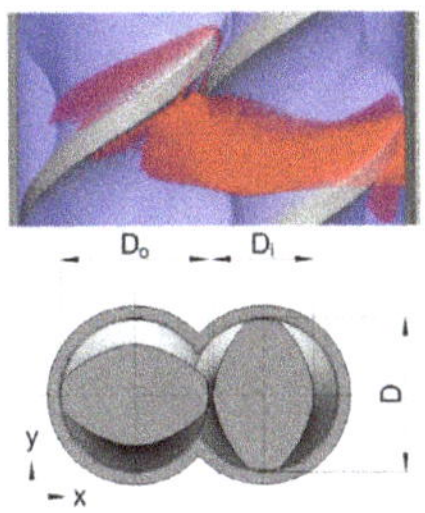 Reprinted with permission from [81,82] (Permission references 5912580169753 and 5912580479271).	- Periodic boundaries in the z-axis ($u_{in} = u_{out}$); - The screw speed was 95 rpm. - The barrel is stationary.	A SPH method considers a mechanistic model that takes into account the mass transfer along the screws and the heat transfer mechanisms.
	BEM *****	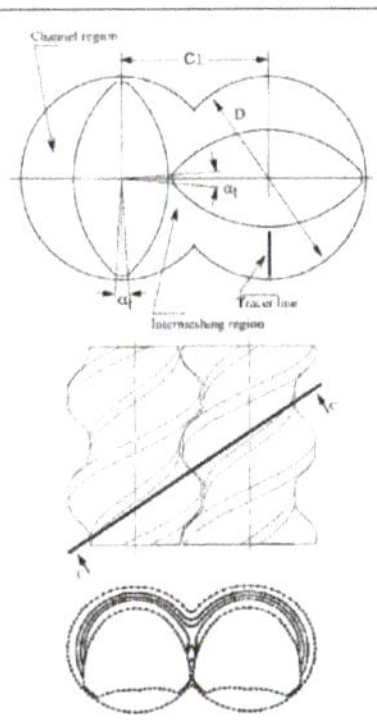 Reprinted from Rios, Gramann [83] (License CC BY).	- The geometry, a cut-long down-channel direction, remains fixed and moves in the axial direction as the screw rotates.	Self-wiping intermeshing corotating model is considered using BEM to study qualitative and quantitative mixing by various screw geometries, includingsingle, double, and triple flighted screws, by computing flow field, pressures, and stress. These simulations consider a flight clearance, a fully filled screw channel, and a Newtonian fluid.
	FEM	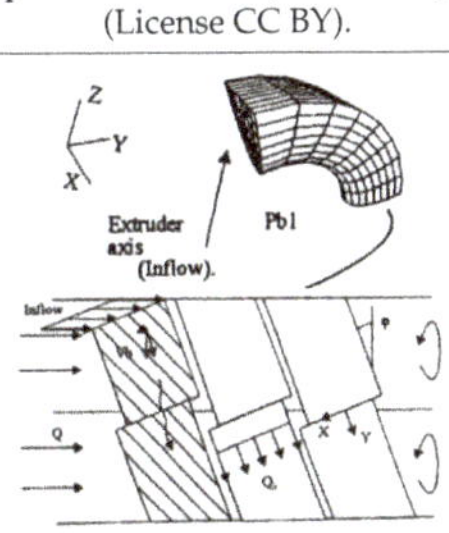 Reprinted with permission from Goffart, Van Der Wal [84] (Permission reference 5913010005059).	- The rotation speed is tangential to the barrel ($u_z = U\cos\theta$); - Inlet flow rate is imposed on the extruder; - The velocity of the screw is $U = 0\,\text{m/s}$; - Free stress condition at the outlet plane.	A 2D CFD model of intermeshing corotating screws is used to analyze the flow profile, pressure build-up, and operational conditions. The fight clearance is neglected. The objective is to determine the throughput of the extruder by studying the effect of the rotation speed and the helix angle of the screw.

Table 3. *Cont.*

Rotation Screw Direction	CFD Method	Geometry of CFD Model	Boundary Conditions	Research Focus
	FEM	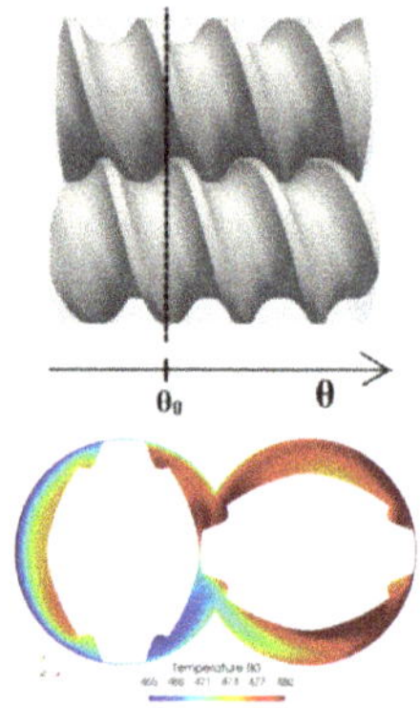 Reprinted with permission from Hinz, Helmig [85] (Permission reference 5913010808006).	- Screw speed 60,120 rpm; - No slip condition at the barrel; - Uniform inflow velocity was imposed at the inlet; - Inflow temperature $T_{inflow} = 473$ K.	Meshing techniques derived from the Snapping Reference Mesh Update Method (SRMUM) and Elliptic Grid Generation were used to simulate the flow in intermeshing corotating extruders using boundary-conforming meshes. The authors generated 2D spline-based geometries for selected screw orientations, which highlights the method's practicality for real-world applications. The density of the working flow is from 1 to 710 kg/m^3.
	FEM	 Reprinted with permission from Helmig, Behr [86] (Permission reference 5913030367101).	- No slip conditions at the barrel; - Dirichlet boundary condition of velocity on the rotating screw; - Inlet pressure is 0.2 MPa.	Authors used the boundary-conforming method to evaluate pressure and temperature distribution in two different screw configurations. The mesh update technique helps to compute the flow in the twin-screw extruder.
	FEM	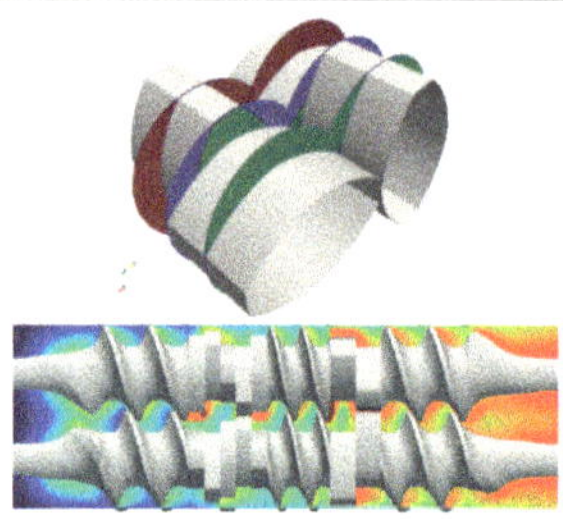 Reprinted with permission from Helmig, Key [87] (License CC BY).	- No slip condition at the barrel; - Atmospheric pressure at the outlet; - Temperature at inlet 473 K; - Adiabatic assumption for screws.	2D and 3D CFD simulation of Newtonian fluid requires the use of finely optimized boundary-conforming meshes within each subdomain, with coupling occurring only at the overlapping boundary interfaces.
	FEM	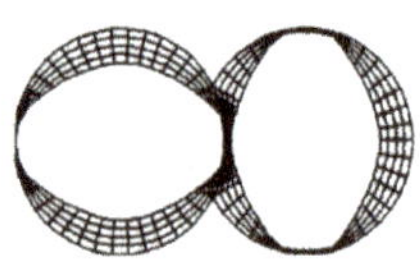 Reprinted with permission from Ishikawa, Kihara [88] (Permission reference 5913051404860).	- Constant flow rate at the inlet; - No slip at the barrel surface - Tangential velocity according to the screw speed for the disc surface; - Constant temperature at the inlet; - Adiabatic condition at disc surface.	A 3D CFD model investigates a simulation by using the FEM method for non-isothermal flow in the kneading disc area. Pressure gradient and temperature distributions at different rotational speeds were performed.

Table 3. *Cont.*

Rotation Screw Direction	CFD Method	Geometry of CFD Model	Boundary Conditions	Research Focus
	SPH	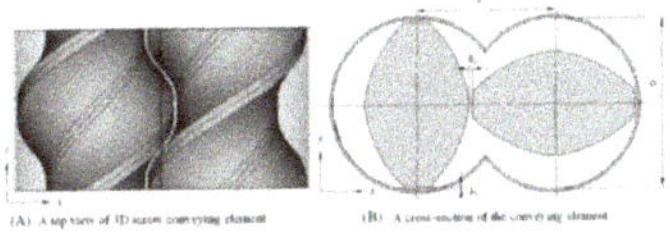 Reprinted with permission from Dong and Wu [89] (Permission reference 5913051404860).	- Periodic condition in z-direction ($u_{in} = u_{out}$); - No slip conditions at the barrel - Initial velocity of the fluid at screw set to zero.	3D CFD model of a conveying region using the SPH method for non-Newtonian fluid in fully and partially filled models. Two flight clearances were studied: 0 and 1 mm.
	SPH	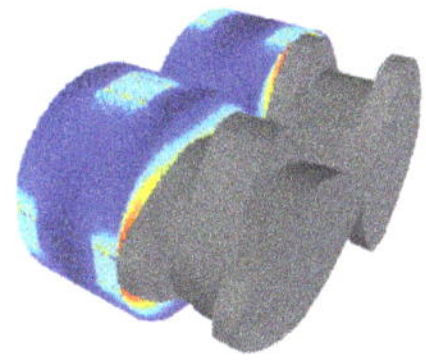 Reprinted with permission from Bauer, Matić [90] (Permission reference 5913100103435).	- No slip conditions at the barrel wall; - Velocity of the screw surface corresponds to rotational speed; - Periodic boundaries between inlet and outlet.	The CFD model evaluates the pressure build-up, power consumption, and mixing behavior of fully filled elements in different operational conditions. This model considers flight clearance.
Counter-rotating Screws	FEM	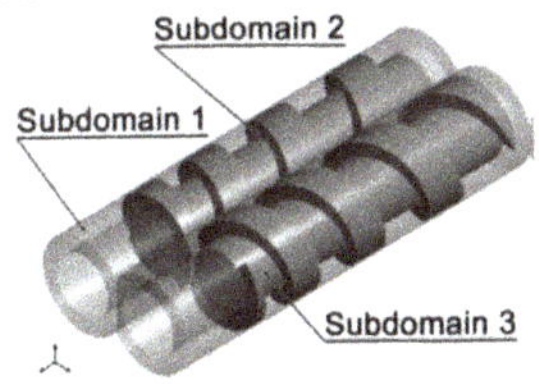 Reprinted with permission from Lewandowski, Wilczyński [91] (Permission reference 5912500224136).	- Flow rates imposed at the inlet were 4, 8, and 14 kg/h; - Screw speeds of 40 and 80 rpm; - Velocity is zero at the barrel wall.	The 3D model for a counter-rotating intermeshing region considers the screws simplification of the kneading block configuration.
	FEM	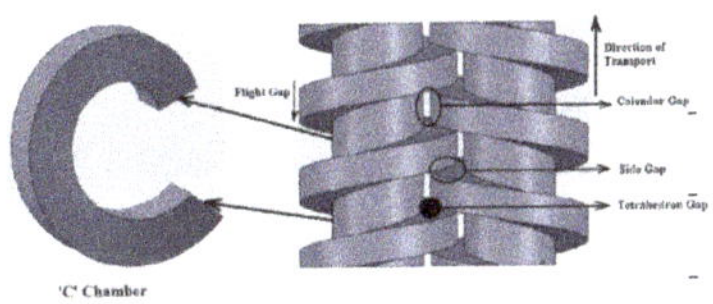 Reprinted with permission from S R, Arumugam [92] (Permission reference 5912970385230).	- Gravity effects were neglected; - Periodic boundary conditions ($u_{in} = u_{out}$); - No slip condition at the barrel wall.	A 3D CFD model using a 3D fully intermeshing counter-rotating extruder optimizes the production process with a minimum dissipation rate of viscosity and specific dimension of the final product. In this model, the Mesh Superposition Technique was used.
	FEM	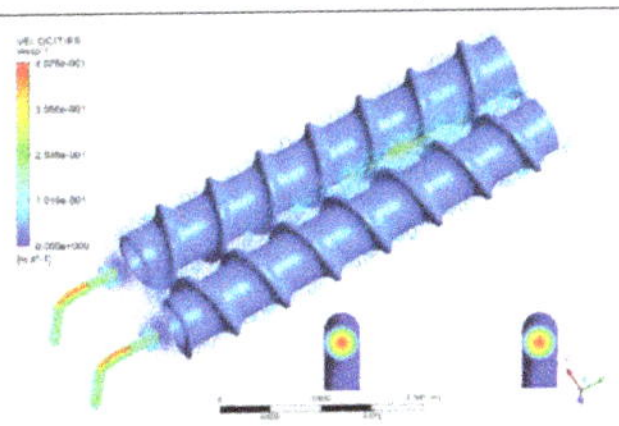 Reprinted from Yuan Zhang [93] (License CC BY).	- Inlet volume flow rate is $6.42 \times 10^{-5} m^3/min$; - Outlet is 6.25×10^6 Pa; - The screw rotational speed was 20 rotations/min.	A 3D model of non-intermeshing counter-rotating predicts a theoretical basis for the design and optimization of the die zone.

**** SPH—Smooth Particle Hydrodynamics. ***** BEM—Boundary Element Method.

3.2.2. Quantification of Mixing

A comprehensive state-of-the-art mixing analysis using CFD simulation is presented here for twin-screw extruders [94–102]. Starting with the state-of-the-art implementation of residence time distribution (RTD) in twin-screw extruders, Bauer and Matić [101] studied the local RTDS in specific regions of the twin-screw extruders. The measurement of local RTDs is a limitation of experiments that only measure the total RTDs in the extruder. The impact of variations in screw speed, throughput, and screw length in local RTDs from-tracer experiments using the SPH method. The CFD simulations helped to understand the impact of the residence time on mixing and predict quantitative mixing per screw revolution. These results were validated from tracer experiments. Robinson and Cleary [103] simulated RTD in twin-screw extruders for particle filling factors (50%) using a Lagrangian method that enabled the tracking of particle positions. Over time, the mixing process improves due to the theoretically infinite residence time resulting from the use of a periodic domain in the axial direction. Mixing efficiency was assessed using the Finite-Time Lyapunov Exponent (FTLE) method and another approach based on the homogeneity of initial color distributions. This approach enabled the quantitative mixing separately in both the cross-sectional and axial directions of the chamber, providing a comprehensive analysis of the process.

Oldemeier and Schöppner [104] investigated the mixing performance of five various configurations of screw elements in a corotating extruder through 3D CFD flow simulations. They combined CFD simulation with particle tracking techniques to achieve both dispersive and distributive mixing, optimizing the extruder mixing section. To obtain distributive mixing, the spatial distribution of particles was evaluated, and a mixing index was determined. To obtain dispersive mixing, the maximum shear stress along the particle paths and the integral of shear stress for each particle's residence time were calculated. They concluded that screw geometry and rotational speed significantly affect both types of mixing.

Alsteens, Legat [99] evaluated the mixing efficiency from the comparison between the residence time and the shear distribution, which are commonly used as metrics for mixing quantification. A set of particles was introduced to the system in kneading blocks. In this study, the stagger angles and number of discs are studied.

Eitzlmayr and Khinast [81] first quantified mixing from the time evolution of the intensity of segregation using the tracer particles for the partially and totally filled regions using the SPH model. On the other hand, the mixing was evaluated using the fitted kinetic laws to determine the mixing rates. This analysis was made separately to analyze the overall mixing and axial mixing and to conclude the axial and cross mixing to the overall mixing rate. Eitzlmayr, Matić [82] evaluated the mixing using the SPH method for different geometries of corotating twin-screw elements, two kneading elements and a mixing element. The mixing analysis was performed based on the time-dependent evolution of the intensity of segregation related to various tracers. The exponential mixing rates describe the decrease in the intensity of segregation per screw revolution. Kwag and Lyu [105] also quantified mixing from the reaction kinetics. The production of polypropylene through a 3D CFD simulation was studied using a corotating extruder using ANSYS Polyflow. The study examined the impact of factors such as screw rotation speed, kneading block stagger angle, and inlet flow rate influenced the mixing and reaction processes. The findings demonstrated that these operational and geometric parameters significantly impact mixing efficiency, residence time distribution, flow temperature, local species concentration, reaction time and rate, and overall conversion rate.

Sarhangi Fard and Anderson [106] implemented a mapping method to analyze and quantify mixing in kneading blocks in a twin-screw extruder. They studied different staggering angles for the kneading blocks. Also, the mixing efficiency of the screw mixing elements was compared with the standard elements of conveying zone. The results demonstrated that the stagger angle has a significantly greater influence on distributive mixing compared to the width of the screw disc.

The impact of different filling factors on both distributive and dispersive mixing characteristics in counter-rotating extruder was evaluated by Ahmed, Poudyal [107]. Although the simulations were conducted in 2D, valuable insights were gained regarding the influence of the filling factor on mixing. The dispersive and distributive mixing were assessed using the mixing index, cumulative distribution of maximum shear stress, cluster distribution index, scale of segregation, and length of stretch calculated from massless particles. Results show that filling factors between 70% and 80% present the best results in terms of distributive and dispersive mixing.

3.2.3. Modelling of Pressure and Temperature Gradients

Temperature and pressure gradients in twin-screw extruders have been simulated using CFD simulations to reach the desired viscosity and homogeneity, facilitating the mixing efficiency and preventing degradation or incomplete melting. The effect of viscosity and velocity on thermal distribution and flow behavior was studied by Ahmed, Poudyal [107]. They concluded that by increasing time, shear results and velocity values rose. This causes an increase in temperature gradient.

Sun, Zhu [108] performed a 3D numerical simulation of a corotating twin-screw extruder to analyze the impact of the screw's rotational speed on the temperature gradient. Results show that the increase in screw rotational speed enhances the increase in temperature. This occurs because a higher rotational speed generates a larger shear rate and, thus, viscous dissipation. However, while the larger pressure drop can speed up material flow, it reduces the residence time of materials, which can negatively impact the efficiency of mixing and polymerization reactions along the axial direction.

Bertrand, Thibault [109] employed a 2D simulation of fluid flow in corotating twin-screw and performed the pressure profiles plotting pressure versus the location of the screws at two different temperatures. The authors showed that when the position of the screw tip is away from the barrel, the pressure value decreases. Moreover, when the position of the screw tip is close to the peak of the nip region of two screws, the pressure increases.

Lewandowski, Wilczyński [91] analyzed the pressure and temperature gradients for a non-Newtonian fluid at different flow rates and screw speeds using 3D simulations in a closely intermeshing counter-rotating extruder. The results showed that the pressure can be increased or decreased in different parts of the screw. At the C-chamber region, pressure decreases and increases at the gap between the flights.

Eitzlmayr and Khinast [110] explore the pressure gradients in Newtonian, isothermal flows within a twin-screw extruder. The pressure gradients (0–35 kPa/m) were validated using the SPH method and shown to scale proportionally with viscosity in creeping flow. However, this simulation assumed isothermal conditions, i.e., the temperature gradients were not considered, limiting the application to real processes. The authors point out that future works must incorporate non-isothermal effects, as neglecting local energy distribution and temperature inhomogeneities could introduce significant errors.

Martelli [111] studied the temperature and pressure gradients in the melt section by considering Newtonian flow in the non-intermeshing counter-rotating extruder. They illustrated two major issues in the conveying zone: the first is how material distribution across different processing zones can be controlled through the use of reverse screw elements, and the second is generating backpressure by these elements.

Nakayama, Takemitsu [112] developed a method to improve mixing characteristics by playing with the pressure gradients using 3D CFD simulation of fully filled, non-isothermal molten polymer in the mixing section. They studied the effect of geometry with both forward and backward configurations of pitched tips of the kneading block on mixing efficiency. Results showed that the geometry of the mixing elements affects the flow's direction and pressure at a certain rotational speed and flow rate throughout the channel. They concluded that the pumping ability was obtained by the disk-stagger angle and tip angle. A positive or a negative amount of the pressure gradient shows that the pressure flow happens on the average of the forward or backward direction of the channel.

Die configuration and the number of dies at the outlet have significant effects on pressure distribution. Yuan Zhang [93] concluded, from CFD simulations, that the pressure gradient by adding two and three dies at the end of a twin-screw extruder. The maximum pressure is in both cases, but the pressure first decreased from the entrance to the die and then increased.

4. Challenges of CFD Models of Reactive Extrusion

This review highlights fundamental studies of mixing mechanisms and flow dynamics in reactive extrusion (REx) using screw extruders, dating back to the 1950s when analytical models were first proposed. These early models provided critical insights and have since been used to simplify geometries in CFD simulations. CFD simulations have been extensively utilized to investigate the effects of various screw element shapes and operating conditions, such as rotation speed and direction, on mixing performance. These studies quantify mixing efficiency using metrics, such as residence time distributions (RTDs) and flow deformation, while also examining the influence of pressure and temperature gradients on mixing efficiency and outlet flow rates. However, REx flows remain highly complex, and accurately simulating the full geometry of extruders is essential for successfully implementing these units in industrial applications.

Modeling reactive extrusion using CFD faces significant challenges due to the intricate interplay of physical and chemical processes. REx involves mixing, heat transfer, and chemical reactions within non-Newtonian fluids subjected to high shear forces and temperature gradients. Capturing these phenomena requires advanced rheological models to account for the material's non-Newtonian behavior and robust kinetic models to describe chemical transformations. Multiphase flows, such as gas generation and particle formation, add further layers of complexity. Additionally, extruder geometries, especially the flight clearance between the screw and barrel, demand high-resolution meshes, making simulations computationally intensive. Finally, validation of these models is also constrained by the limited availability of in situ measurements and the challenges of replicating industrial conditions at a laboratory scale.

Future research should aim to improve the accuracy and efficiency of CFD simulations to better capture complex phenomena such as mixing, heat transfer, and chemical reactions. Addressing multiphase flow modeling, including gas generation and particle interactions, is critical; it can be performed using a global model that combines DEM/CFD simulations, as proposed by Wilczyński, Nastaj [5]. Moreover, incorporating real-time sensor data during extrusion processes could enable dynamic and adaptive modeling, paving the way for more reliable industrial applications.

5. Conclusions

CFD models have been intensively explored for REx, offering valuable insights for industrial applications. These CFD models have evolved from mere geometric conceptualizations of extruders to actual geometric simulations. This work presents a literature review on single and twin-screw extruders, primarily focusing on the geometries of CFD models, studies on reactive mixing, and modeling of pressure and temperature gradients. The influence of geometrical configurations (e.g., mixing element shapes) and operating conditions (such as rotation speed, relative screw direction in twin extruders, and pressure and temperature gradients) has been extensively investigated. Additionally, the analytical expression that describes the outlet flow rate was validated here using CFD simulations of a helical geometry to calculate the outlet flow rate. However, the results reveal discrepancies between the model, which assumes simplified flow conditions, and the CFD results, particularly due to flow restrictions at the outlet caused by the presence of a die.

While numerous studies have explored various geometries and flow constrictions in reactive extrusion using CFD, significant challenges remain due to the complex interactions between physical and chemical processes. REx involves mixing, heat transfer, and chemical reactions in non-Newtonian fluids under high shear and temperature gradients, requiring detailed rheological and kinetic models. The presence of multiphase flows, such as gas generation and particle formation, further complicates simulations. Additionally, the extruder's geometry, particularly flight clearance, necessitates high-resolution meshes, increasing computational demands. Validation is also hindered by limited access to in situ data and challenges in replicating industrial conditions at the lab scale. Future efforts should improve simulation accuracy and efficiency, address multiphase flow modeling, and incorporate real-time sensor data to enable more dynamic and adaptive modeling approaches.

Author Contributions: Conceptualization, E.D., I.O., M.-F.B., A.S.-E., C.G.S., R.J.S. and M.S.C.A.B.; methodology and formal analysis, E.D., I.O. and M.S.C.A.B.; writing—original draft preparation, E.D., I.O. and M.S.C.A.B.; writing—review and editing, R.J.S., C.G.S. and M.S.C.A.B.; supervision, M.-F.B., A.S.-E., C.G.S., M.S.C.A.B. and R.J.S.; project administration, M.-F.B., C.G.S., R.J.S., All authors have read and agreed to the published version of the manuscript.

Funding: This work was supported by BioShoes4All (PRR–000011), supported by PRR—Plano de Recuperação e Resiliência—Bioeconomia. This research was also supported by national funds through FCT/MCTES (PIDDAC): LSRE-LCM-UIDB/50020/2020 (DOI: 10.54499/UIDB/50020/2020) and UIDP/50020/2020 (DOI: 10.54499/UIDP/50020/2020); and ALiCE—LA/P/0045/2020 (DOI: 10.54499/LA/P/0045/2020), CIMO UIDB/00690/2020 (DOI:10.54499/UIDB/00690/2020) and UIDP/00690/2020 (DOI:10.54499/UIDP/00690/2020); SusTEC LA/P/0007/2020 (DOI:10.54499/LA/P/0007/2020). Elham Delvar acknowledges her PhD scholarship supported by Fundação para a Ciência e Tecnologia (FCT) (2023.01245.BD). Margarida S.C.A. Brito acknowledges her contract (2023.06416.CEECIND) supported by the FCT Individual Call to Scientific Employment Stimulus—6th Edition.

Conflicts of Interest: The authors declare no conflict of interest.

List of Variables

A	channel's cross-sectional area [m^2]
A_b	barrel's area [m^2]
A_f	filled channel's cross-sectional area [m^2]
A_p	area between the barrel and the screw [m^2]
A_s	screw's area [m^2]
C	centerline distance [m]
D	diameter of the extruder [m]
e	flight's width [m]
F	force's work [N]
f	degree of fill
H	channel's height [m]
N	screw velocity [rpm]
n	number of tips
p	pressure [Pa]

Q	flowrate [$m^3\ s^{-1}$]
Q_{max}	maximum flowrate [$m^3\ s^{-1}$]
Q_c	calendaring flowrate [$m^3\ s^{-1}$]
Q_f	flight flowrate [$m^3\ s^{-1}$]
Q_s	side flowrate [$m^3\ s^{-1}$]
Q_t	tetrahedron flowrate [$m^3\ s^{-1}$]
$R_{clearance}$	distance between the rotation axis and the flight's clearance surface [m]
R_{screw}	distance between the rotation axis and the screw surface [m]
R_s	screw's radius [m]
t	time [s]
U	velocity [$m\ s^{-1}$]
u	velocity profile [$m\ s^{-1}$]
$\mathbf{u}$	velocity vector field [$m\ s^{-1}$]
U_x	velocity in the x component [$m\ s^{-1}$]
u_x	velocity profile in the x component [$m\ s^{-1}$]
U_x	velocity in the z component [$m\ s^{-1}$]
u_z	velocity profile in the z component [$m\ s^{-1}$]
W	channel's width [m]
$\dot{W}$	power consumption [m]
x	space coordinate [m]
y	space coordinate [m]
Z	helical length [m]
z	space coordinate [m]
Greek letters	
α_i	volume fraction of species i
β_i	intermesh angle [°]
β_t	tip angle [°]
δ	flight's clearance [m]
θ	helix's angle [°]
μ	dynamic viscosity [Pa s]
ρ	density [$kg\ m^{-3}$]
φ	circumferential angle [°]
ω	angular velocity [$rad\ s^{-1}$]
ϕ	pressure-to-drag flow ratio
VIC	variable intermeshing clearance

References

1. Scott, H.G.; Humphries, J.F. Modern Pelastic. Patent WO2004009356A1, March 1973.
2. Hopmann, C.; Adamy, M.; Cohnen, A. *Introduction to Reactive Extrusion: Principles and Applications*; Wiley-VCH Verlag GmbH & Co. KGaA: Hoboken, NJ, USA, 2017; pp. 1–10.
3. Biermann, L.; Brepohl, E.; Eichert, C.; Paschetag, M.; Watts, M.; Scholl, S. Development of a continuous PET depolymerization process as a basis for a back-to-monomer recycling method. *Green Process. Synth.* **2021**, *10*, 361–373. [CrossRef]
4. Misiura, D.; Majka, T.M. An overview on obtaining foamed PET by reactive extrusion. *Tech. Trans.* **2018**, *115*, 97–102.
5. Wilczyński, K.; Nastaj, A.; Lewandowski, A.; Wilczyński, K.J.; Buziak, K. Fundamentals of Global Modeling for Polymer Extrusion. *Polymers* **2019**, *11*, 2106. [CrossRef]
6. Lewandowski, A.; Wilczyński, K. Modeling of Twin Screw Extrusion of Polymeric Materials. *Polymers* **2022**, *14*, 274. [CrossRef] [PubMed]
7. Cassagnau, P.; Bounor-Legaré, V.; Fenouillot, F. Reactive Processing of Thermoplastic Polymers: A Review of the Fundamental Aspects. *Int. Polym. Process.* **2007**, *22*, 218–258. [CrossRef]
8. Martin, C. Twin Screw Extruders as Continuous Mixers for Thermal Processing: A Technical and Historical Perspective. *AAPS Pharmscitech* **2016**, *17*, 3–19. [CrossRef]
9. Tsao, T.F.; Harper, J.M.; Repholz, K.M. The effects of screw geometry on extruder operational characteristics. *AIChE Symp* **1978**, *74*, 142–147.

10. Kousemaker, T.M.; Druetta, P.; Picchioni, F.; Vakis, A.I. Modelling supercritical CO_2 flow in a co-rotating twin screw extruder using the level-set method. *Chem. Eng. Res. Des.* **2024**, *205*, 569–577. [CrossRef]
11. Rowell, H.; Finlayson, D. Screw viscosity pumps. *Engineering* **1928**, *126*, 639.
12. Maillefer, C. Etude théorique et expérimentale sur le fonctionnement des boudineuses. Doctoral thesis, Available online, 1952. Available online: https://infoscience.epfl.ch/entities/publication/50539f50-40fd-4810-b473-37fceb10d636 (accessed on 1 November 2024).
13. Carley, J.F.; Mallouk, R.S.; McKelvey, J.M. Simplified Flow Theory for Screw Extruders. *Ind. Eng. Chem.* **1953**, *45*, 974–978. [CrossRef]
14. Mohr, W.D.; Saxton, R.L.; Jepson, C.H. Theory of Mixing in the Single-Screw Extruder. *Ind. Eng. Chem.* **1957**, *49*, 1857–1862. [CrossRef]
15. Li, Y.; Hsieh, F. Modeling of flow in a single screw extruder. *J. Food Eng.* **1996**, *27*, 353–375. [CrossRef]
16. Carley, J.F.; Strub, R.A. Basic Concepts of Extrusion. *Ind. Eng. Chem.* **1953**, *45*, 970–973. [CrossRef]
17. Middleman, S. Extrusion. In *Fundamentals of Polymer Processing*; McGraw-Hill College: New York, NY, USA, 1997; pp. 123–169.
18. Erdmenger, R. Mehrwellen-Schnecken in der Verfahrenstechnik. *Chem. Ing. Tech.* **1964**, *36*, 175–185. [CrossRef]
19. Booy, M.L. Geometry of fully wiped twin-screw equipment. *Polym. Eng. Sci.* **1978**, *18*, 973–984. [CrossRef]
20. Denson, C.D.; Hwang, B.K. The influence of the axial pressure gradient on flow rate for Newtonian liquids in a self wiping, co-rotating twin screw extruder. *Polym. Eng. Sci.* **1980**, *20*, 965–971. [CrossRef]
21. Booy, M.L. Isothermal flow of viscous liquids in corotating twin screw devices. *Polym. Eng. Sci.* **1980**, *20*, 1220–1228. [CrossRef]
22. Szydlowski, W.; White, J.L. An improved theory of metering in an intermeshing corotating twin-screw extruder. *Adv. Polym. Technol.* **1987**, *7*, 177–183. [CrossRef]
23. White, J.L.; Szydlowski, W. Composite models of modular intermeshing corotating and tangential counter-rotating twin screw extruders. *Adv. Polym. Technol.* **1987**, *7*, 419–426. [CrossRef]
24. Tayeb, J.; Vergnes, B.; Valle, G.D. A Basic Model for a Twin-Screw Extruder. *J. Food Sci.* **1989**, *54*, 1047–1056. [CrossRef]
25. Baird, D.G.; Collias, D.I. Extruders. In *Polymer Processing*; Brenner, H., Ed.; Butterworth-Heinemann: Oxford, UK, 1995; pp. 235–240.
26. Kiesskalt, S. Untersuchungen an einer Kapsel Pumpe. *VDI Zeitschr* **1927**, *71*, 453.
27. Schenkel, G. *Kunststoff-Extrudertechnik: Konstruktionsgrundlagen und Betriebstechnik der Kunststoff-Schneckenpressen*; Carl Hanser Verlag: Munich, Germany, 1963.
28. Kaplan, A.; Tadmor, Z. Theoretical model for non-intermeshing twin screw extruders. *Polym. Eng. Sci.* **1974**, *14*, 58–66. [CrossRef]
29. Kim, M.H.; Szydlowski, W.; White, J.L.; Min, K. A new model of flow in a tangential counter-rotating twin screw extruder. *Adv. Polym. Technol.* **1989**, *9*, 87–95. [CrossRef]
30. Doboczky, Z.E. Doppelschnecke. *Plastverarbeiter* **1965**, *16*, 57–67.
31. Doboczky, Z. Theoretische and wirkliche Ausstoßleistung der Doppelschnecken Extruder. *Plastverarbeiter* **1965**, *16*, 395–400.
32. Janssen, L. *Twin Screw Extrusion*; Elseviers Sci. Publ. Company: Amsterdam, The Netherlands, 1978.
33. Janssen, L.P.B.; Lphrm, M.; Jm, S. A Model for the Output from the Pump Zone of the Double-screw Processor or Extruder. *Plast. Polym.* **1975**, *43*, 93–98.
34. Nastaj, A.; Wilczyński, K. Scaling-Up for the Counter-Rotating Twin Screw Extrusion of Polymers. *Polymers* **2024**, *16*, 2720. [CrossRef]
35. Klein, I.; Tadmor, Z. The simulation of the plasticating screw extrusion process with a computer programmed theoretical model. *Polym. Eng. Sci.* **1969**, *9*, 11–21. [CrossRef]
36. Acur, E.E.; Vlachopoulos, J. Numerical simulation of a single-screw plasticating extruder. *Polym. Eng. Sci.* **1982**, *22*, 1084–1094. [CrossRef]
37. Amellal, K.; Lafleur, P.G.; Arpin, B. A Computer Model for the Simulation of Conventional and Barrier Screws. *Polym.-Plast. Technol. Eng.* **1991**, *30*, 655–664. [CrossRef]
38. Macgregor, A.; Vlachopoulos, J.; Vlcek, J. Computer Simulation of Conventional and Barrier Screw Extruders. *J. Reinf. Plast. Compos.* **1997**, *16*, 1270–1280. [CrossRef]
39. Aigner, M.; Köpplmayr, T.; Kneidinger, C.; Miethlinger, J. A network-analysis-based comparative study of the throughput behavior of polymer melts in barrier screw geometries. *AIP Conf. Proc.* **2014**, *1593*, 606–610.
40. Stueker, D.; Schoeppner, V. Simplified numerical calculation of the isothermal, three-dimensional, Non-Newtonian flow characteristics of single-screw melt-extruder. *AIP Conf. Proc.* **2023**, *2607*, 060008.
41. Hube, S.; Behr, M.; Elgeti, S.; Schön, M.; Sasse, J.; Hopmann, C. Numerical design of distributive mixing elements. *Finite Elem. Anal. Des.* **2022**, *204*, 103733. [CrossRef]
42. Marschik, C.; Roland, W. Predicting the pumping capability of single-screw extruders: A comparison of two- and three-dimensional modeling approaches. *AIP Conf. Proc.* **2023**, *2607*, 020002.

43. Ershov, S.V.; Trufanova, N.M. Numerical studies of the polymer melt flow in the extruder screw channel and the forming tool. *IOP Conf. Ser. Mater. Sci. Eng.* **2017**, *208*, 012018. [CrossRef]
44. Kadyirov, A.; Gataullin, R.; Karaeva, J. Numerical Simulation of Polymer Solutions in a Single-Screw Extruder. *Appl. Sci.* **2019**, *9*, 5423. [CrossRef]
45. Lim, K.H.; Hwang, W.R.; Kim, S.J. A finite-element technique for flows in the single screw extruder using a partial periodic unit. *Korea-Aust. Rheol. J.* **2019**, *31*, 59–67. [CrossRef]
46. Vachagina, E.K.; Kadyirov, A.I.; Karaeva, J.V. Simulation of Giesekus fluid flow in extruder using helical coordinate system. *IOP Conf. Ser. Mater. Sci. Eng.* **2020**, *733*, 012033. [CrossRef]
47. Herzog, D.; Roland, W.; Marschik, C.; Berger-Weber, G.R. Generalized predictions of the pumping characteristics and viscous dissipation of single-screw extruders including three-dimensional curvature effects. *Polym. Eng. Sci.* **2024**, *64*, 5566–5587. [CrossRef]
48. Janßen, M.; Schiffers, R. Predicting the rotation speed of mixing sleeves in single-screw extrusion by means of CFD-simulation. *AIP Conf. Proc.* **2023**, *2607*, 020003.
49. Schöppner, V.; Schadomsky, M.; Hopmann, C.; Lemke, F. Investigations of the mixing behaviour of pin-type rubber extruders. *AIP Conf. Proc.* **2016**, *1713*, 130003.
50. Kimura, K.; Nakayama, Y.; Kajiwara, T. Distributive mixing characteristics of a Dulmage-type screw for a single-screw extruder: Experimental and numerical evaluation. *Chem. Eng. J. Adv.* **2021**, *7*, 100137. [CrossRef]
51. Erb, T.; Celik, O.; Bonten, C. Three-dimensional characterization of dispersive mixing in single screw extruders. *AIP Conf. Proc.* **2016**, *1779*, 050001.
52. Frank, M.; Hollenhorst, V.; Schöppner, V. Investigation of the pressure-throughput behaviour of different faceted mixer geometries on an analytical and simulative basis. *AIP Conf. Proc.* **2023**, *2607*, 060010.
53. Wang, Z.; Pan, Y.; Liu, Y.; Huang, J.; Wang, N.; Hu, X. Investigations of the Elongational Deformation Induced by Pins in Pin-Barrel Cold-Feed Extruders. *Adv. Polym. Technol.* **2022**, *2022*, 3513804. [CrossRef]
54. Marschik, C.; Osswald, T.A.; Roland, W.; Albrecht, H.; Skrabala, O.; Miethlinger, J. Numerical analysis of mixing in block-head mixing screws. *Polym. Eng. Sci.* **2019**, *59*, E88–E104. [CrossRef]
55. Motzigemba, M.; Broecker, H.C.; Prüss, J.; Bothe, D.; Warnecke, H.J. Chapter 37—A contribution to simulation of mixing in screw extruders employing commercial CFD-software. In *10th European Conference on Mixing*; van den Akker, H.E.A., Derksen, J.J., Eds.; Elsevier Science: Amsterdam, The Netherlands, 2000; pp. 297–304.
56. Straka, K.; Praher, B.; Steinbichler, G. Analyzing melt homogeneity in a single screw plasticizing unit of an injection molding machine. *AIP Conf. Proc.* **2013**, *1558*, 94–97.
57. Alemaskin, K.; Manas-Zloczower, I.; Kaufman, M. Color mixing in the metering zone of a single screw extruder: Numerical simulations and experimental validation. *Polym. Eng. Sci.* **2005**, *45*, 1011–1020. [CrossRef]
58. Domingues, N.; Gaspar-Cunha, A.; Covas, J.A. Computational and Experimental Study of Mixing in a Single Screw Extruder. *AIP Conf. Proc.* **2007**, *907*, 867–872.
59. Frank, M.; Schöppner, V. Investigation of the homogenization performance of various faceted mixers and optimization with regard to mixing as well as pressure throughput behavior. *AIP Conf. Proc.* **2023**, *2884*, 090003.
60. Liu, L.; Meng, Z.; Zhang, Y.; Sun, Y. Simulation of High-Viscosity Generalized Newtonian Fluid Flows in the Mixing Section of a Screw Extruder Using the Lattice Boltzmann Model. *ACS Omega* **2023**, *8*, 47991–48018. [CrossRef]
61. Pandey, V.; Maia, J.M. Extension-dominated improved dispersive mixing in single-screw extrusion. Part 1: Computational and experimental validation. *J. Appl. Polym. Sci.* **2021**, *138*, 49716. [CrossRef]
62. Pandey, V.; Maia, J.M. Comparative computational analysis of dispersive mixing in extension-dominated mixers for single-screw extruders. *Polym. Eng. Sci.* **2020**, *60*, 2390–2402. [CrossRef]
63. Wang, W.; Manas-Zloczower, I. Temporal distributions: The basis for the development of mixing indexes for scale-up of polymer processing equipment. *Polym. Eng. Sci.* **2001**, *41*, 1068–1077. [CrossRef]
64. Schulz, L.; Vorjohann, F.; Janßen, M.; Schiffers, R. Evaluating distributive mixing potential by mean distance of streamlines and a passive scalar. *AIP Conf. Proc.* **2023**, *2884*, 110004.
65. Kubik, P.; Zatloukal, M.; Vlcek, J.; Womer, T. Three-dimensional finite element method simulation study of fusion screw geometry. *Plast. Rubber Compos.* **2015**, *44*, 11–18. [CrossRef]
66. Moritzer, E.; Wittke, M.; Heiderich, G. Numerical simulation of residence time, shear rates and throughput in single screw extruders considering wall shear rates. *AIP Conf. Proc.* **2019**, *2065*, 030034.
67. Kim, S.J. Dimensionless analysis of three-dimensional residence time distribution in single-screw extrusion processes. *Korea-Aust. Rheol. J.* **2018**, *30*, 179–188. [CrossRef]
68. Olofsson, E.H.; Roland, M.; Spangenberg, J.; Jokil, N.H.; Hattel, J.H. A CFD model with free surface tracking: Predicting fill level and residence time in a starve-fed single-screw extruder. *Int. J. Adv. Manuf. Technol.* **2023**, *126*, 3579–3591. [CrossRef]

69. Mateboer, T.J.; Hummel, C.; van Dijk, D.J.; Buist, J. To quantify mixing quality in a single screw extruder simulation. In Proceedings of the 14th International Conference on CFD in 6 Oil & Gas, Metallurgical and Process Industries SINTEF, Trondheim, Norway, 12–14 October 2020.
70. Hopmann, C.; Schön, M.; Reul, M.M.; Facklam, M. A Method for the Validation of Simulated Mixing Characteristics of Two Dynamic Mixers in Single-Screw Extrusion. *Polymers* **2020**, *12*, 2234. [CrossRef]
71. Strutt, D.; Tzoganakis, C.; Duever, T.A. Mixing analysis of reactive polymer flow in a single-screw extruder channel. *Polym. Eng. Sci.* **2000**, *40*, 992–1003. [CrossRef]
72. Roy, S.; Lawal, A. Isothermal Pseudo-2D Analysis of Reactive Extrusion in Single-Screw Extruders. *J. Reinf. Plast. Compos.* **2004**, *23*, 685–706. [CrossRef]
73. Habla, F.; Obermeier, S.; Dietsche, L.; Kintzel, O.; Hinrichsen, O. CFD Analysis of the Frame Invariance of the Melt Temperature Rise in a Single-screw Extruder. *Int. Polym. Process.* **2013**, *28*, 463–469. [CrossRef]
74. Manohar, B.; Periasamy, C. Computational fluid dynamics simulation of single screw extruders in cable industries. *IJRET Int. J. Res. Eng. Technol.* **2016**, *5*, 5.
75. Casari, N.; Fadiga, E.; Pinelli, M.; Suman, A.; Ziviani, D. CFD Simulations of Single- and Twin-Screw Machines with OpenFOAM. *Designs* **2020**, *4*, 2. [CrossRef]
76. Roland, W.; Marschik, C.; Kommenda, M.; Haghofer, A.; Dorl, S.; Winkler, S. Predicting the Non-Linear Conveying Behavior in Single-Screw Extrusion: A Comparison of Various Data-Based Modeling Approaches used with CFD Simulations. *Int. Polym. Process.* **2021**, *36*, 529–544. [CrossRef]
77. Abdel-Ghany, W.E.; Ebeid, S.J.; Fikry, I. Mechanical design aspects of single screw extruders using finite element analysis. *Int. J. Eng. Tech. Res. (IJETR)* **2015**, *3*, 2454–4698.
78. Marschik, C.; Roland, W.; Miethlinger, J. A Network-Theory-Based Comparative Study of Melt-Conveying Models in Single-Screw Extrusion: A. Isothermal Flow. *Polymers* **2018**, *10*, 929. [CrossRef]
79. Stritzinger, U.; Roland, W.; Berger-Weber, G.; Steinbichler, G. Modeling melt conveying and power consumption of co-rotating twin-screw extruder kneading blocks: Part A. Data generation. *Polym. Eng. Sci.* **2022**, *62*, 3721–3745. [CrossRef]
80. Van Der Wal, D.; Goffart, D.; Klomp, E.; Hoogstraten, H.; Janssen, L. Three-dimensional flow modeling of a self-wiping corotating twin-screw extruder. Part II: The kneading section. *Polym. Eng. Sci.* **1996**, *36*, 912–924. [CrossRef]
81. Eitzlmayr, A.; Khinast, J. Co-rotating twin-screw extruders: Detailed analysis of conveying elements based on smoothed particle hydrodynamics. Part 2: Mixing. *Chem. Eng. Sci.* **2015**, *134*, 880–886. [CrossRef]
82. Eitzlmayr, A.; Matić, J.; Khinast, J. Analysis of Flow and Mixing in Screw Elements of Co-Rotating Twin-Screw Extruders via SPH. *AIChE J.* **2016**, *63*, 2451–2463. [CrossRef]
83. Rios, A.C.; Gramann, P.J.; Osswald, T.A. Comparative study of mixing in corotating twin screw extruders using computer simulation. *Adv. Polym. Technol.* **1998**, *17*, 107–113. [CrossRef]
84. Goffart, D.; Van Der Wal, D.; Klomp, E.; Hoogstraten, H.; Janssen, L.; Breysse, L.; Trolez, Y. Three-dimensional flow modeling of a self-wiping corotating twin-screw extruder. Part I: The transporting section. *Polym. Eng. Sci.* **1996**, *36*, 901–911. [CrossRef]
85. Hinz, J.; Helmig, J.; Möller, M.; Elgeti, S. Boundary-conforming finite element methods for twin-screw extruders using spline-based parameterization techniques. *Comput. Methods Appl. Mech. Eng.* **2020**, *361*, 112740. [CrossRef]
86. Helmig, J.; Behr, M.; Elgeti, S. Boundary-conforming finite element methods for twin-screw extruders: Unsteady-temperature-dependent-non-Newtonian simulations. *Comput. Fluids* **2019**, *190*, 322–336. [CrossRef]
87. Helmig, J.; Key, F.; Behr, M.; Elgeti, S. Combining boundary-conforming finite element meshes on moving domains using a sliding mesh approach. *Int. J. Numer. Methods Fluids* **2021**, *93*, 1053–1073. [CrossRef]
88. Ishikawa, T.; Kihara, S.I.; Funatsu, K. 3-D numerical simulations of nonisothermal flow in co-rotating twin screw extruders. *Polym. Eng. Sci.* **2000**, *40*, 357–364. [CrossRef]
89. Dong, T.; Wu, J. Simulation of non-Newtonian flows in a partially filled twin-screw extruder by smoothed particle hydrodynamics. *Polym. Eng. Sci.* **2022**, *62*, 802–814. [CrossRef]
90. Bauer, H.; Matić, J.; Khinast, J. Characteristic parameters and process maps for fully-filled twin-screw extruder elements. *Chem. Eng. Sci.* **2021**, *230*, 116202. [CrossRef]
91. Lewandowski, A.; Wilczyński, K.J.; Nastaj, A.; Wilczyński, K. A composite model for an intermeshing counter-rotating twin-screw extruder and its experimental verification. *Polym. Eng. Sci.* **2015**, *55*, 2838–2848. [CrossRef]
92. Prashanth, S.R.; Arumugam, S.K.; Gangradey, R.; Mukherjee, S.; Kasthurirengan, S.; Behera, U.; Pabbineedi, G.; Mugilan, M. CFD modelling and performance analysis of a twin screw hydrogen extruder. *Fusion Eng. Des.* **2019**, *138*, 151–158.
93. Zhang, Y.; Jiang, X.; Fan, H.; Li, X. Optimization and Numerical Simulation of Outlet of Twin Screw Extruder. *MATEC Web Conf.* **2018**, *153*, 05004. [CrossRef]
94. Ilinca, F.; Hétu, J.-F. Three-dimensional finite element solution of the flow in single and twin-screw extruders. *Int. Polym. Process.* **2010**, *25*, 275–286. [CrossRef]

95. Hétu, J.-F.; Ilinca, F. Immersed boundary finite elements for 3D flow simulations in twin-screw extruders. *Comput. Fluids* **2013**, *87*, 2–11. [CrossRef]
96. Malik, M.; Kalyon, D.; Golba, J. Simulation of co-rotating twin screw extrusion process subject to pressure-dependent wall slip at barrel and screw surfaces: 3D FEM analysis for combinations of forward-and reverse-conveying screw elements. *Int. Polym. Process.* **2014**, *29*, 51–62. [CrossRef]
97. Valette, R.; Vergnes, B.; Coupez, T. Multiscale simulation of mixing processes using 3D-parallel, fluid-structure interaction techniques. *Int. J. Mater. Form.* **2008**, *1*, 1131–1134. [CrossRef]
98. Avalosse, T.; Rubin, Y.; Fondin, L. Non-isothermal modeling of co-rotating and contra-rotating twin screw extruders. *J. Reinf. Plast. Compos.* **2002**, *21*, 419–429. [CrossRef]
99. Alsteens, B.; Legat, V.; Avalosse, T. Parametric study of the mixing efficiency in a kneading block section of a twin-screw extruder. *Int. Polym. Process.* **2004**, *19*, 207–217. [CrossRef]
100. Xu, B.; Liu, Y.; He, L.; Chen, J.; Turng, L.S. Numerical study of mixing dynamics inside the novel elements of a corotating nontwin screw extruder. *Adv. Polym. Technol.* **2018**, *37*, 2478–2496. [CrossRef]
101. Bauer, H.; Matić, J.; Evans, R.C.; Gryczke, A.; Ketterhagen, W.; Sinha, K.; Khinast, J. Determining local residence time distributions in twin-screw extruder elements via smoothed particle hydrodynamics. *Chem. Eng. Sci.* **2022**, *247*, 117029. [CrossRef]
102. Valette, R.; Coupez, T.; David, C.; Vergnes, B. A Direct 3D Numerical Simulation Code for Extrusion and Mixing Processes. *Int. Polym. Process.* **2009**, *24*, 141–147. [CrossRef]
103. Robinson, M.; Cleary, P.W. Effect of geometry and fill level on the transport and mixing behaviour of a co-rotating twin screw extruder. *Comput. Part. Mech.* **2019**, *6*, 227–247. [CrossRef]
104. Oldemeier, J.P.; Schöppner, V. Analysis of the Dispersive and Distributive Mixing Effect of Screw Elements on the Co-Rotating Twin-Screw Extruder with Particle Tracking. *Polymers* **2024**, *16*, 2952. [CrossRef] [PubMed]
105. Kwag, D.S.; Lyu, M.Y.; Kim, W.S. A 3D Numerical Study of Fluid Flow and Heat Transfer in a Single Screw Extruder. *Int. Polym. Process.* **2002**, *17*, 95–101. [CrossRef]
106. Sarhangi Fard, A.; Anderson, P.D. Simulation of distributive mixing inside mixing elements of co-rotating twin-screw extruders. *Comput. Fluids* **2013**, *87*, 79–91. [CrossRef]
107. Ahmed, I.; Poudyal, H.; Chandy, A.J. Fill Factor Effects in Highly-Viscous Non-Isothermal Rubber Mixing Simulations. *Int. Polym. Process.* **2019**, *34*, 182–194. [CrossRef]
108. Sun, D.; Zhu, X.; Gao, M. 3D Numerical Simulation of Reactive Extrusion Processes for Preparing PP/TiO_2 Nanocomposites in a Corotating Twin Screw Extruder. *Materials* **2019**, *12*, 671. [CrossRef] [PubMed]
109. Bertrand, F.; Thibault, F.; Delamare, L.; Tanguy, P.A. Adaptive finite element simulations of fluid flow in twin-screw extruders. *Comput. Chem. Eng.* **2003**, *27*, 491–500. [CrossRef]
110. Eitzlmayr, A.; Khinast, J. Co-rotating twin-screw extruders: Detailed analysis of conveying elements based on smoothed particle hydrodynamics. Part 1: Hydrodynamics. *Chem. Eng. Sci.* **2015**, *134*, 861–879. [CrossRef]
111. Martelli, F.G. *Twin Screw Extruders: A Basic Understanding*; Springer: Berlin/Heidelberg, Germany, 1983.
112. Nakayama, Y.; Takemitsu, H.; Kajiwara, T.; Kimura, K.; Takeuchi, T.; Tomiyama, H. Improving mixing characteristics with a pitched tip in kneading elements in twin-screw extrusion. *AIChE J.* **2018**, *64*, 1424–1434. [CrossRef]

fluids

MDPI

Article

Analysis of the Effect of Cut Sweep Ratio of Lily Impeller on the Distribution of Dissolved Oxygen

Mohammad Tauviqirrahman *, Eflita Yohana, Jourdy Cakranegara , Jamari and Budi Setiyana

Department of Mechanical Engineering, Faculty of Engineering, Diponegoro University, Jl. Prof. Soedharto SH, Tembalang, Semarang 50275, Indonesia
* Correspondence: mtauviq99@lecturer.undip.ac.id

Abstract: The aquaculture industry encounters substantial obstacles, including organic pollution, oxygen insufficiency, and elevated levels of ammonia and carbon dioxide. Aeration systems are employed to enhance the process of oxygen transfer and promote circulation. The Lily impeller, a newly developed technology, has demonstrated reduced energy consumption in comparison to conventional impeller designs. The objective of this study is to examine how changes in the cut sweep ratio impact the distribution of dissolved oxygen in shrimp ponds, using computational fluid dynamics (CFD) simulation. A user-defined function (UDF) was utilized to incorporate a dissolved oxygen model into the pond. Five designs of Lily impellers were analyzed and compared with each other. This study demonstrated that alterations in the cut sweep ratio significantly affected the distribution of dissolved oxygen, dynamic pressure, and flow velocity in the pond. The "no cut" variant exhibited the highest average dissolved oxygen value of 0.00385 kg/m^3, along with a maximum dynamic pressure of 11.5 Pa and a maximum flow velocity of 0.96 m/s, resulting in the most significant outcomes. This study determined that only the immediate area surrounding the aerator possesses dissolved oxygen levels that are sufficiently elevated to support the survival of shrimp. Consequently, the installation of additional aerators is necessary to guarantee the presence of adequate dissolved oxygen throughout the entire pond.

Keywords: aeration; CFD (computational fluid dynamics); cut sweep ratio; dissolved oxygen; Lily impeller

1. Introduction

Indonesia is a maritime country—around two-thirds of its territory is water, both sea and river. Therefore, Indonesia has abundant water resource potential. Based on data from the Indonesian Ministry of Maritime Affairs and Fisheries (KKP), shrimp is the largest export commodity compared to other fisheries commodities. The export volume reached 239.28 million kg with a value of USD 2.04 billion in 2020 [1]. With increasing market demand, Indonesia is in a good position to develop shrimp production. In intensive cultivation systems, ponds are filled with many organisms and food supplies are high. As a result, many problems occur, such as organic pollution, lack of oxygen, increased levels of carbon dioxide, and ammonia [2–4].

Oxygen depletion in freshwater ponds is a dangerous and limiting factor because it causes hypoxia, which impacts production. Dissolved oxygen is one of the most important parameters and the main limiting factor that controls the growth and survival of fish [5]. Aquatic ecosystems have dissolved oxygen. There are two main sources of oxygen in ponds: oxygen diffusion from the atmosphere to the pond surface and photosynthesis by phytoplankton. Photosynthesis requires sunlight, so the pond surface produces oxygen during the day. The bottom layer lacks oxygen because sunlight cannot reach it. At night, the respiration of phytoplankton and fish consumes oxygen, thereby reducing the overall dissolved oxygen concentration of the pond [6].

An efficient aeration system can help increase the amount of dissolved oxygen in the water. Aeration is the process of introducing and distributing oxygen throughout the pond. To improve oxygen transport and circulation, aquaculture ponds often use aeration systems such as paddle wheels, diffusers, and impeller aerators [7]. The aeration system channels air into the pond, increasing oxygen levels and expanding DO dispersion. An indirect impact of aeration is water mixing, which is largely determined by the equipment used to aerate the pond [8]. Fish become more active and grow faster when given enough oxygen. Chen et al. [9] indicated that aerated ponds can yield 47% greater productivity compared to non-aerated ponds. Enhancing water quality will enhance the well-being of fish, consequently boosting their survival and growth rate. This can be achieved by implementing adequate aeration procedures [10].

Many aeration systems are available. Each option has pros and cons that must be considered in light of the environment. Wind-powered aeration (WPA) could be used in developing-country aquaculture ponds. WPA's vertical-axis wind turbine is directly connected to a subsurface impeller [6]. The impeller type and design greatly affect aeration system performance. The impeller design must meet three criteria, making it difficult. The impeller's leading and trailing blade edges must have good relative velocity distribution, reducing flow separation and performance degradation. The blade shape must also be precision-made and affordable. Engineered blades must sustain pressure at a safe level to eliminate operation malfunction [11]. Li and Chen [12] conducted a comparative analysis of impeller pitch angles of 30°, 45°, and 60° in their research. The outcome is that a pitch angle of 30° results in the highest oxygen transfer rate, but it necessitates the most power. In contrast, pitch angles of 45° and 60° result in lower speeds, but they also produce less power. Later, Zheng et al. [13] investigated concave-shaped impeller designs in comparison to the Rushton and Baker types. This study demonstrated that concave impellers yield superior oxygen-transfer efficiency while necessitating less power than Rushton and Baker-type impellers. Three types of impellers were examined in the research conducted by Karimi et al. [14], i.e., impellers with six blades, four blades, and two blades. At speeds ranging from 400 rpm to 800 rpm, impellers with six blades exhibit an oxygen mass transfer coefficient of 23% to 77%, which is higher than that of other impeller configurations. De Jesus et al. [15] analyzed the oxygen mass transfer coefficient between the radial and axial impellers at different velocities. This study revealed that the radial impeller yielded an oxygen mass transfer coefficient threefold greater than the axial impeller at speeds under 400 rpm. Jamshidzadeh et al. [16] evaluated the impact of impeller combinations on the oxygen mass transfer coefficient. The amalgamation utilized a Rushton-type impeller, a pitched blade impeller, and an elephant ear-type impeller. Pairing two Rushton-type impellers yields the highest oxygen mass transfer coefficient among all the combinations tested. Despite the numerous studies exploring different impeller configurations, the existing research has predominantly focused on traditional impeller designs, leaving a gap in understanding the influence of more unconventional parameters, such as the cut sweep ratio, particularly in aquatic environments like shrimp ponds.

While previous studies have extensively analyzed factors like impeller type, rotational speed, and tank geometry, the cut sweep ratio—a parameter that modifies the impeller blade's profile to influence the fluid flow and mixing efficiency—remains underexplored. Currently, researchers are exploring various types of impeller shapes, sizes, and operational parameters to improve the performance of aerators in aquaculture ponds. The purpose of the study is to ascertain the impact of the cut sweep ratio of the Lily impeller on the dissolved oxygen distribution using computational fluid dynamics (CFD). This research will examine five novel variations of the cut sweep ratio parameter on the Lily impeller.

2. Materials and Methods

2.1. Governing Equations

Momentum and continuity equations are fundamental principles in fluid dynamics that describe the behavior of flowing fluids. The continuity equation, on the other hand, expresses the conservation of mass within a fluid system. The continuity equation as expressed in Equation (2) states that the mass flow rate entering a control volume must equal the mass flow rate leaving it, ensuring that mass is neither created nor destroyed. The momentum equation, derived from Newton's second law, relates the rate of change of momentum of a fluid element to the forces acting on it. This equation is expressed in Equation (2).

$$\nabla \cdot \vec{u} = 0 \tag{1}$$

$$\rho\left(\frac{\partial \overline{u}_i}{\partial t} + \overline{u}_j \frac{\partial \overline{u}_i}{\partial x_j}\right) = \rho g + \frac{\partial}{\partial x_j}\left[-p\delta_{ij} + \mu\left(\frac{\partial \overline{u}_i}{\partial x_j} + \frac{\partial \overline{u}_j}{\partial x_i}\right) - \rho\overline{u_i' u_j'}\right] \tag{2}$$

where subindexes i and j are the summation of cartesian components, $\vec{u}$ is fluid velocity, $\overline{u}$ is the mean flow velocity, u' is the fluctuating component, ρ is the fluid density, p is the pressure, μ is the fluid dynamic viscosity, δ_{ij} is the Kronecker delta, and $\rho\overline{u_i' u_j'}$ are the Reynolds stresses.

The turbulence model used in this research is the Renormalization Group (RNG) k-ε turbulence model. This model has a sub-model called swirl-dominated flow, which makes this model more accurate and reliable for this research than the standard k-ε model [6]. The k-ε turbulence model consists of two transport equations. Equation (3) shows the transport equation for turbulent kinetic energy k and Equation (4) depicts the transport equation for turbulent dissipation rate ε.

$$\frac{\partial(\rho k)}{\partial t} + \frac{\partial}{\partial x_i}(\rho u_i k) = \frac{\partial}{\partial x_j}\left(\frac{\rho c_\mu k^2}{\varepsilon \sigma_k}\frac{\partial k}{\partial x_i}\right) + P_k - \rho\varepsilon \tag{3}$$

$$\frac{\partial(\rho \varepsilon)}{\partial t} + \frac{\partial}{\partial x_i}(\rho u_i \varepsilon) = \frac{\partial}{\partial x_j}\left(\frac{\rho c_\mu k^2}{\varepsilon \sigma_k}\frac{\partial \varepsilon}{\partial x_i}\right) + c_1 P_k \frac{\varepsilon}{k} - c_2 \rho \frac{\varepsilon^2}{k} - R_\varepsilon \tag{4}$$

where k is the turbulent kinetic energy, ε is the turbulent dissipation rate, P_k is the turbulent kinetic energy generation, c_μ, c_1, c_2 are the model coefficients, and R_ε is an additional term for RNG k-ε model.

2.2. Aeration

Aeration systems can be classified into two fundamental groups. The destratification aeration system employs compressed air to generate vertical water movement and mix the pond's water column, while the hypolimnetic aeration system introduces oxygen to the pond without disrupting stratification [17]. The inability to recognize and manage stratification may result in diminished oxygen levels and mortality in shrimp [18]. There are many factors that can affect aeration, such as temperature and pH. As water temperature increases, DO concentration at saturation decreases [19]. Meanwhile, aqueous systems have been shown to have lower mass transfer coefficients than pure systems (distilled water) [20]. Microorganisms and organisms inside the pond consume oxygen during respiration. A source term was created to take into account the biological uptake of oxygen based on known metabolic rates by Vinatea et al. [21]. This research focuses on vaname shrimp, an aquatic organism residing at the pond's bottom, necessitating the installation of an aerator to ensure sufficient dissolved oxygen levels for its growth. A dissolved oxygen model was developed to replicate the aeration process in ponds, based on the work of Wu et al. [22], utilizing a user-defined function (UDF). This model employs the transient convection–diffusion equation as the mass transfer equation, as illustrated in Equation (5):

$$\frac{\partial \rho \phi}{\partial t} + \frac{\partial}{\partial x_j}\left(\vec{\psi}\phi - \Gamma\frac{\partial \phi}{\partial x_j}\right) = S_\phi \tag{5}$$

where ϕ is the mass concentration, which in this case is the dissolved oxygen concentration, $\vec{\psi}$ is the mass flux, Γ is the mass diffusivity which is $\Gamma = 3.36 \times 10^{-4}$ m^2/s and S_ϕ is the mass source item. The value of S_ϕ is calculated by Equation (6):

$$S_\phi = a(C_S - C) \tag{6}$$

where C_S is the saturated DO concentration, C is the local DO concentration, and a is the source item coefficient, which is called the reaeration coefficient. The initial value of DO or local DO concentration is 0.003 kg/m^3. The saturated DO concentration value is 0.0085 kg/m^3, and the value for a is a = 22.34/s, which was obtained using data fitting from experimental data [22]. The saturated DO concentration is determined by considering the temperature and pH level of the pond water using equations based on the study by Akinwole et al. [7].

2.3. Lily Impeller

The Lily impeller is a spiral mixer intended for insertion into a water tank to generate a rotational motion in the contained water. The design draws inspiration from the spiral form of the Lily flower and seeks to emulate the dynamics of ocean eddies. The Lily impeller emulates the natural spiral flow patterns found in whirlpools and tornadoes, hence enhancing the performance and energy efficiency of mixed water tanks [22]. The Lily impeller delivers a greater mass flow rate per unit of energy input compared to the traditional hydrofoil impeller [6]. The Lily impeller possesses multiple geometric parameters. In this instance, the cut sweep ratio will serve as the primary parameter of the investigation. The ratio of the impeller's upper and lower radii is known as the cut sweep ratio. The shape of the impeller is important for maximizing mass transfer efficiency, energy efficiency, improving yields, and sustainability in shrimp ponds. Optimizing the cut sweep ratio will not only enhance the performance of the impeller but also contribute to the long-term viability and sustainability of aquaculture systems [23]. Examples of models and their parameters are illustrated in Figure 1 and Table 1.

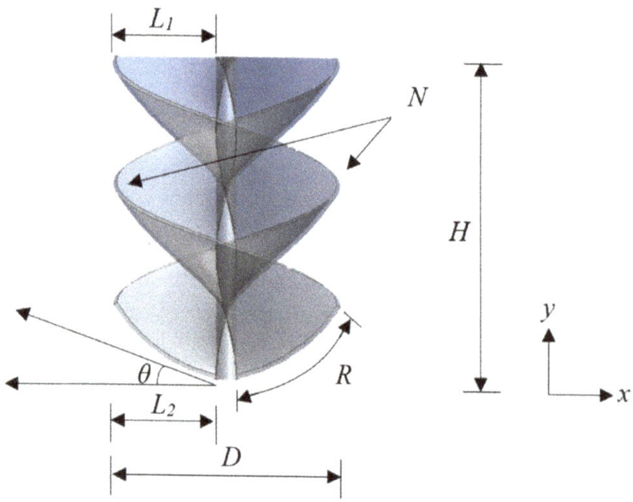

Figure 1. Geometric parameters of the Lily impeller.

Table 1. Geometric parameters of Figure 1.

Parameter	Symbol	Figure 1	No Cut	Cut 3/4	Cut 2/4	Cut 1/4	Full Cut	Validation	Simulation
Scale	-	-	-	-	-	-	-	1	5
Angle between the start of the blade and the horizontal line	θ	25°	25°	25°	25°	25°	25°	25°	25°
Cut sweep ratio (L1/L2)	α	1	1	3/4	2/4	1/4	0	1 (no cut)	1 (no cut)
Blade curvature radius	R	40 mm	40 mm	40 mm	40 mm	40 mm	40 mm	40 mm	200 mm
Number of blades	N	2	2	2	2	2	2	2	2
Impeller height	H	75 mm	75 mm	75 mm	75 mm	75 mm	75 mm	75 mm	375 mm
Radius of the blade at the top of the impeller	L_1	25 mm	25 mm	18.75 mm	12.5 mm	6.25 mm	0 mm	25 mm	125 mm
Radius of the blade at the bottom of the impeller	L_2	25 mm	25 mm	25 mm	25 mm	25 mm	25 mm	25 mm	125 mm
Impeller diameter	D	55 mm	55 mm	55 mm	55 mm	55 mm	55 mm	55 mm	275 mm

In this study, five cut sweep ratios are analyzed: no cut, cut 3/4, cut 2/4, cut 1/4, and full cut. All these variations and their respective parameter values can be seen in Figure 2 and Table 1.

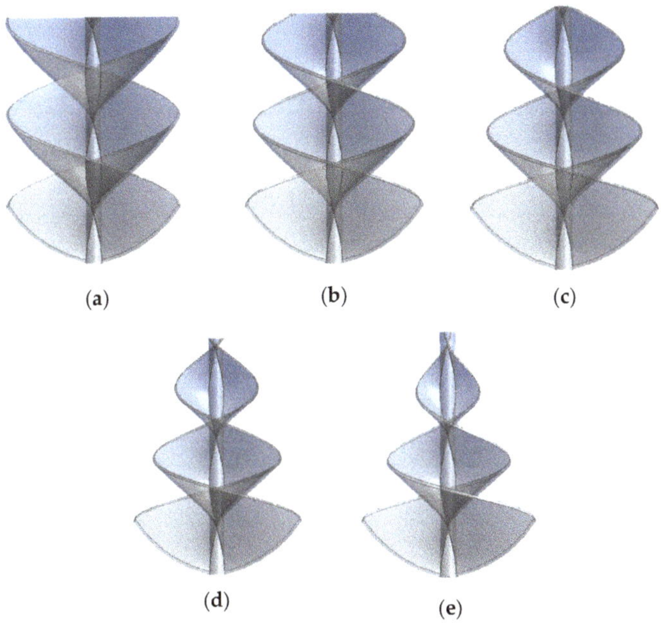

Figure 2. Lily impeller with the variations of cut sweep ratio: (**a**) no cut, (**b**) cut 3/4, (**c**) cut 2/4, (**d**) cut 1/4, (**e**) full cut.

The five impellers are then enlarged using the scale feature in DesignModeler. This enlargement is carried out based on the height ratio of the validation pond and the simulation pond, which is 1:5. Table 1 shows the results of the scale-enlargement measures.

In this simulation study, a pond with dimensions of 25 m in length and width, and a depth of 1.2 m, serves as the model for analysis. This pond is designed to replicate a controlled aquatic environment for the simulation, which aims to assess various physical and environmental factors. Figure 3 offers a visual representation of the pond model, allowing for a clearer understanding of its structure and layout.

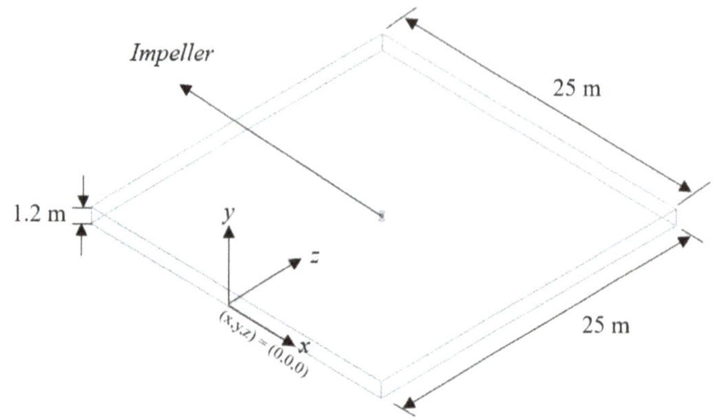

Figure 3. Geometric parameters of pond.

2.4. Cell Zone Conditions

In this study, Boolean operations are used to obtain two fluid zones for the computational fluid domain. The first is the rotating zone that includes the impeller and the water around it, and the second is the remaining water in the pond. Impeller rotation is simulated using the Frame Motion feature in Fluent. A rotation speed of 50 rpm is given to the rotating zone, while the rest of the pond is set in a stationary state. Figure 4 shows the computational fluid domain. The model consists of two distinct domains: a stationary domain representing the outer region of the pond and a rotating domain where the agitation occurs. The rotating domain, is illustrated in the right-hand section of the image as a cylindrical area containing a Lily impeller structure. This setup enables the simulation to capture the interaction between the stationary fluid and the rotating agitation, which is critical for analyzing flow patterns and other dynamic behaviors within the pond environment. Each side of the domain is subjected to wall boundary conditions with no-slip boundary conditions, with the exception of the top surface, where shear stress is set to zero. This simulation is conducted using the pressure implicit with splitting of operators (PISO) algorithm.

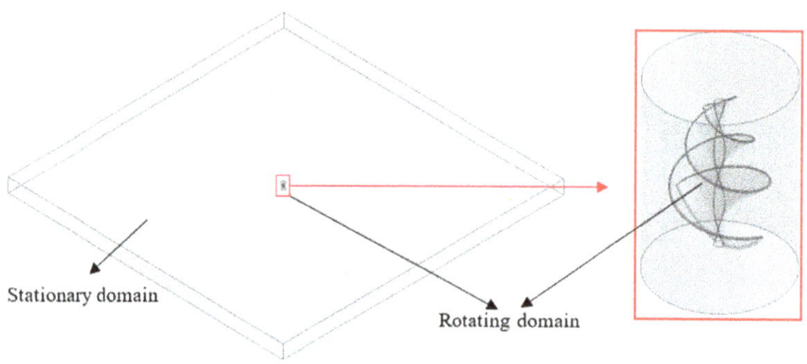

Figure 4. Fluid domain of the simulation.

2.5. Meshing

In this simulation, meshing is performed using an automatic method to ensure that the intricate geometries of the system, particularly around the impeller, are accurately captured. The mesh is composed of tetrahedral elements, which are well-suited for the modeling complex: curved surfaces like those seen in the impeller design. As depicted in Figure 5, the overall mesh comprises approximately 1.4 million elements, offering a high level of detail and precision in areas of both high and low curvature. The figure shows the progression of the mesh at various magnifications, starting from a wider view of the entire system down to the localized region around the impeller. The closer views highlight the fine mesh used to capture the precise contours of the impeller blades, ensuring that flow dynamics in this region are accurately represented in the simulation. This detailed meshing is critical for resolving the complex interactions between the rotating impeller and the surrounding fluid, allowing for a more accurate analysis of fluid velocity, turbulence, and pressure distribution in the system. The balance between mesh resolution and computational efficiency has been optimized to deliver reliable results without unnecessary computational effort.

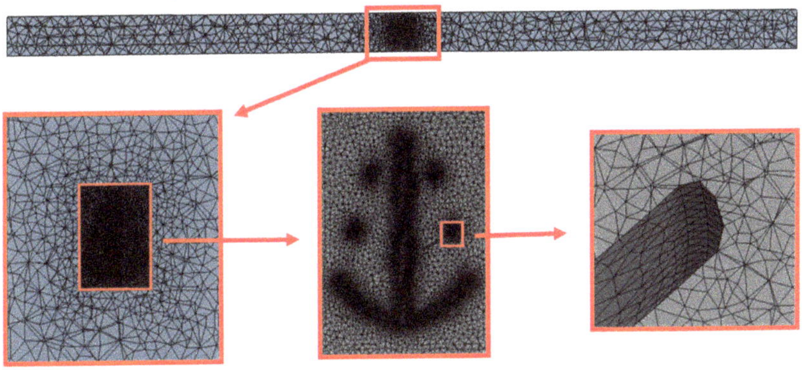

Figure 5. Meshing of the computational domain.

The meshing results are evaluated based on two key quality parameters: skewness and orthogonal quality, both of which are crucial for ensuring accurate and reliable computational fluid dynamics (CFD) simulations. Skewness refers to how distorted the mesh elements are concerning an ideal shape, with lower skewness values indicating better quality. In this case, the maximum skewness obtained is 0.83, which is within the acceptable range for tetrahedral elements. Orthogonal quality, on the other hand, measures the angular alignment of the mesh elements relative to each other and is a direct indicator of the element's shape quality. The minimum orthogonal quality is 0.16, which, although on the lower side, is still considered acceptable for this type of simulation. These quality metrics ensure that the mesh is capable of accurately capturing the flow patterns and interactions, especially around critical areas such as the impeller. By meeting acceptable skewness and orthogonal quality thresholds, the mesh is considered reliable for the simulation, ensuring that the results will be valid while balancing computational efficiency.

3. Results and Discussion

3.1. Grid Independence Test

The grid independence test is carried out to ensure that the simulation results are not influenced by the meshing resolution. This test is very important because the accuracy of the results can vary depending on the quality of the meshing. Poor meshing quality can cause incorrect simulation results. Figure 6 shows the independence grid results with the mass flow rate on the y-axis and the number of elements on the x-axis. The mass flow rate in Figure 6 begins to converge at 1.1 million elements, indicating that there is no significant difference in the result as the mesh number increases. Consequently, these elements will be employed in the meshing options for all subsequent simulations.

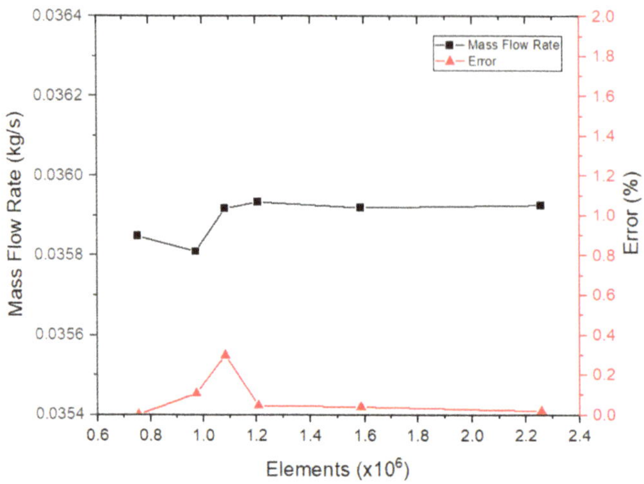

Figure 6. Grid independence result.

3.2. Validation

Validation is an essential process in assessing the reliability of a model to yield accurate and valuable insights. Validation was conducted by comparing the simulation results with data from Mahmudov et al. [6]. The reference data indicated a mass flow rate of 7.96 kg/s. The Lily impeller model used in this study was constructed based on the parameters provided in the reference. The simulation yielded a mass flow rate of 7.84 kg/s, resulting in a deviation of 1.5%. Figure 7 illustrates the comparison between the findings from the reference study and this research. From this comparison, it can be concluded that the developed method is sufficiently accurate.

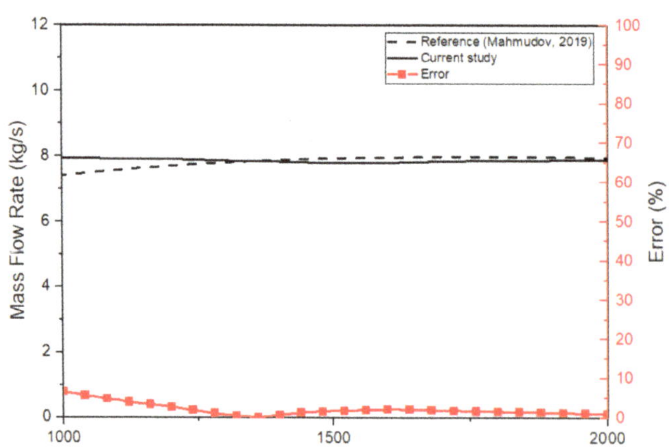

Figure 7. Comparison of mass flow rate between the current simulation and reference data from Mahmudov et al. [6].

3.3. Velocity Contours

The aerator generates a rotating motion, which induces water circulation throughout the pond. This circulation is crucial for ensuring the even distribution of dissolved oxygen, which is vital for aquatic life. The simulation results are visualized through a flow streamline in Figure 8, while the velocity pathline on the horizontal plane is depicted in Figure 9.

Each variation in the simulation influences the velocity contours differently, demonstrating distinct flow behaviors. The highest velocity was observed in the "no cut" variation, reaching a peak speed of 0.96 m/s. This indicates that the "no cut" configuration optimizes the flow rate and enhances circulation, making it the most effective setup in terms of speed and distribution efficiency within the pond. These findings highlight the importance of optimizing aerator design for improved water movement and particle management.

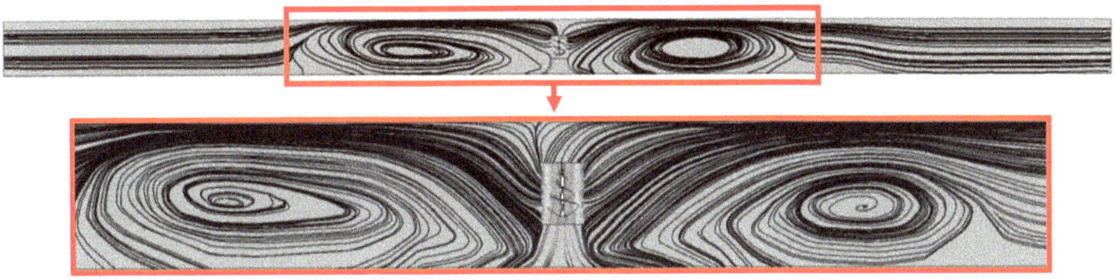

Figure 8. Flow streamline.

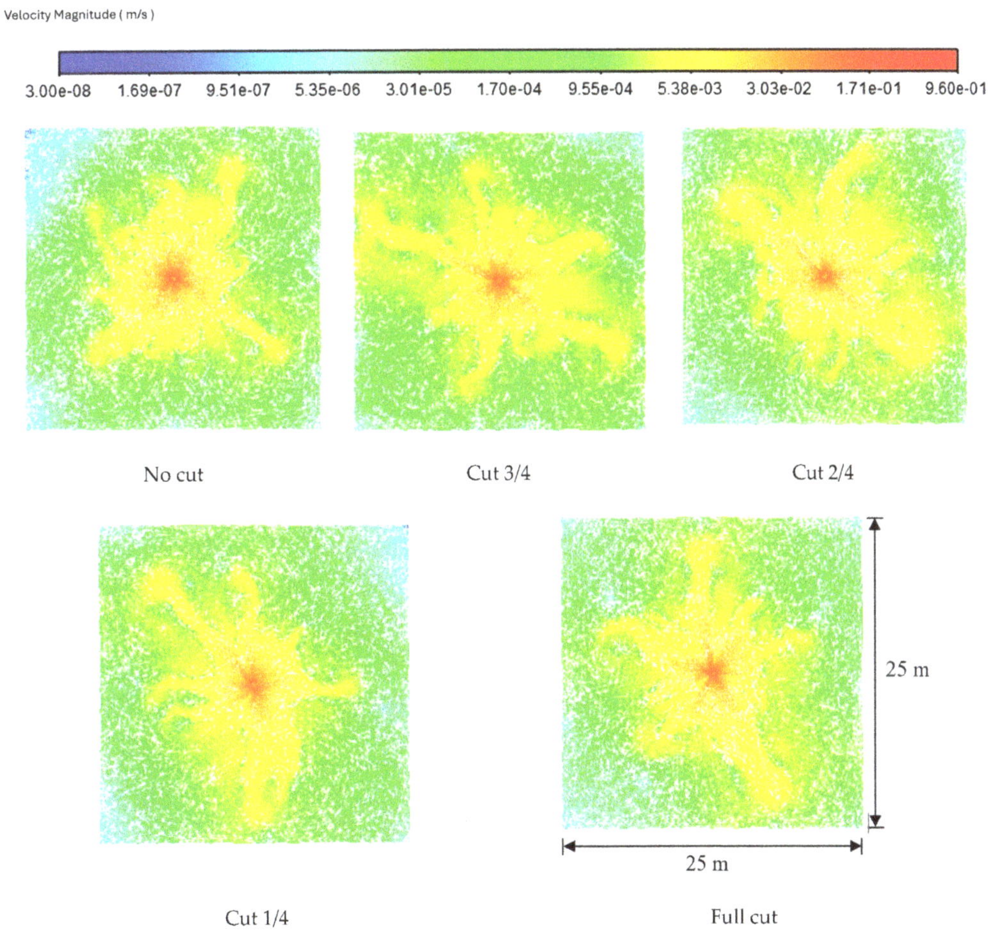

Figure 9. Velocity pathline and varying patterns (Note: the results are evaluated at y = 0 m).

Figure 10 illustrates the flow velocity pathline at $y = 1.2$ m, corresponding to the pond surface. The maximum speed achieved by the "no cut" is 0.89 m/s. The velocity of each variation on the pond's surface is comparatively minimal relative to that at the pond's bottom. This occurs because the aerator is oriented towards the pond's bottom, where the shrimp reside. The surface velocity converges towards the center of the pond due to the aerator's influence, in contrast to Figure 9, where the water disperses from the center.

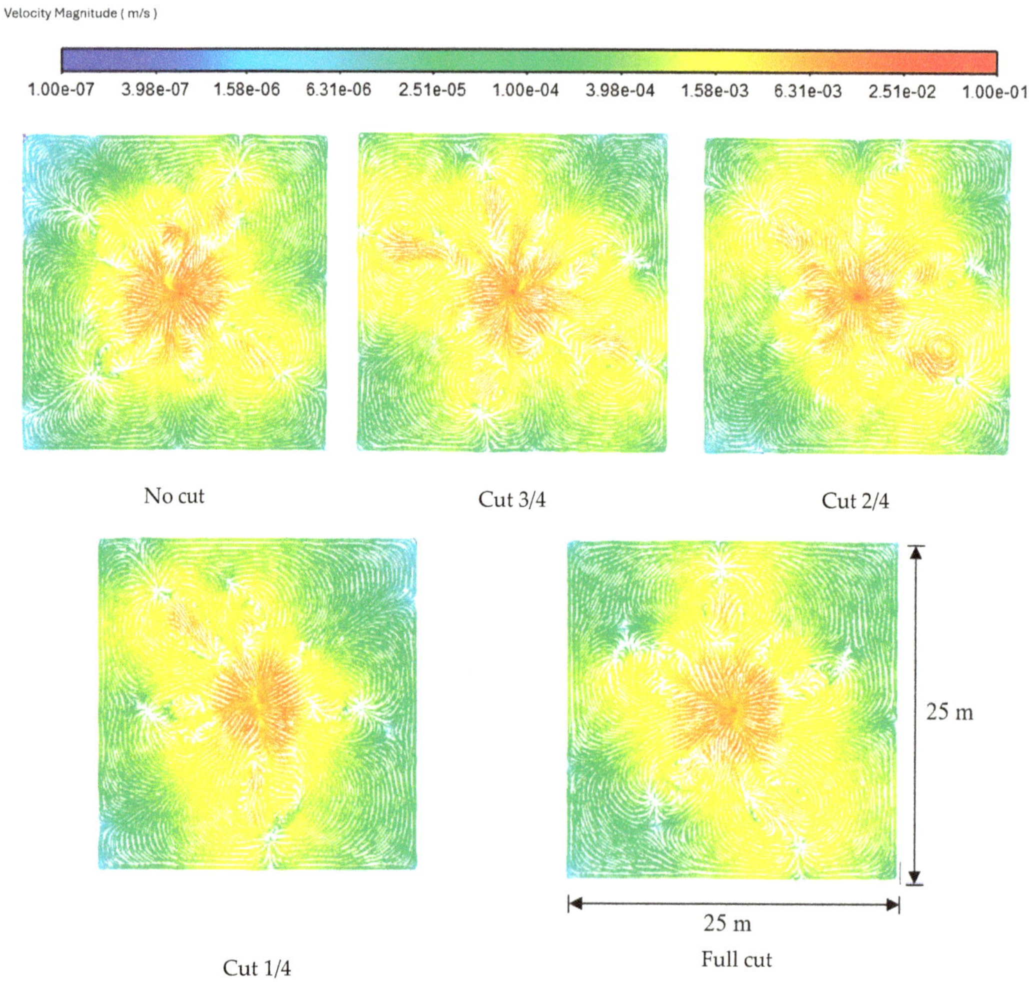

Figure 10. Velocity pathline at $y = 1.2$ m.

3.4. Pressure Contours

In this study, a strong correlation was observed between pressure and velocity, where dynamic pressure increases with higher fluid velocity. The pressure contour results in the vertical plane at $y = 0$ m are presented in Figure 11, showing how different variations affect pressure distribution. The highest dynamic pressure recorded in this analysis was for the "no cut" variation, with a peak value of 7.89 Pa. Additionally, the maximum dynamic pressure across the entire simulation was 11.5 Pa, also observed in the "no cut" variation, measured at $z = 12.5$ m. These findings highlight that as fluid velocity increases, dynamic pressure rises significantly, with the "no cut" configuration consistently producing the highest pressure values, demonstrating its effectiveness in optimizing fluid flow within the system.

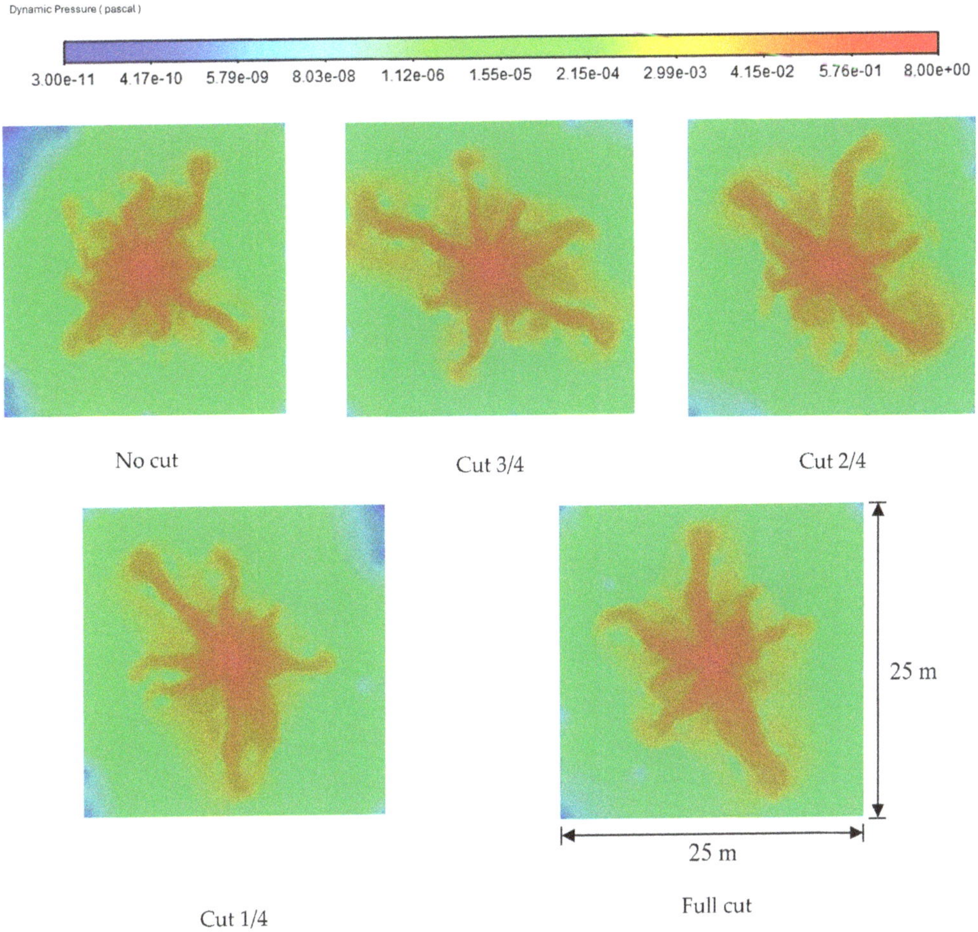

Figure 11. Pressure contour and varying impeller patterns (Note: the results are evaluated at $y = 0$ m).

Figure 12 presents the pressure contour at $y = 1.2$ m, representing the surface of the pond. The "no cut" variation achieved a maximum pressure of 0.24 Pa at this level. Similar to the trends discussed in Figures 10 and 11, the pressure at the surface of the pond is significantly lower compared to the bottom. This is due to the aerator's design, which directs water flow and pressure toward the pond's bottom, where aquatic life, such as shrimp, resides, ensuring optimal oxygenation and circulation in that region. The lower surface pressure reflects the reduced intensity of flow dynamics near the surface, where water is primarily drawn inward toward the aerator, contrasting with the more vigorous flow and higher pressures near the bottom.

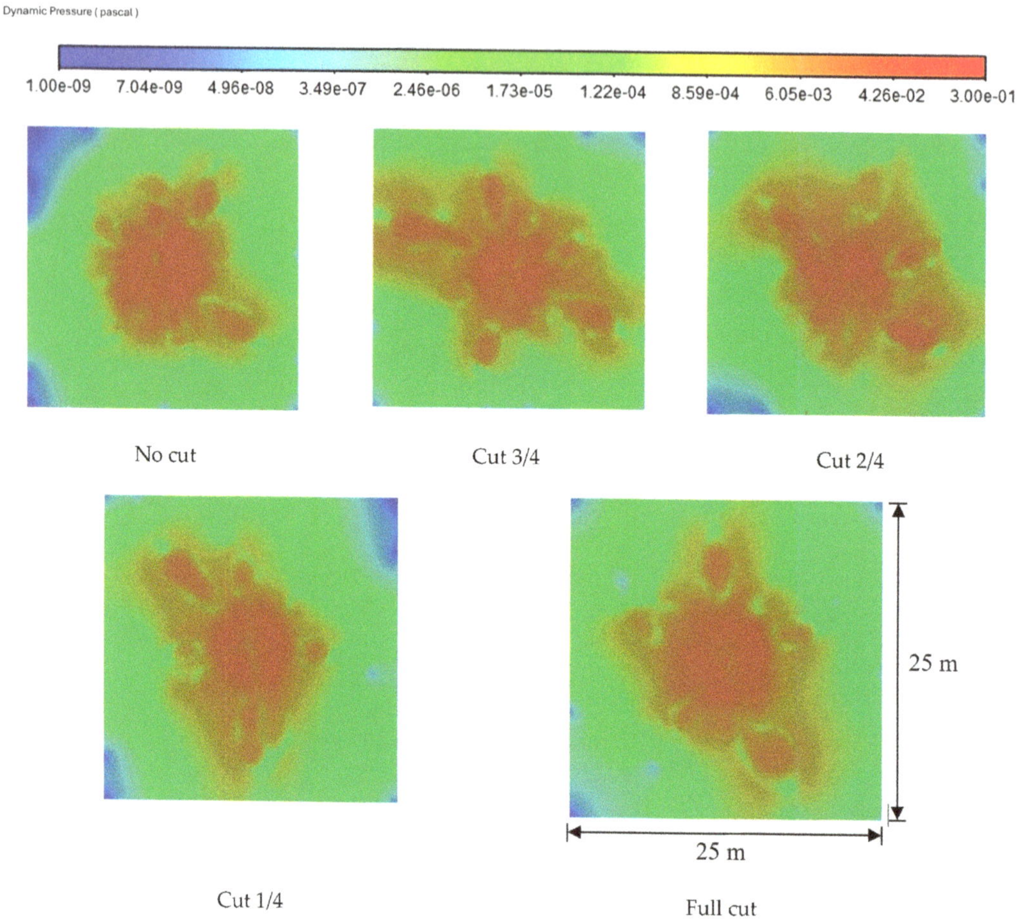

Figure 12. Pressure contour and varying impeller configurations (Note: the results are evaluated at y = 1.2 m).

3.5. *Dissolved Oxygen*

Figure 13 shows the pathline of dissolved oxygen (DO) concentration on the horizontal plane at y = 0 m, varying the cut configurations of the aerator system. The color gradient, as indicated by the scale, ranges from a low concentration (dark blue) to a high concentration (yellow-orange), representing dissolved oxygen values between 0.00300 kg/m^3 and 0.00600 kg/m^3. From Figure 13, it is clear that the "cut 3/4" pattern yields the highest dissolved oxygen concentration, with a peak value reaching 0.00561 kg/m^3. The yellow regions in each of the subfigures demonstrate that the areas where dissolved oxygen levels are sufficient to sustain shrimp life are primarily concentrated near the aerator, while the other configurations reflect lower concentrations. These regions form a distinct, star-like pattern around the aerator, diminishing rapidly as distance from the aerator increases. The pattern across all variations suggests that while there is a localized high oxygen concentration, the overall dispersion diminishes significantly as one moves farther from the aerator. This observation highlights that only specific regions immediately adjacent to the aerator contain oxygen levels that are favorable for shrimp habitation. As the cuts increase (moving towards a full-cut configuration), the oxygen distribution becomes less effective, reinforcing the critical role of the aerator's configuration in maintaining a suitable dissolved oxygen environment for aquatic life, particularly for shrimp-farming applications.

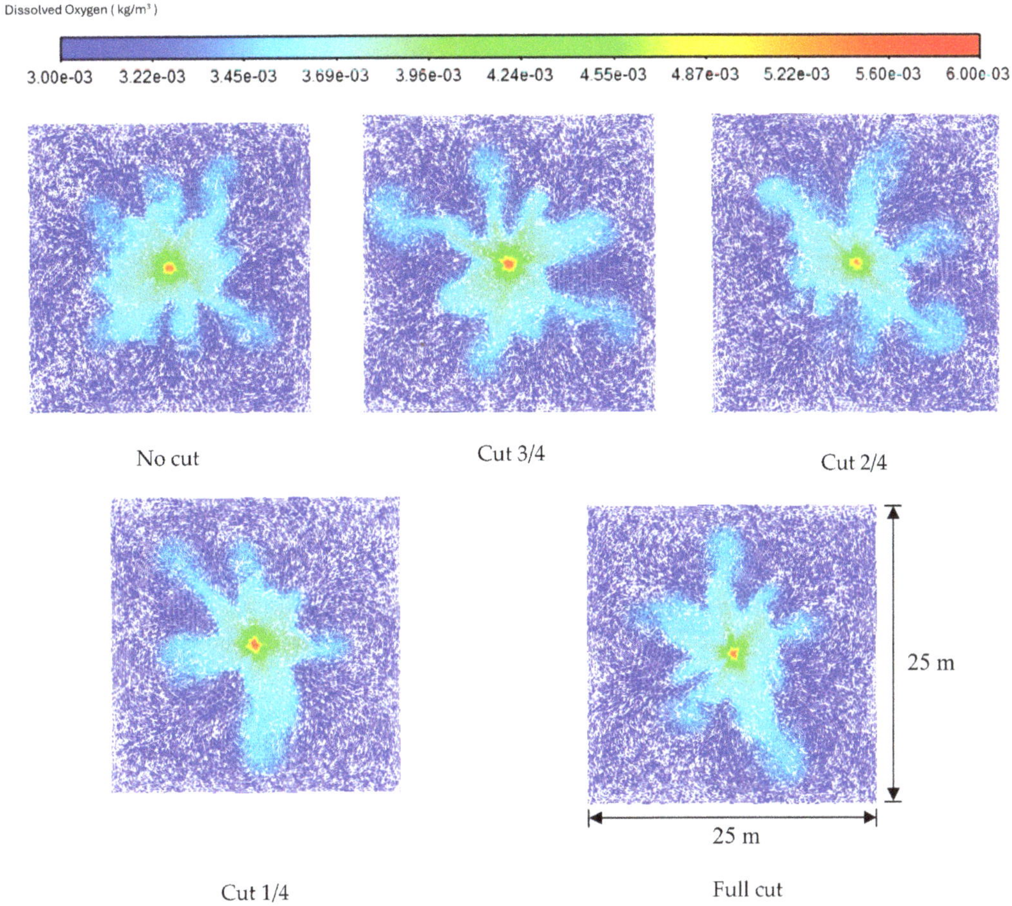

Figure 13. Pathline of dissolved oxygen (DO) distribution at $y = 0$ m.

In Figure 14, the pathline of dissolved oxygen (DO) concentrations at a depth of $y = 1.2$ m in the pond is shown. The results show that only minimal amounts of dissolved oxygen are able to reach the surface of the pond. This is due to the design of the aerator, which directs the oxygen flow towards the bottom of the pond, thus limiting its upward movement. Among the different aerator configurations, the "cut 3/4" variation again stands out as the most effective, achieving the highest dissolved oxygen concentration at this depth, with a peak value of 0.00361 kg/m^3. However, the overall concentration values are lower than those observed at shallower depths (e.g., $y = 0$ m), suggesting that the oxygen transfer to the upper layers of the pond is still restricted. The oxygen distribution across all variations forms a star-like pattern around the aerator, similar to the patterns observed in earlier figures at $y = 0$ m. However, these patterns are less pronounced and more diffuse at this depth. The central regions near the aerator still show relatively higher concentrations of dissolved oxygen, while the outer regions exhibit significantly lower values, as indicated by the dominant blue color in the contour map. The idea that the aerator's efficiency decreases with increasing depth is supported by this.

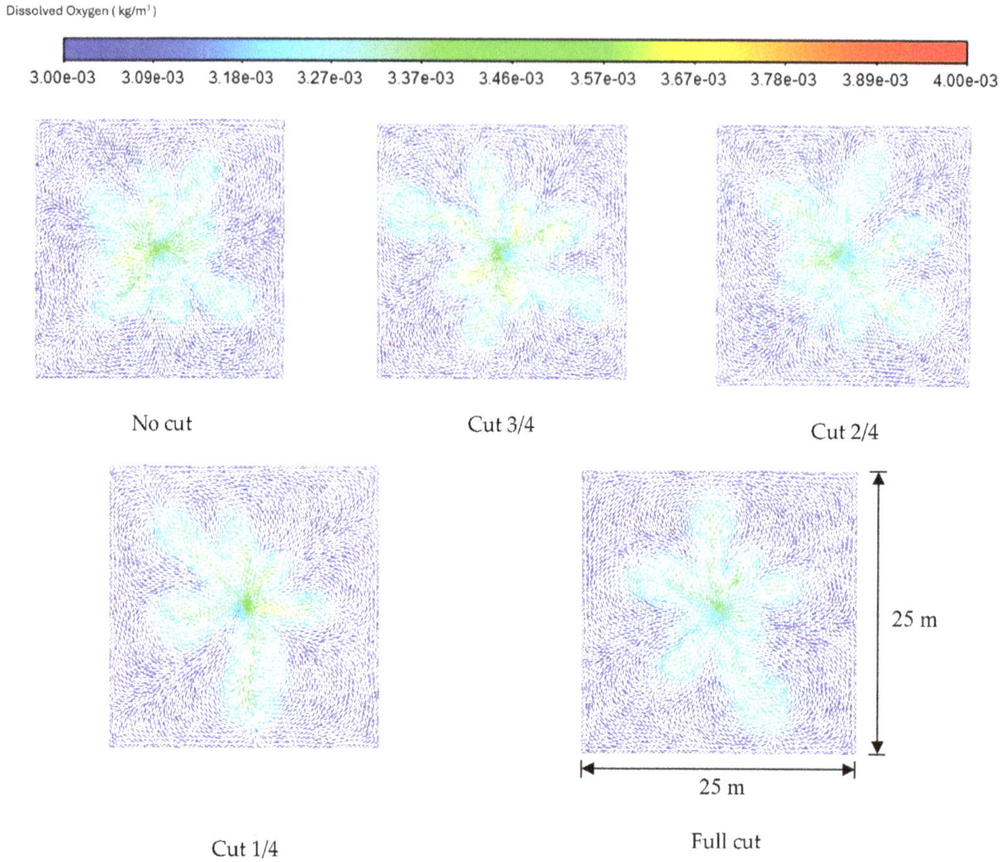

Figure 14. Pathline of dissolved oxygen (DO) distribution at $y = 1.2$ m.

Figure 15 presents the average dissolved oxygen (DO) concentration at the pond's bottom for 3000 s for five distinct aerator configurations. The results demonstrate that the initial dissolved oxygen concentration for all variants begins at a relatively high level, approximately 0.0045 kg/m^3, before gradually stabilizing over time. Over time, the dissolved oxygen levels for all variations stabilize, while subtle differences arise. The "no cut" design proves to be the most effective in sustaining the highest average dissolved oxygen concentration during the simulation. After 3000 s, the dissolved oxygen content for the "no cut" variation stabilizes at around 0.00345 kg/m^3, somewhat above that of the other variations. Notably, the "cut 3/4" variation, which had demonstrated the highest peak dissolved oxygen concentration in prior figures, now displays a slightly reduced average concentration in this particular context of bottom oxygenation. The remaining variations, i.e., "cut 2/4", "cut ¼", and "full cut" exhibit like patterns but with somewhat reduced average DO values. It can be noted that the inset in Figure 15 focuses on the stabilization period, providing a detailed look at the dissolved oxygen variations between 2000 and 3000 s. This segment illustrates slight fluctuations in DO concentration across all configurations, with the "no cut" variation consistently exhibiting higher levels than the others. This figure also highlights that whereas specific aerator designs, such as "cut 3/4," may excel in particular regions, the "no cut" configuration yields a more consistent and elevated average dissolved oxygen level at the pond's bottom over prolonged durations.

This is particularly pertinent in situations where consistent oxygen distribution and stable oxygen levels are essential for benthic organisms or for preserving long-term water quality.

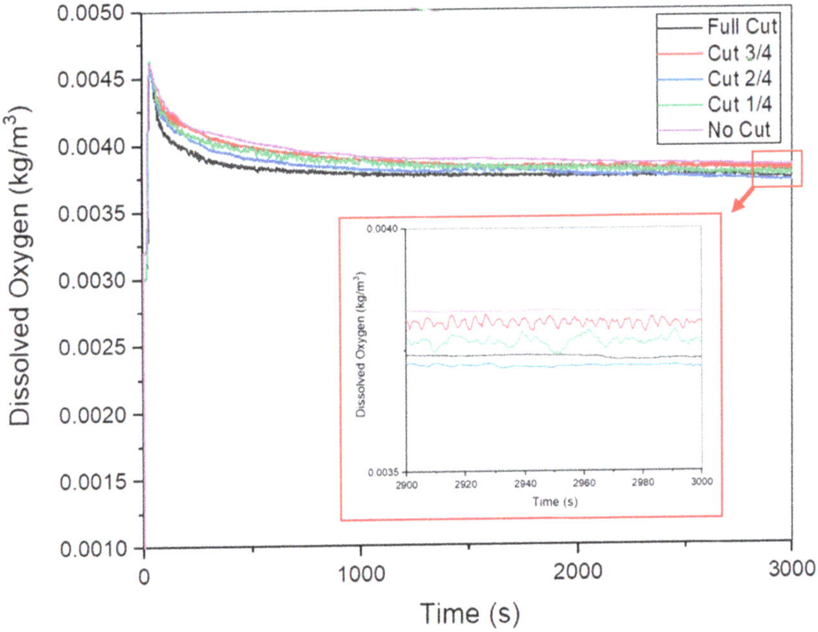

Figure 15. Average value of dissolved oxygen (DO) at the bottom of the pool.

Analysis of the data presented in Figures 9–15 indicates that the "no cut" variation yields the most extensive distribution, as the DO distribution nearly reaches the pond's perimeter and generates the most significant circulation within the pond, despite the DO levels at the edge being inadequate to sustain white vaname shrimp. Dissolved oxygen levels of 3.5–7.5 mg/L are necessary. This research indicates that dissolved oxygen distribution adequate for sustaining shrimp life is exclusively located near the aerator. It can be concluded that additional aerators are required for the optimal utilization of the existing ponds for vaname shrimp cultivation.

4. Conclusions

The study conducted a comprehensive analysis of aerator performance utilizing the Lily impeller through the CFD approach, emphasizing the impact of cut sweep ratio on dissolved oxygen (DO) distribution in shrimp ponds. Five designs of Lily impellers were examined and compared against one another. The key findings from the research are as follows:

1. The "no cut" type produces the highest maximum flow velocity in the pond, reaching 0.96 m/s, while the "Cut 3/4" type provides the lowest maximum flow speed of 0.77 m/s.
2. The "no cut" variation also exhibits the widest DO distribution, providing the best circulation of dissolved oxygen throughout the pond. This is evident from the contour images, which show that this variation enhances oxygen coverage across a larger area.
3. In terms of hydrodynamic pressure, the "no cut" variation achieves the highest maximum dynamic pressure at 11.5 Pa, whereas the "Cut 3/4" variation results in the lowest maximum dynamic pressure at 6.87 Pa.

4. Despite the "no cut" variation improving overall circulation and oxygen distribution, the DO levels sufficient to support vaname shrimp (which require 3.5–7.5 mg/L of oxygen) are still confined to areas near the aerator. Thus, more aerators are necessary to extend oxygen distribution to areas farther from the aerator.

Author Contributions: Conceptualization, J.C.; methodology, M.T.; software, B.S. and J.; validation, J.C.; formal analysis, M.T. and E.Y.; investigation, J.C.; resources, M.T. and J.; data curation, B.S.; writing—original draft preparation, M.T. and J.C; writing—review and editing, E.Y. and M.T.; visualization, J.C. and E.Y.; supervision, J.; project administration, B.S.; funding acquisition, J. and M.T. All authors have read and agreed to the published version of the manuscript.

Funding: This research is fully funded by Diponegoro University through WCRU-A Grant, No. 357-29/UN7.D2/PP/IV/2024.

Data Availability Statement: The data presented in this study are available upon request from the corresponding author.

Acknowledgments: The authors acknowledge the Institute for Research and Community Services (LPPM)—Diponegoro University, Indonesia for the approved fund that makes this important research viable and effective.

Conflicts of Interest: The authors declare no conflicts of interest.

References

1. Statistik KKP. Available online: https://statistik.kkp.go.id/home.php (accessed on 9 August 2024).
2. Chakravarty, M.S.; Ganesh, P.R.C.; Amarnath, D.; Sudha, S.; Srinu Babu, T.; Ganesh, P. Spatial Variation of Water Quality Parameters of Shrimp (*Litopenaeus vannamei*) Culture Ponds at Narsapurapupeta, Kajuluru and Kaikavolu Villages of East Godavari District, Andhra Pradesh. *IJFAS* **2016**, *4*, 390–395.
3. Sultana, T.; Haque, M.M.; Salam, M.A.; Alam, M.M. Effect of Aeration on Growth and Production of Fish in Intensive Aquaculture System in Earthen Ponds. *J. Bangladesh Agril. Univ.* **2017**, *15*, 113–122. [CrossRef]
4. Das, P.C.; Ayyappan, S.; Jena, J.K.; Singh, S.K.; Patamajhi, P.; Muduli, H.K. Effect of Aeration on Production and Water Quality Changes in Intensive Carp Culture. *Indian J. Fish.* **2004**, *51*, 173–183.
5. Qayyum, A.; Ayub, M.; Tabinda, A.B. Effect of Aeration on Water Quality, Fish Growth and Survival in Aquaculture Ponds. *Pakistan J. Zool.* **2005**, *37*, 75–80.
6. Mahmudov, K.; Mahmoud, A.; Sur, S.; Cruz, F.C.; Bilton, A.M. Feasibility of a Wind-Powered Aeration System for Small-Scale Aquaculture in Developing Countries. *Energy Sustain. Dev.* **2019**, *51*, 40–49. [CrossRef]
7. Akinwole, A.; Dauda, A.; Oredein, T. Effect of Aeration on Selected Water Quality Parameters in Freshwater Fish Culture. *Int. J. Biol. Chem. Sci.* **2015**, *8*, 2858. [CrossRef]
8. Avnimelech, Y.; Mozes, N.; Weber, B. Effects of Aeration and Mixing on Nitrogen and Organic Matter Transformations in Simulated Fish Ponds. *Aquac. Eng.* **1992**, *11*, 157–169. [CrossRef]
9. Chen, H.; Hu, B. Chinese Integrated Fish Farming: A Comparative Bioeconomic Analysis. *Aquac. Res.* **1995**, *26*, 81–94. [CrossRef]
10. Izel-Silva, J.; Ono, E.A.; de Queiroz, M.N.; dos Santos, R.B.; Affonso, E.G. Aeration Strategy in the Intensive Culture of Tambaqui, Colossoma Macropomum, in the Tropics. *Aquaculture* **2020**, *529*, 735644. [CrossRef]
11. Jansen, W.; Kirschner, A.M. Impeller Blade Design Method for Centrifugal Compressors. *North. Res. Eng. Corp.* **1974**, *304*, 537–563.
12. Li, D.; Chen, W. Effects of Impeller Types on Gas-Liquid Mixing and Oxygen Mass Transfer in Aerated Stirred Reactors. *Process Saf. Environ. Prot.* **2022**, *158*, 360–373. [CrossRef]
13. Zheng, Z.; Sun, D.; Li, J.; Zhan, X.; Gao, M. Improving Oxygen Transfer Efficiency by Developing a Novel Energy-Saving Impeller. *Chem. Eng. Res. Des.* **2018**, *130*, 199–207. [CrossRef]
14. Karimi, A.; Golbabaei, F.; Mehrnia, M.R.; Neghab, M.; Mohammad, K.; Nikpey, A.; Pourmand, M.R. Oxygen Mass Transfer in a Stirred Tank Bioreactor Using Different Impeller Configurations for Environmental Purposes. *Iranian J. Environ. Health Sci. Eng.* **2013**, *10*, 6. [CrossRef] [PubMed]
15. de Jesus, S.S.; Moreira Neto, J.; Santana, A.; Maciel Filho, R. Influence of Impeller Type on Hydrodynamics and Gas-Liquid Mass-Transfer in Stirred Airlift Bioreactor. *AIChE J.* **2015**, *61*, 3159–3171. [CrossRef]
16. Jamshidzadeh, M.; Ein-Mozaffari, F.; Lohi, A. Local Distribution of Oxygen Mass Transfer Coefficient in CMC Solutions in Bioreactors Furnished with Different Types of Coaxial Mixers. *Chem. Eng. Res. Des.* **2021**, *174*, 213–224. [CrossRef]
17. Nordin, R.N.; McKean, C.J.P. *Review of Lake Aeration as a Technique for Water Quality Improvement*; Province of British Columbia, Ministry of Environment, Assessment and Planning Division: Victoria, BA, Canada, 1982; Volume 22.
18. Hasan, K.; Alam, K.; Chowdhury, S.A. The Use of an Aeration System to Prevent Thermal Stratification of Water Bodies: Pond, Lake and Water Supply Reservoir. *Appl. Ecol. Environ. Sci.* **2013**, *2*, 1–7. [CrossRef]

19. Boyd, C.E. Pond Water Aeration Systems. *Aquac. Eng.* **1998**, *18*, 9–40. [CrossRef]
20. Ferreira, A.; Cardoso, P.; Teixeira, J.A.; Rocha, F. Ph influence on oxygen mass transfer coefficient in a bubble column. individual characterization of KL and a. *Chem. Eng. Sci.* **2013**, *100*, 145–152. [CrossRef]
21. Vinatea, L.; Muedas, W.; Arantes, R. The impact of oxygen consumption by the shrimp litopenaeus vannamei according to body weight, temperature, salinity and stocking density on pond aeration: A simulation. *Acta Sci. Biol. Sci.* **2011**, *33*, 125–132. [CrossRef]
22. Wu, G.L.; Zhu, L.S.; Li, F.C. Numerical Simulation of Dissolved Oxygen Transfer in an Aerated Pond. *Appl. Mech. Mater.* **2014**, *587–589*, 588–593. [CrossRef]
23. Weidner, B.V.; Nagel, J.; Weber, H.J. Facilitation Method for the Translation of Biological Systems to Technical Design Solutions. *Int. J. Des. Creat. Innov.* **2018**, *6*, 211–234. [CrossRef]

fluids

MDPI

Article

Two-Stage Multi-Objective Optimization for Improving the Aerodynamic Characteristics of High-Speed Train Nose Shape

Suhwan Yun [1,*], Minho Kwak [2], Hyungmin Kang [3], Wonhee Park [1] and Taesoo Kwon [1]

[1] Railroad Safety Division, Korea Railroad Research Institute, Uiwang 16105, Republic of Korea
[2] Innovative Transportation & Logistics Research Center, Korea Railroad Research Institute, Uiwang 16105, Republic of Korea
[3] Department of Mechanical Engineering, Dongyang Mirae University, Seoul 08221, Republic of Korea
* Correspondence: shyun@krri.re.kr

Abstract: An optimal high-speed train nose design was formulated to reduce aerodynamic drag, mitigate tunnel micro-pressure waves, and improve crosswind safety. Using the vehicle modeling function, the EMU-320 nose shape was derived and applied as the base model for the optimal design. Following the initial optimization stage aimed at reducing aerodynamic drag, the second stage of optimization was conducted, using the results of the first stage as constraints, to further reduce tunnel micro-pressure waves and enhance crosswind safety. In the first stage, the aerodynamic drag was reduced by 8.7% due to the pointed nose shape. In the second stage, a Pareto front of optimal shapes was derived to reduce tunnel micro-pressure waves and improve crosswind safety, ensuring that the aerodynamic drag did not exceed the optimum achieved in the first stage by more than 1.5%. This two-stage multi-objective optimization is expected to be effectively utilized in the optimal design of high-speed train shapes that meet the intended purposes.

Keywords: high-speed train; nose shape; two-stage optimization; multi-objective; aerodynamic characteristics

1. Introduction

High-speed trains are among the fastest and most efficient modes of transportation, having established themselves as essential in modern society, and their speed and efficiency continue to improve. However, this increase in speed brings various challenges, such as an energy consumption increase, noise generation, and operational safety reduction. These issues are predominantly closely related to the aerodynamic characteristics of high-speed trains, making aerodynamic performance improvements crucial to their resolution.

The aerodynamic performance of high-speed trains is not defined by a single factor but manifests in various forms under different conditions. Among these, aerodynamic drag constitutes the majority of running resistance at speeds exceeding 300 km/h, and an increase in drag leads to higher energy consumption [1]. When high-speed trains pass through tunnels, the micro-pressure waves generated at the tunnel exit can damage surrounding structures, which is related to passenger comfort within the tunnel. Due to the elongated shape of high-speed trains relative to their cross-sectional area, crosswinds can have severe impacts, such as causing derailments, making crosswind stability crucial for the operational safety of the trains. These external shapes, especially the nose shapes of the leading and trailing car, can largely influence the aerodynamic characteristics of high-speed trains. The optimization of the nose shape of high-speed trains is therefore considered the best method to enhance aerodynamic performance [2].

In the research on the optimal aerodynamic design of high-speed trains, previous studies have primarily focused on single-objective optimization. However, recent studies have increasingly aimed to enhance multi-objective aerodynamic performance by optimizing train nose shapes. Kwon et al. optimized an axisymmetric nose shape to reduce aerodynamic drag and micro-pressure waves [3]. Kwak et al. achieved an optimal design to reduce the overall aerodynamic drag of a front–rear symmetric train based on cross-sectional distribution for reducing micro-pressure waves [4]. He et al. proposed an improved optimization algorithm and derived an optimal design to minimize the total aerodynamic drag and the absolute value of the lift on the trailing car [5]. Sun et al. utilized an LSF-based FFD technique to create an optimal design that reduces aerodynamic drag and aero-acoustic noise [6]. Wang et al. employed PDE-based parametric modeling to conduct a multi-objective nose optimization that minimizes the total aerodynamic drag and the absolute value of the lift on the rear car [7]. Munoz-Paniagua and Garcia optimized the nose shape of high-speed trains concerning lateral stability under crosswind and passing-by conditions [8]. Zhang et al. optimized the nose shape of a high-speed train with the objective functions of minimizing total aerodynamic drag, lift, and side force on the leading car under crosswind conditions [9].

However, there are still few design cases that optimize energy efficiency, such as reducing vehicle resistance, considering actual railway operating conditions. In particular, in South Korea, approximately 63% of the land is mountainous, resulting in a higher proportion of tunnels in high-speed rail operation sections compared to other countries. Various aerodynamic issues that may arise when running through tunnels, such as micro-pressure waves and pressure fluctuations inside tunnels, must be considered when designing high-speed trains. Additionally, due to the climatic characteristics of South Korea, numerous typhoons pass through the Korean Peninsula during the summer, often accompanied by heavy rain and strong winds. Crosswind safety must therefore be considered in high-speed train design to mitigate the risk of a train overturning due to strong winds.

2. Methodology

2.1. Vehicle Modeling Function

The vehicle modeling function that simulates the nose shape of high-speed trains is based on the fundamental curve expressed by Equation (1), as shown in Figure 1. Given the starting point (x_p, z_p), width (L), and height (H), the shape is determined by coefficients that adjust the curvature. By combining multiple fundamental curves from the vehicle modeling function, it is possible to simulate the 2D cross-sectional shapes of a high-speed train nose, and based on these 2D cross-sectional shapes, the 3D shape can be constructed [10,11].

$$Z(x) = \frac{H}{L^{A_1+A_2}}\left(x-x_p\right)^{A_1}\left[2L-\left(x-x_p\right)\right]^{A_2}+z_p \quad (1)$$

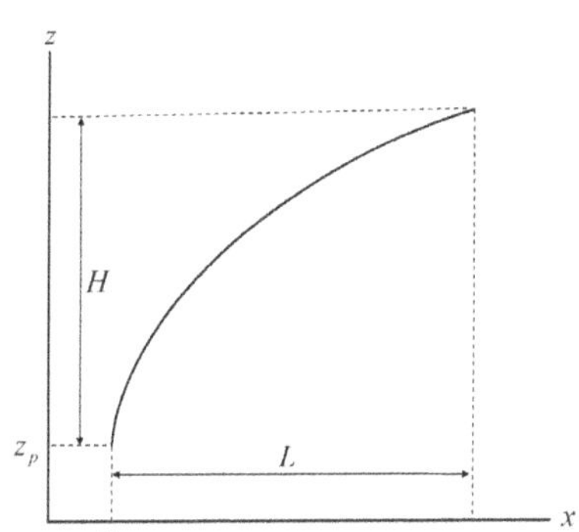

Figure 1. Basic curve of vehicle modeling function.

The process of constructing the 3D shape of the EMU-320 train using the vehicle modeling function is shown in Figure 2. By varying the coefficients of the train shape function and performing iterative calculations, the coefficients that best simulate the actual 3D shape are derived. These derived coefficients are then used to construct the 3D shape, simplifying and parameterizing the EMU-320 shape. From an overall isometric projection perspective, the derived 3D basic shape relatively similarly simulates the actual shape. Due to the nature of parameterization, it is challenging to represent every geometric irregularity; so, the simulation is performed from an engineering perspective.

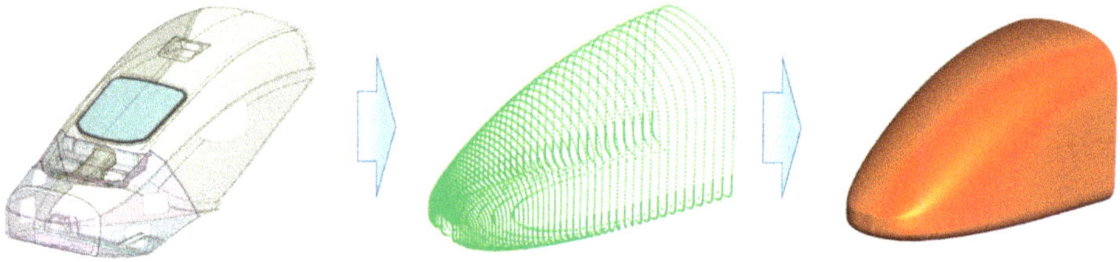

Figure 2. Forming process of EMU-320 by vehicle modeling function.

2.2. Numerical Method

The analysis model for evaluating aerodynamic drag and crosswind effects is a streamlined model of the EMU-320, consisting of three cars without bogies and pantographs, as shown in Figure 3. The nose shape derived using the vehicle modeling function was applied to both the leading and trailing cars. To maximize the impact of the nose shape, other influencing factors were removed, and a streamlined model was used as the analysis model. The train model has a width of 3.14 m, a height of 3.67 m, and a total length of 75 m. The gap between the train model and the ground is 0.445 m, which is the sum of the distance between the train under floor and the rail upper surface plus the rail height, based on the Korean railway standard [12].

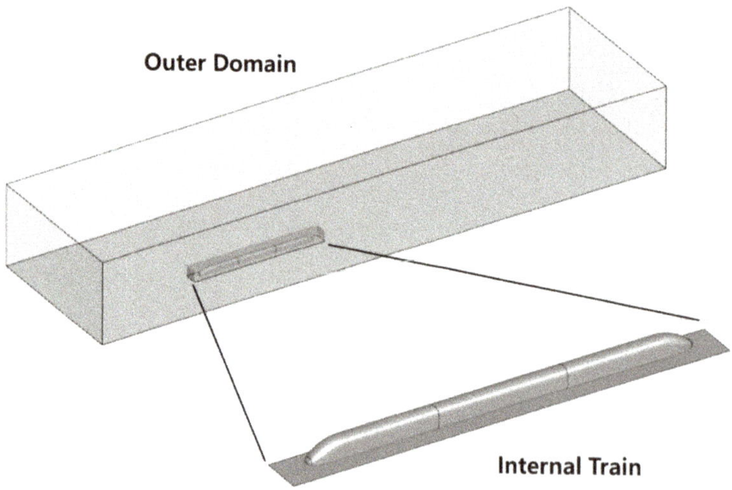

Figure 3. The computational domain via 3D analysis.

Figure 4 shows the mesh around the surface of the train model with the smallest grid on the nose surface measuring 30 mm. It also shows the mesh around the near field and far field for 3D numerical analysis. To improve the accuracy of the calculations, a dense mesh is required around the vehicle, while the far-field region, where flow changes are minimal, requires a mesh that balances efficiency and economy. In this study, the computational domain was thus divided into near-field and far-field regions, as shown in Figure 4. For the near-field region, triangular meshes were used with the height of the first mesh layer from the train surface set at 0.2 mm, forming a total of 15 boundary prism layers. Subsequently, tetrahedral meshes were densely distributed in the near-field region. For the far-field region, a structured hexahedral mesh was used to generate the 3D mesh. The total number of 3D mesh elements constructed for each numerical case ranged from approximately 13 million to 14 million. y+ is at the O(1) level.

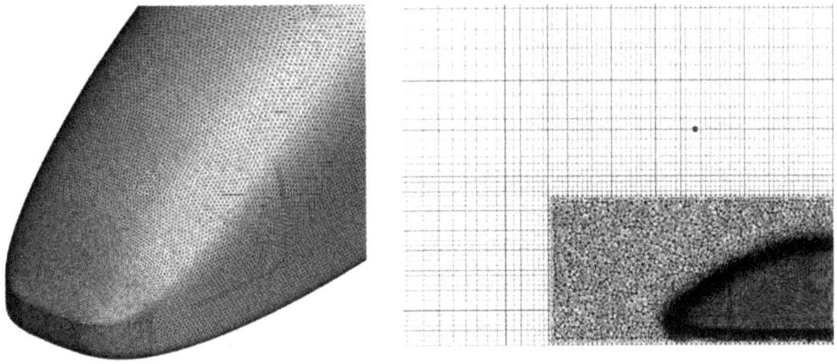

Figure 4. Mesh construction via 3D analysis.

The commercial CFD solver ANSYS Fluent 2022 R2 is used for aerodynamic drag analysis and crosswind analysis. The governing equations are the three-dimensional compressible Navier–Stokes equations, as shown in Equation (2).

$$\frac{\partial}{\partial t}\int_V \vec{W} dV + \oint \left[\vec{F} - \vec{G}\right] \cdot dA = 0, \tag{2}$$

where $\vec{W} = \begin{bmatrix} \rho \\ \rho u \\ \rho v \\ \rho w \\ \rho E \end{bmatrix} \vec{F} = \begin{bmatrix} \rho \vec{v} \\ \rho \vec{v} u + p\hat{i} \\ \rho \vec{v} v + p\hat{j} \\ \rho \vec{v} w + p\hat{k} \\ \rho \vec{v} E + p\hat{v} \end{bmatrix} \vec{G} = \begin{bmatrix} 0 \\ \tau_{xi} \\ \tau_{yi} \\ \tau_{zi} \\ \tau_{xi} v_j + \vec{q} \end{bmatrix}$.

The turbulence model used was the $\mathrm{k}-\omega$SST model, known for effectively combining the advantages of the $\mathrm{k}-\varepsilon$ model and $\mathrm{k}-\omega$ models [13]. Numerical simulations were conducted at the design maximum speed of the EMU-320, which is 352 km/h. The Reynolds number (Re) based on the train's speed (V = 352 km/h) and height (h = 3.67 m) is 2.37×10^7. For the comparison of aerodynamic drag, the drag coefficient, as shown in Equation (3), was used with a reference cross-sectional area of 10 m^2.

$$C_D = \frac{F}{\frac{1}{2}\rho V^2 A_{ref}} \tag{3}$$

The grid convergence test was conducted as shown in Figure 5. The C_D value remained constant regardless of the number of cells. Therefore, subsequent analyses were performed using the current grid.

To enhance the efficiency of numerical analysis under crosswind conditions, the flow was assumed to be the vector sum of the inlet flow velocity of 352 km/h (97.8 m/s) in the train's running direction and a crosswind at an angle of 90 degrees with a speed of 45 m/s, which was determined considering the high-speed rail operation regulations and typhoon speeds. To compare crosswind safety, the side force coefficient, as shown in Equation (4), was used, with a reference cross-sectional area of 263 m^2.

$$C_S = \frac{F_S}{\frac{1}{2}\rho V^2 A_{Sref}} \tag{4}$$

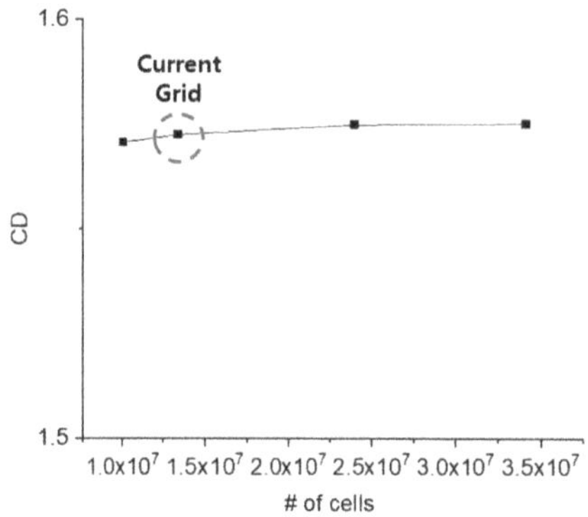

Figure 5. Grid convergence test.

The numerical analysis for predicting micro-pressure waves was conducted using the method proposed by Kwon et al., employing the axisymmetric compressible Navier–Stokes equations derived in a general curvilinear coordinate system, as shown in Equation (5). This axisymmetric compressible analysis provides results close to those of three-dimensional numerical analysis for a high-speed train running in tunnel while offering superior computational efficiency. To enhance computational efficiency, the analysis was therefore performed using axisymmetric numerical methods [14].

$$\frac{\partial Q}{\partial t} + \frac{\partial E}{\partial x} + \frac{\partial F}{\partial y} + H = \frac{1}{Re_c}\left[\frac{\partial E_v}{\partial x} + \frac{\partial F_v}{\partial y} + H_v\right] \tag{5}$$

Q represents the conserved quantity vector, E and F are the flux vectors, and H denotes the axisymmetric source term. Figure 6 shows the grid system for the axisymmetric analysis. Zone 1 is the mesh around the train, which moves through Zone 2, representing the tunnel. Zones 3 and 4 are the meshes for the entrance and exit regions of the tunnel. The train mesh was applied with the cross-sectional distribution and length of the EMU-320 three-car model used in the 3D analysis. The tunnel mesh is configured with a length of 1 km and a cross-sectional area of 96.7 m^2, based on the Honam high-speed railway route specifications. From pressure fluctuation history at a point 25 m away from the tunnel exit, the first maximum value is extracted and it is used as a micro-pressure wave.

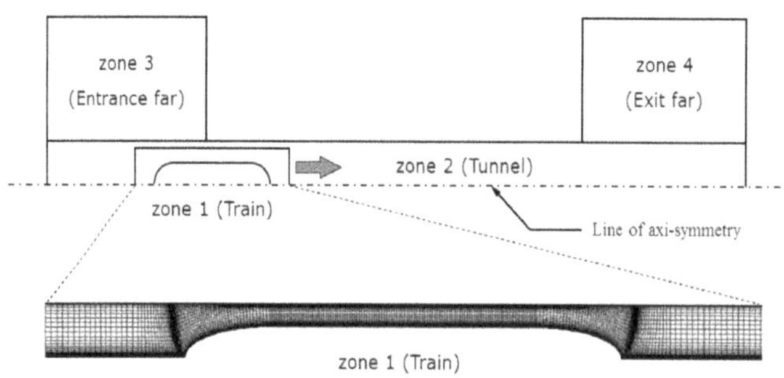

Figure 6. The computational domain and the grid system of the train and the tunnel at axisymmetric analysis.

2.3. *Two-Stage Optimization Process*

By analyzing the basic shape, five design variables that significantly impact aerodynamic performance were selected: nose length, nose tip height, upper and lower surface curvature from a side view, and side surface curvature from a top view. Since the design process involves the nose shape of an actual operating train, the design space was set within the limits that do not violate the spatial allowance for internal equipment and the driver's field of view. The variations within the design space for each design variable are shown in Table 1, and due to the constraints of the actual operational vehicle shape, the design space is relatively small.

Table 1. Shape deformations according to variations in the design variables.

Design Variables	2D Shape Variation	Lower Limit	Base Model	Upper Limit
1. Nose length	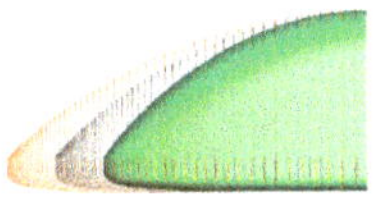	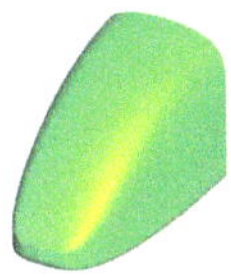		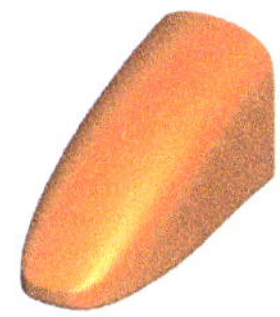
2. Nose tip height				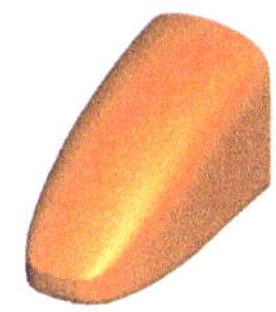
3. Upper surface curvature	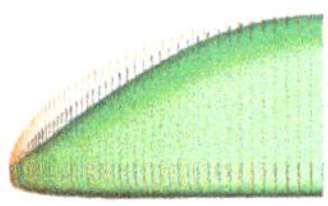			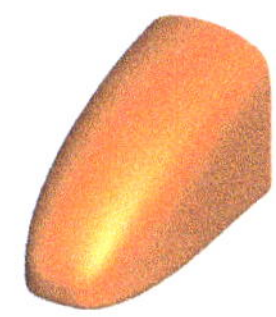
4. Lower surface curvature				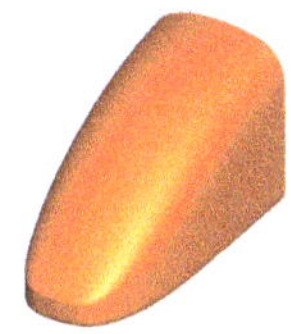
5. Side surface curvature				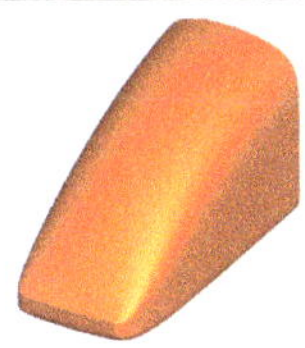

To derive the experimental points that constitute the design space, the Maximin Latin Hypercube method was used. The Maximin Latin Hypercube method is advantageous for experimental design using computational analysis, as it is known for distributing experimental points across the entire design space without overlap [15]. Thirty experimental points were extracted to construct the design space.

The effects of each design variable on the corresponding objective functions are shown in Figure 7. In design variable #5, the side surface curvature significantly affects aerodynamic drag; in design variable #1, the nose length predominantly influences micro-pressure waves; and in design variable #2, the nose tip height has a substantial impact on the side force coefficient. These factors are considered to play a dominant role in deriving the optimal shape.

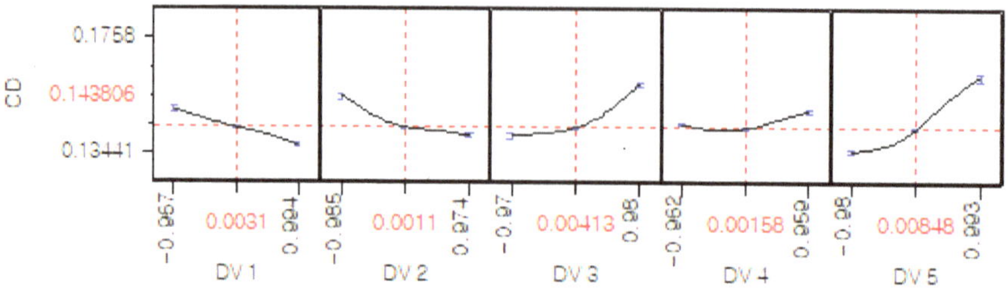

(**a**) Effect on aerodynamic drag coefficient of each design variable.

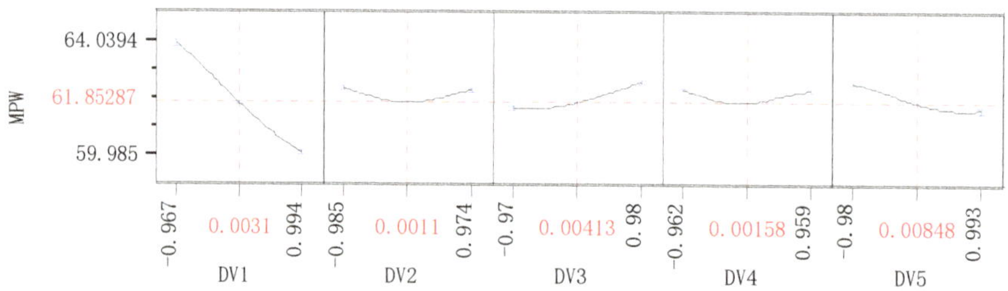

(**b**) Effect on micro-pressure wave of each design variable.

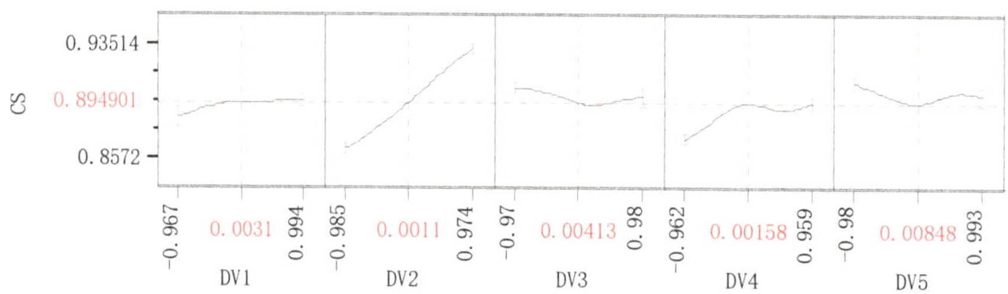

(**c**) Effect on side force coefficient of each design variable.

Figure 7. Effect on each objective function of each design variable.

The design optimization was conducted in two stages, as shown in Figure 8. Among the three aerodynamic performance objectives, the reduction in aerodynamic drag, which is considered the most critical in the current train operation environment in South Korea, was prioritized and optimized in the first stage [16]. Subsequently, using the result from the first-stage optimization as a constraint, a multi-objective optimization was performed in the second stage to reduce tunnel micro-pressure waves and improve crosswind safety. In the first stage of optimization, an artificial neural network model was used to construct a surrogate model [17]. For the five design variables, a single-layer structure with eight hidden nodes was used. The Broyden–Fletcher–Goldfarb–Shanno (BFGS) algorithm was applied for the first-stage optimization method [18]. In the second stage of optimization, a surrogate model was also constructed using an artificial neural network model. The same five design variables were used, with a two-layer structure consisting of 100 and 20 hidden nodes. The NSGA-II algorithm was applied for the optimization method at the second stage.

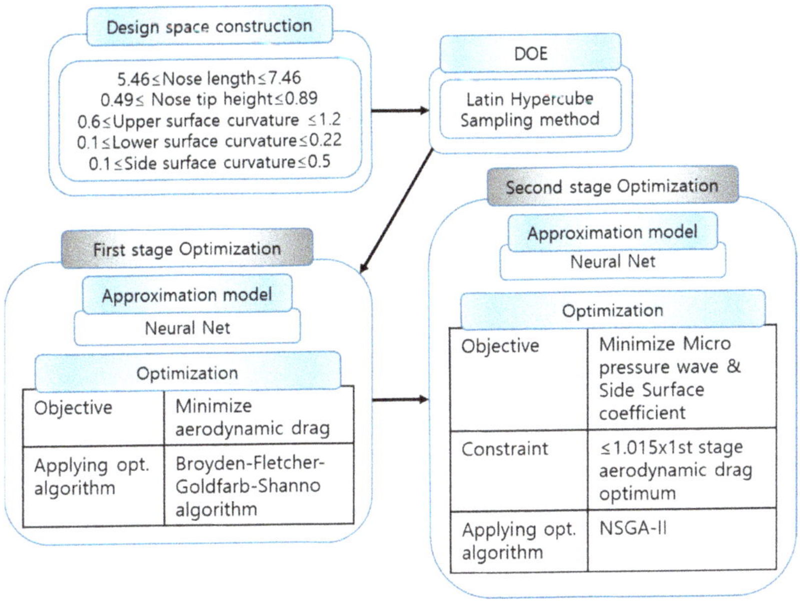

Figure 8. Two-stage optimization process.

3. Results and Discussion

The aerodynamic drag coefficient of the optimized shape derived from the first-stage optimization was 0.1311, representing an approximate 8.7% reduction compared to the base model of 0.1436. Figures 9 and 10 compare the pressure contours around the first and last cars. Around the first car, the transition from a blunt of the base model to a pointed front shape of the optimized shape resulted in a reduction in high-pressure regions, thereby decreasing aerodynamic drag. For the last car, a reduction in low-pressure regions was observed, leading to relatively faster pressure recovery. Additionally, the size of the vortices behind the train was found to be smaller compared to that of the base model.

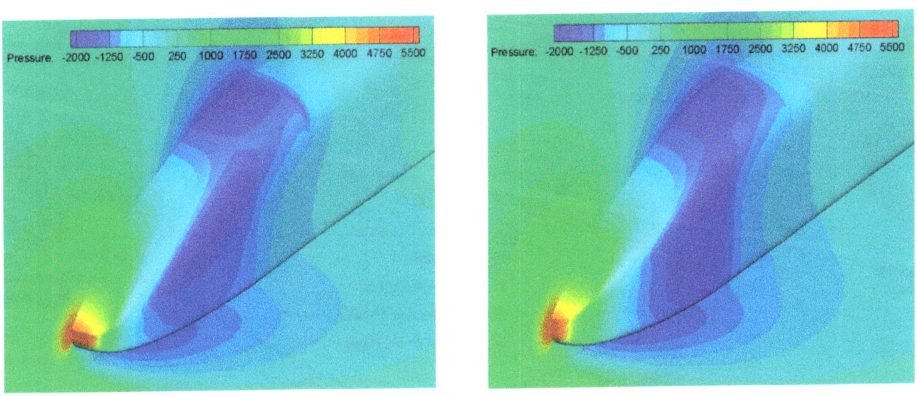

(**a**) Base model. (**b**) Optimized shape at first stage.

Figure 9. Pressure contour around the first car.

Table 3. *Cont.*

Model	Base Model	Stage 2 Optimized Shape #1	Stage 2 Optimized Shape #2	Stage 2 Optimized Shape #3
3D shape				
Pressure obtained via aerodynamic drag analysis	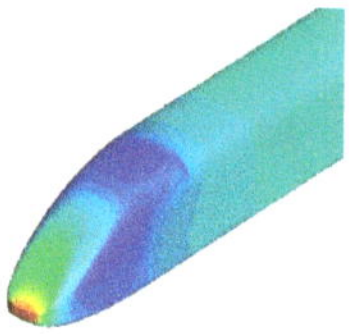	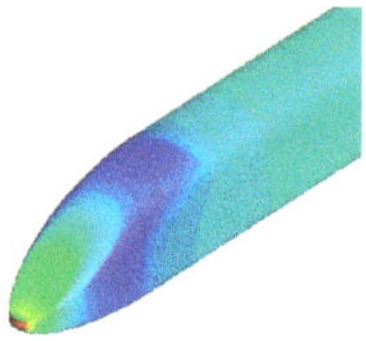	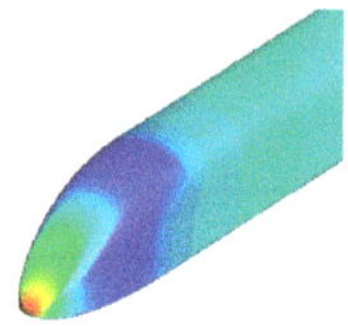	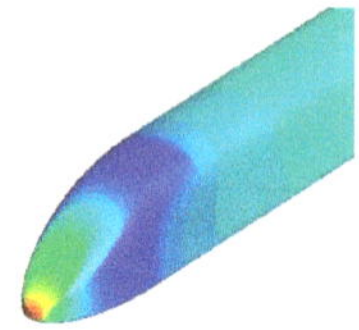
Micro-pressure wave	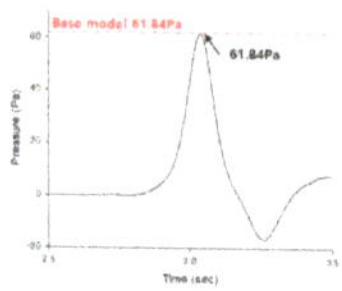	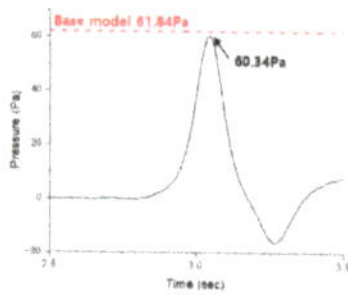	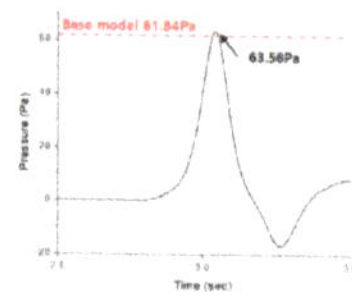	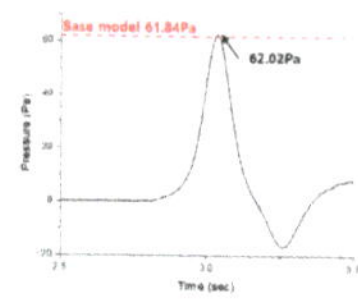
Pressure obtained via side wind analysis	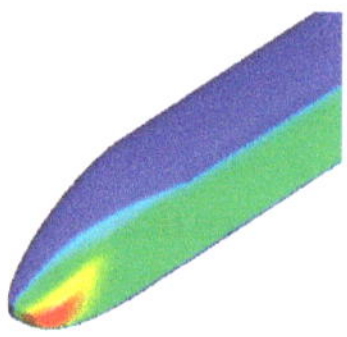	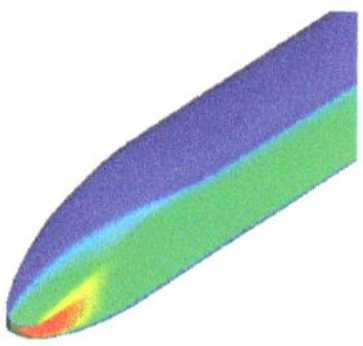	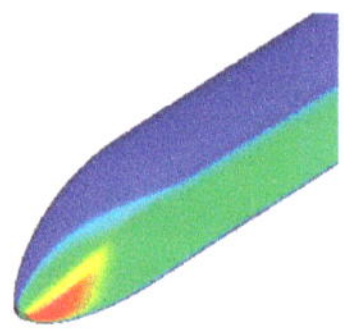	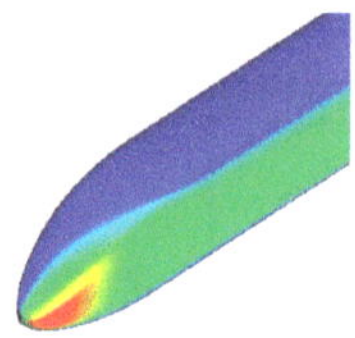

Comparing the surface pressure in the aerodynamic drag analysis, all three optimized shapes exhibited reduced high-pressure regions due to their pointed forms, resulting in decreased aerodynamic drag compared to the base model. Similarly to the optimal shape derived from the first-stage optimization, the pointed shape led to a reduction in the high-pressure region around the first car and faster pressure recovery around the last car. Additionally, the size of the vortices generated behind the last car was reduced, contributing to the overall reduction in aerodynamic drag.

4. Conclusions

Considering the operational environment characteristics of high-speed trains in South Korea, an optimal train nose shape design was formulated with the objectives of reducing aerodynamic drag, minimizing tunnel micro-pressure waves, and enhancing crosswind safety. The EMU-320 nose shape was derived and applied as the base model for optimal design using a vehicle modeling function capable of appropriately simulating the three-dimensional shape of a high-speed train. Following the initial optimization stage aimed at reducing aerodynamic drag, the second stage of optimization was conducted, using the results of the first stage as constraints, to further reduce tunnel micro-pressure waves and enhance crosswind safety.

In the first-stage optimization, the aerodynamic drag was reduced by 8.7% due to the pointed nose shape. In the second-stage optimization, a Pareto front of optimal shapes was derived to reduce tunnel micro-pressure waves and improve crosswind safety, ensuring that the aerodynamic drag did not exceed the optimum achieved in the first step by more than 1.5%.

This demonstrates that the two-stage multi-objective optimization is effectively applicable for the optimal design of high-speed train shapes to meet the specified objectives.

Author Contributions: Conceptualization, S.Y.; Methodology, S.Y.; Validation, W.P.; Formal analysis, M.K. and H.K.; Writing—original draft, M.K. and H.K.; Supervision, S.Y.; Project administration, S.Y.; Funding acquisition, T.K. All authors have read and agreed to the published version of the manuscript.

Funding: This research was supported by a grant from the R&D Program of the Ministry of Land, Infrastructure, and Transport (Development of frontal hatch system for power distributed high-speed rail vehicles, RS-2022-KA165880), Republic of Korea.

Data Availability Statement: The original contributions presented in this study are included in the article. Further inquiries can be directed to the corresponding author.

Conflicts of Interest: The authors declare no conflict of interest.

References

1. Peters, J.L. Optimizing aerodynamics to raise IC performance. In *Railway Gazette International*; DVV Media International Limited: London, UK, 1982; pp. 817–819.
2. Kim, K.M.; Han, S.W. The changes of Shinkansen vehicles' nose shape. *J. Korean Soc. Railw.* **2007**, *10*, 124–130.
3. Kwon, H.B.; Kim, Y.S.; Lee, D.H.; Kim, M.S. Nose Shape Optimization of the High-speed Train to Reduce the Aerodynamic drag and Micro-pressure Wave. In Proceedings of the KSME Conference, Jeju, Republic of Korea, 27–29 June 2001; pp. 373–379.
4. Kwak, M.H.; Yun, S.H.; Lee, Y.B.; Kwon, H.B.; Kim, K.H.; Lee, D.H. Optimum nose shape of a front-rear symmetric train for the reduction of the total aerodynamic drag. *J. Mech. Sci. Technol.* **2013**, *27*, 3733–3743. [CrossRef]
5. He, Z.; Liu, T.; Liu, H. Improved particle swarm optimization algorithms for aerodynamic shape optimization of high-speed train. *Adv. Eng. Softw.* **2022**, *173*, 103242. [CrossRef]
6. Sun, Z.; Zhang, Y.; Yang, G. Surrogate Based Optimization of Aerodynamic Noise for Streamlined Shape of High Speed Trains. *Appl. Sci.* **2017**, *7*, 196. [CrossRef]
7. Wang, S.; Wang, R.; Xia, Y.; Sun, Z.; You, L.; Zhang, J. Multi-objective aerodynamic optimization of high-speed train heads based on the PDE parametric modeling. *Struct. Multidiscip. Optim.* **2021**, *64*, 1285–1304. [CrossRef]
8. Munoz-Paniagua, J.; Garcia, J. Aerodynamic surrogate-based optimization of the nose shape of a high-speed train for crosswind and passing-by scenarios. *J. Wind. Eng. Ind. Aerodyn.* **2019**, *184*, 139–152. [CrossRef]
9. Zhang, L.; Zhang, J.; Li, T.; Zhang, Y. A multi objective aerodynamic optimization design of a high-speed train head under crosswinds. *Proc. Inst. Mech. Eng. Part F J. Rail Rapid Transit* **2018**, *232*, 895–912. [CrossRef]
10. Kulfan, B.M. A Universal Parametric Geometry Representation Method–CST. In Proceedings of the 45th AIAA Aerospace Sciences Meeting and Exhibit, Reno, NV, USA, 8–11 January 2007.
11. Ku, Y.C.; Rho, J.H.; Yun, S.H.; Kwak, M.H.; Lee, D.H. A Study of Design Method for Train Nose Shape Using Configuration Function. In Proceedings of the 2008 Spring Conference of Korean Society for Railway, Daegu, Republic of Korea, 12 June 2008.
12. *KRS TR 0001-12®*; Korean Railway Standards. Korea Railroad Research Institute: Uiwang, Republic of Korea, 2012.
13. Rogers, S.E.; Kwak, D.C. Upwind differencing scheme for the time-accurate incompressible Navier-Stokes equations. *AIAA J.* **1990**, *28*, 253–262. [CrossRef]
14. Kwon, H.B.; Kim, T.Y.; Lee, D.H.; Kim, M.S. Numerical simulation of unsteady compressible flows induced by a high-speed train passing through a tunnel. *J. Rail Rapid Transit* **2003**, *217*, 111–124. [CrossRef]
15. Johnson, M.E.; Moore, L.M.; Ylvisaker, D. Minimax and maximin distance designs. *J. Stat. Plan. Inference* **1990**, *26*, 131–148. [CrossRef]
16. Yun, S.H.; Kwak, M.H.; Park, C.S. ; Study of Shape Optimization for Aerodynamic Drag Reduction of High-speed. *J. Korean Soc. Railw.* **2016**, *19*, 709–716. [CrossRef]
17. Jun, S.O.; Jeon, Y.H.; Kim, J.H.; Lee, D.H. Application of the robust and reliability-based design optimization to the aircraft wind design. *J. Korean Soc. Aeronaut. Space Sci.* **2006**, *34*, 24–32.
18. Vanderplaats, G.N. *Numerical Optimization Techniques for Engineering Design*; Vanderplaats Research & Development Inc.: Colorado Springs, CO, USA, 1999.

et al. [26] carried out numerical simulations to study fluid flow and thermal transfer over a rotating cylinder at low rotation, demonstrating that increased rotational speeds resulted in lower mean Nusselt numbers, thereby affecting heat transfer efficiency. Brahim et al. [27] investigated the thermal and hydrodynamic behavior of a non-Newtonian nanofluid in a stirred tank, examining various anchor impeller configurations. Their study highlighted enhanced heat transfer at higher Reynolds numbers and nanoparticle fractions, with the arm-bladed anchor impeller demonstrating superior performance. Youcefi et al. [28]. aimed to present a thorough analysis utilizing CFD, focusing on the thermal, energetic, and hydrodynamic behaviors of a Newtonian fluid stirred by a modified two-blade impeller made up of two adjacent sub-blades. Their investigation assessed the effects of the rotational speed of the two-blade impeller, various design configurations, and the thermal fluid's characteristics on these behaviors. Moreover, Mokhfi [29]. numerically analyzed the hydrodynamic and isothermal characteristics of a viscoplastic fluid for food substance, in a mechanically stirred vessel equipped with an inclined anchor. Their results demonstrated that the inclined anchor design enhances the axial flow inside the vessel compared to the classical anchor and maintaining a predominant tangential flow.

The literature review highlights the significant complexities involved in agitating non-Newtonian fluids, which necessitate the use of advanced and well-designed impellers. Effective heat transfer in stirred vessels is crucial for enhancing mixture quality and accelerating chemical reactions. However, research specifically addressing the thermal properties of viscoplastic fluids in these systems remains limited. Standard anchor impellers, primarily designed to scrape the vessel's lateral surfaces, often do not suffice in improving efficiency, particularly in the inter-blade space (low-shear regions) where fluid dynamics stagnate. Therefore, there is a recognized need for innovative impeller designs. The main objective of this paper is to evaluate the energy efficiency and thermal performance of various anchor impeller configurations employed for agitating yield stress fluids, with chamfered angles of 22.5°, 45°, and 67.5°, comparing their performance to that of a classical anchor impeller.

The remainder of this paper is organized as follows: Section 2 delves into the mathematical formulation, providing a comprehensive exploration of the governing equations, fluid rheology, and the various configurations of mechanical agitation systems analyzed. Section 3 presents a thorough analysis and discussion of the simulation results, highlighting key findings and their implications for the field. Finally, Section 4 summarizes the main conclusions drawn from this research.

2. Numerical Modeling and Computational Settings

2.1. Agitated System Presentation

This study investigates a novel anchor impeller design with chamfered angles (α) of 22.5°, 45°, and 67.5°, comparing their performance to that of a classical anchor impeller with α = 0°, ensuring that the total surface area of the blades in contact with the fluid remains consistent across all configurations. The geometries of the stirring systems are depicted in Figure 1. The system consists of a cylindrical vessel with a flat bottom and no baffles, having a diameter D_v = 300 mm and a height $H_v/D_v = 1$. The impellers, each with a diameter of d = 288 mm, are mounted on a shaft with a diameter ratio d_a/D_v = 0.023. The blades feature a length ratio $L/D_v = 0.067$ and a thickness of $e/D_v = 0.027$, with geometrical parameters detailed in Table 1.

Mixing fluids that exhibit Bingham behavior typically occurs within the laminar and transient regimes, as highlighted by Bertrand et al. [30]. In these systems, heat transfer is primarily influenced by forced or mixed convection, particularly at lower rotational speeds, while natural convection effects act on the free surface of the fluid. Previous studies [23–25] often assume an adiabatic free surface due to thermal insulation; however, in this study, the free surface is directly exposed to ambient air, as examined in [28,29]. This exposure results in a distinct thermal profile, which may significantly impact the mixing and heat transfer dynamics within the system.

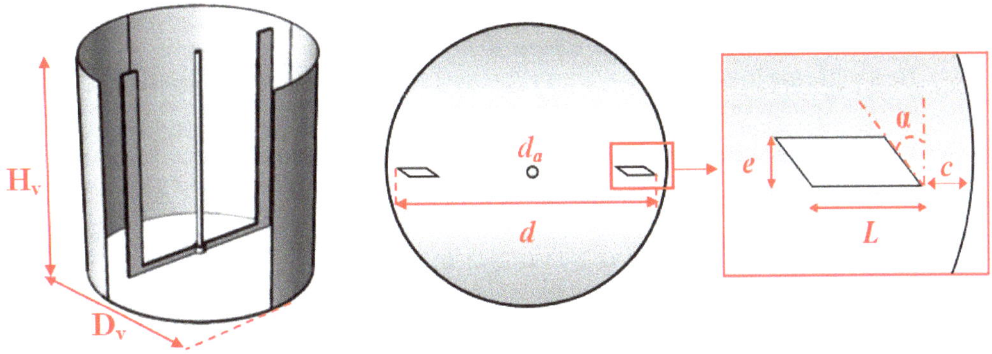

Figure 1. Geometry and presentation of the simulated system using a chamfered anchor impeller.

Table 1. Geometrical dimensions of the agitated system.

D_v	d	d_a/D_v	e/D_v	L/D_v	c/D_v
300 mm	288 mm	0.023	0.027	0.067	0.02

To simplify the numerical analysis, the following assumptions are made:

- The thermo-physical properties of the fluid are considered constant, with the exception of the apparent viscosity, which is shear-rate dependent as described by the Bingham model.
- Given the rotational speed and the predominance of tangential flow generated by the anchor impeller, free surface deformation is assumed to be negligible. Therefore, heat transfer at the free surface can be treated as natural convection.
- Effects of viscous dissipation and radiative heat transfer are considered negligible in the energy balance.

2.2. Governing Equations

The fluid is considered a non-Newtonian incompressible type. The flow dynamics are described by the mass conservation equation, along with the filtered momentum and energy equations:

$$\nabla \cdot V = 0 \tag{1}$$

$$\rho\frac{\partial V}{\partial t} + \rho(V\nabla)V - \eta_{ap}\Delta V - 2\overline{D} \cdot \nabla\eta_{ap} + \nabla P = 0 \tag{2}$$

$$\frac{\partial T}{\partial t} + V.\nabla T = k\Delta^2 T, \tag{3}$$

where k is the thermal diffusivity defined by:

$$k = \frac{\lambda}{\rho C_p}. \tag{4}$$

Here Cp, λ and ρ refer to the specific heat, thermal conductivity and density of the fluid, respectively.

2.3. Fluid Behavior

The behavior of the fluid follows a regularized version of Bingham model proposed by Bercovier and Engelman [31], which exhibits distinct flow behavior, remaining stationary until the applied shear stress exceeds a critical yield point, defined by a minimum yield stress (τ_0), which there is no flow below. It is expressed as:

$$\dot{D} = 0 \text{ in } \|\overline{\overline{\tau}}\| \leq \tau_0 \tag{5}$$

- On the vessel

$$v_t = -\pi N D_v \text{ and } v_r = v_z = 0.$$

Temperature conditions are:

- On the anchor impeller and the bottom of the vessel

$$\frac{\partial \Theta}{\partial n} = 0,$$

- On the lateral wall of tank

$$\Theta = 1.$$

At the fluid free surface:

$$\frac{\partial v_t}{\partial z} = \frac{\partial v_r}{\partial z} = \frac{\partial v_z}{\partial z} = 0 \text{ and } \left(-k\frac{\partial T}{\partial z}\right)_{z=H} = h_{air}(T - T_c) = 0$$

3. Results and Discussions

3.1. Mesh Independence Test and Model Validation

To ensure accurate calculations and reliable results, a mesh independence test was conducted using four different grid resolutions: M1, M2, M3 and M4. These meshes comprise varying cell counts—244,848, 465,210, 790,858, and 1,257,460 cells, respectively. Table 2 summarizes the effect of these meshes on the evolution of the stirring power number (Np). As shown, the difference in Np values between M3 and M4 is negligible, indicating that M3 provides a sufficient resolution for accurate simulations. Consequently, all simulations were performed using the M3 mesh, which consists of 790858 cells in the computational domain.

Table 2. Variation in the power number with the number of mesh elements.

Mesh	M1	M2	M3	M4
Elements	244,848	465,210	790,858	1,257,460
Np	29.863	28.931	28.681	28.679

To validate the accuracy of our numerical simulation, we compared our results with those documented by Marouche et al. [34], who performed numerical simulations under identical rheological and geometrical conditions. These parameters involved a standard anchor impeller and a Bingham fluid characterized by a viscosity of μ = 0.1 Pa·s and a yield stress of τ_0 = 1 Pa. We assessed the tangential velocity profile at the impeller plane and the power number across various Reynolds numbers. The findings, illustrated in Figure 3, demonstrate a strong agreement between our computed results and those of Marouche et al. [31]. Additionally, in Figure 4, we compare our results with experimental data from Anne-Archard et al. [35] for a Newtonian fluid, where τ_0 = 0 Pa in our model. The agreement between the numerical and experimental results further reinforces the accuracy of our model in capturing the behavior of both Bingham and Newtonian fluids.

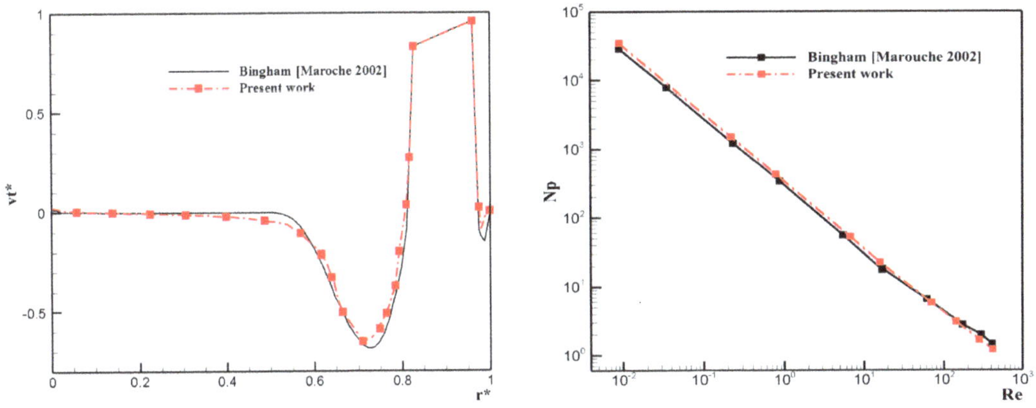

(a) (b)

Figure 3. Comparison of the tangential velocity on the impeller line (**a**) and power number (**b**) of the current CFD model and results of Marouche et al. [34].

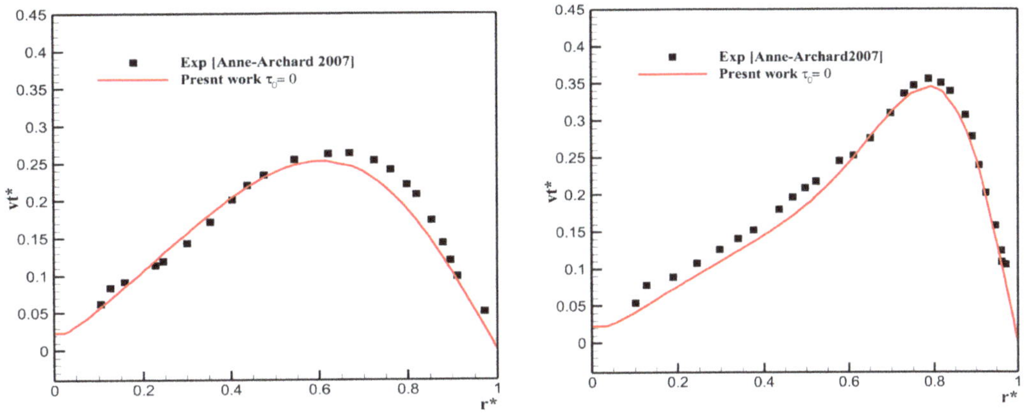

(a) (b)

Figure 4. Comparison of the tangential velocity on the median line (**a**) Re = 14; (**b**) Re = 70 of the current CFD model and results of Anne-Archard et al. [35].

3.2. Effect of Inertia

The analysis of flow patterns in stirred vessels is a key factor in evaluating the flow dynamics within the system. These dynamics are influenced not only by the impeller design but also by the operational conditions, particularly it's rotational speed. Figure 5 illustrates the streamline patterns within the stirred tank for different Reynolds numbers (13.8, 140 and 415) with τ_0 = 1 Pa and β = 0.001, showing both three-dimensional and two-dimensional views. These Reynolds numbers represent the transition from viscous-dominated to inertial-dominated flow, providing insight into the complex interplay between shear and inertial forces in stirred vessels. The chosen values—13.8, 140, and 415 were selected based on their relevance to the flow regimes typically observed in stirred tank systems and are consistent with similar studies in the literature, particularly those by of Marouche et al. [34] and Anne-Archard et al. [35].

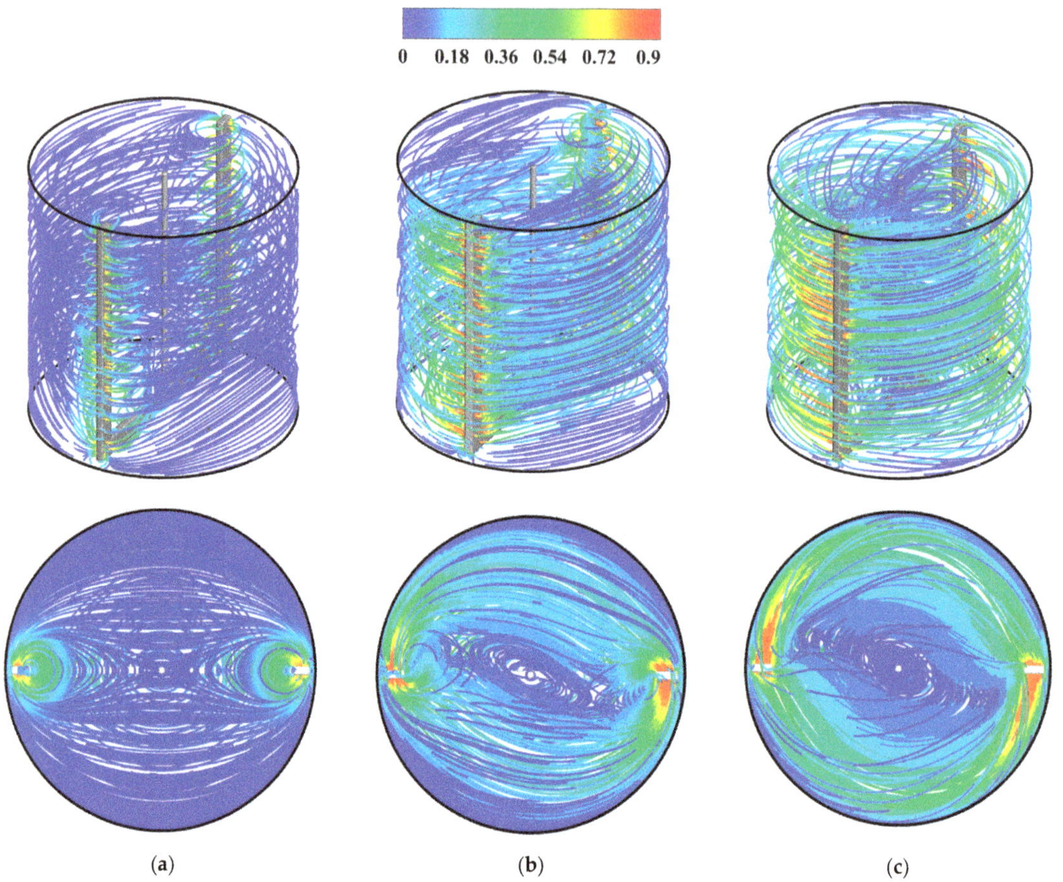

Figure 5. Three-dimensional streamlines for different Reynolds numbers: (**a**) Re = 13.8; (**b**) Re = 140; (**c**) Re = 415 for $\tau_0 = 1$ Pa.

For Re = 13.8, the streamlines in the horizontal section of the vessel exhibit remarkable uniformity, highlighting the dominance of tangential velocity over axial flow in stirred tanks with low axial velocities. The formation of recirculation zones near the impeller blades indicates a significant reduction in fluid movement, which is a hallmark of inefficient mixing. The flow in this regime is characterized by sluggish material transport and poor energy dissipation, leading to stagnant regions that limit the effective contact between the impeller and the bulk fluid. As the Reynolds numbers increase to 140 and 415, a significant change in the streamline distribution is observed, with the streamlines showing a more pronounced axial inclination. This shift indicates a change in flow dynamics, where higher Reynolds numbers intensify inertial effects in the system, leadingto the effective elimination of the dead zone region and more swept area.

3.3. Effect of Rehology

Figure 6 illustrates the effect of yield stress (τ_0) on the velocity distribution and 2D streamlines for a classical anchor impeller. The analysis includes three yield stress values: 0, 1, and 30 Pa, which correspond to Hedström number (He) equal to 0, 2640 and 79,205, respectively, where a yield stress of zero represents the Newtonian fluid case. As the yield stress increases, the velocity near the blades decreases, and large regions within the vessel remain stationary, resulting in expanded immobilized zones and smaller active shear regions near the blades. This leads to larger regions of immobilized fluid and narrower shear zones near the blades. Under these conditions, the flow's resistance increases due to the higher apparent viscosity in most parts of the vessel, which stabilizes the flow structure. The presence of yield stress fundamentally alters the hydrodynamics by reducing or even nullifying motion across large portions of the vessel, causing the fluid to exhibit solid-like behavior.

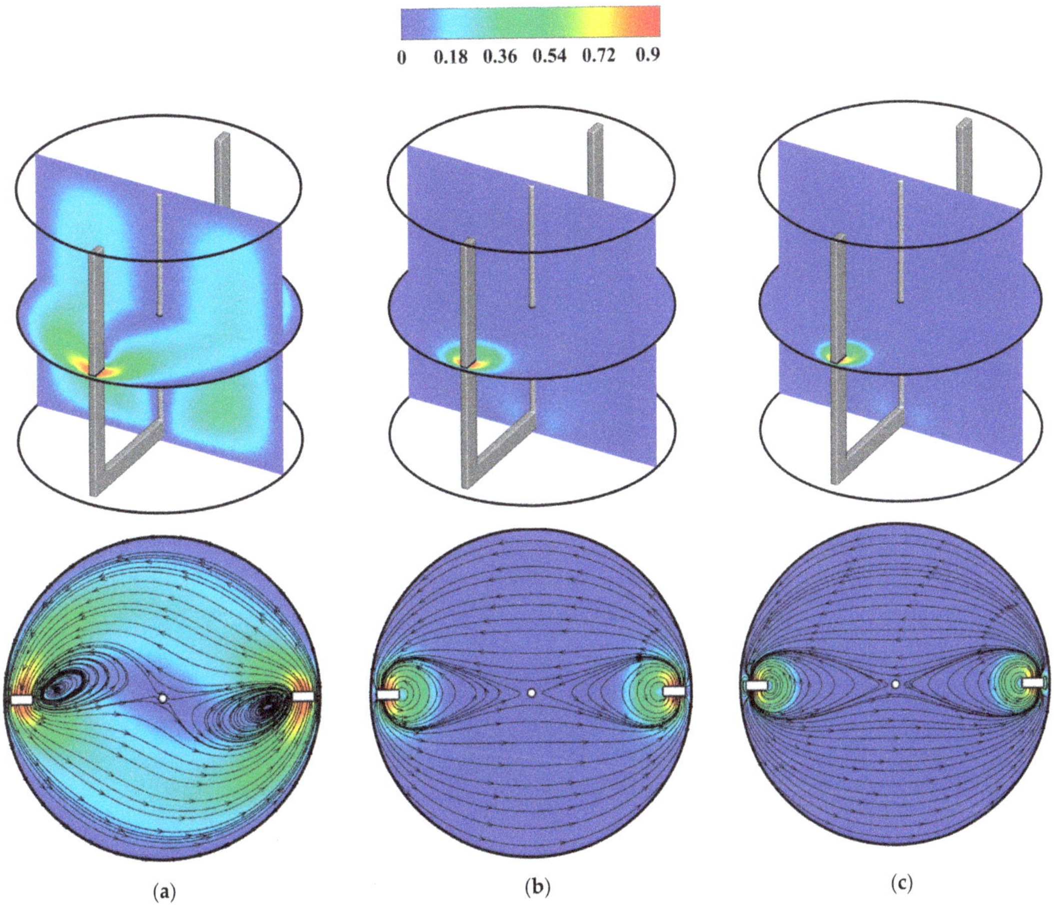

Figure 6. Velocity magnitude contour comparison for different yield stress: (**a**) $\tau_0 = 0$ Pa; (**b**) $\tau_0 = 1$ Pa; (**c**) $\tau_0 = 30$ Pa for Re = 13.8.

3.4. *Effect of Chamfered Angle Design*

Standard anchor impellers effectively scrape the vessel's lateral surfaces but often do not adequately tackle the challenges posed by low-shear regions between the blades, where fluid dynamics can stagnate. Therefore, we analyze the hydrodynamic and thermal performance of various chamfered impeller configurations with angles of 22.5°, 45°, and 67.5°. The performance of these new designs is compared to that of a classical anchor impeller to assess their effectiveness in agitating yield stress fluids.

Figures 7 and 8 present dimensionless velocity magnitude contours for chamfered anchor impellers with angles of 0°, 22.5°, 45°, and 67.5° at Reynolds numbers of Re = 10 and Re = 415. In Figure 7, at Re = 10, the contours show minimal fluid circulation across all configurations, with flow concentrated near the blades and a stagnant zone forming at the vessel's center, limiting mixing efficiency. Although increasing the chamfered angle from 0° to 67.5° improves fluid movement near the blades, circulation remains weak due to the low Reynolds number, resulting in inadequate mixing.

In contrast, at Re = 415 (Figure 8), fluid circulation improves significantly across all configurations. The higher Reynolds number increases fluid velocities, enhancing mixing. The chamfered impeller with a 67.5° angle exhibits the most extensive region of active fluid motion, suggesting superior pumping and circulation efficiency. Larger chamfered angles promote better flow patterns around the blades, making them more effective at enhancing circulation under high Reynolds number conditions. Optimizing the chamfered angle is thus crucial for improving mixing performance.

Figures 9 and 10 present the effect of chamfered anchor impellers on the tangential velocity profiles, focusing on two critical lines within the fluid domain: the impeller line and the median line, respectively.

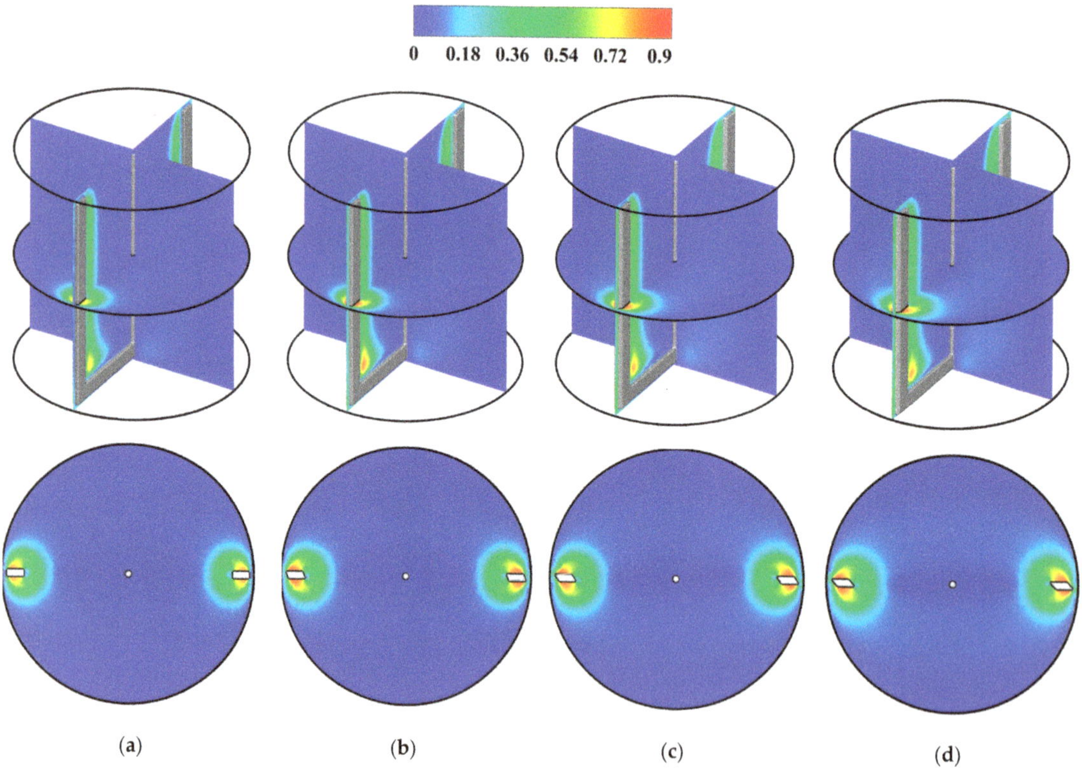

Figure 7. Effect of chamfered angles on the velocity magnitude contour: (**a**) $\alpha = 0°$; (**b**) $\alpha = 22.5°$; (**c**) $\alpha = 45°$; (**d**) $\alpha = 67.5°$ for Re = 13.8 and $\tau_0 = 1$ Pa.

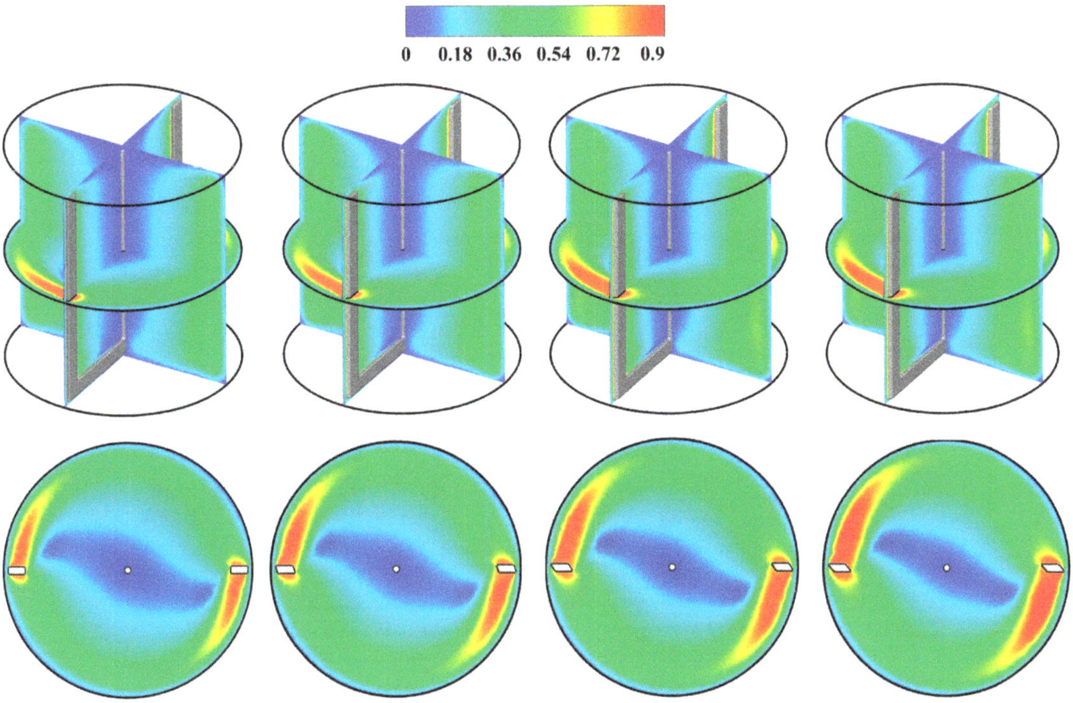

(**a**) (**b**) (**c**) (**d**)

Figure 8. Effect of blade chamfered angles on the velocity magnitude contour: (**a**) $\alpha = 0°$; (**b**) $\alpha = 22.5°$; (**c**) $\alpha = 45°$; (**d**) $\alpha = 67.5°$ for Re = 415 and $\tau_0 = 1$ Pa.

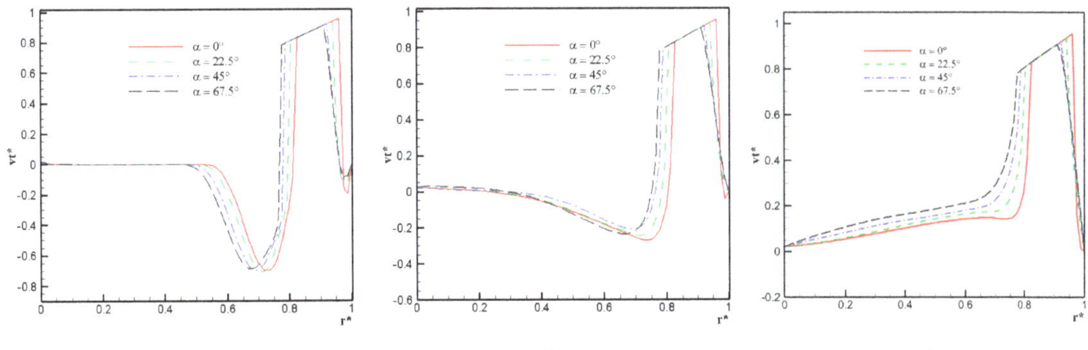

(**a**) (**b**) (**c**)

Figure 9. Effect of blade chamfered angles on tangential velocity profiles along the impeller line for different Reynolds numbers: (**a**) Re = 13.8; (**b**) Re = 140; (**c**) Re = 415 for $\tau_0 = 1$ Pa.

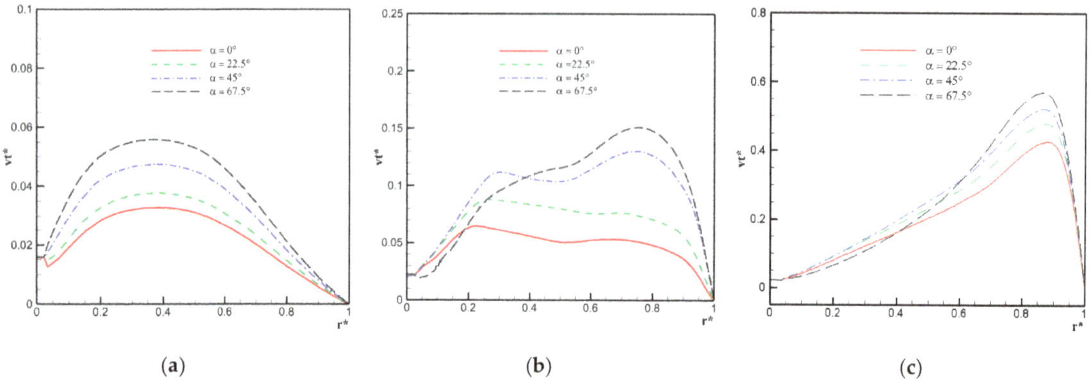

Figure 10. Effect of blade chamfered angles on tangential velocity profiles along the median line for different Reynolds numbers: (**a**) Re = 13.8; (**b**) Re = 140; (**c**) Re = 415 for $\tau_0 = 1$ Pa.

In Figure 9, at a lower Reynolds number of Re = 13.8, the velocity profiles reveal that fluid movement is concentrated primarily near the impeller blades, where it peaks at the tips of the blades and gradually diminishes to zero along the walls of the vessel. The profile for the 0° angle shows the least fluid movement, with negative velocities observed close to the blades, indicating poor mixing efficiency. As the chamfered angle increases, there is a slight enhancement in fluid movement near the blades, with progressively higher velocities recorded at a radial position of r^* = 0.7. However, despite this slight improvement, the overall circulation remains weak, and a significant stagnant region is evident at the center of the vessel.

At a higher Reynolds number of Re = 415, the situation markedly changes. The velocity contours in the impeller plane demonstrate significantly increased fluid velocities across all configurations. The 67.5° chamfered impeller exhibits the most extensive area of active fluid motion, indicating a superior pumping action and enhanced circulation efficiency. The chamfered angles of 22.5° and 45° also show improved flow patterns compared to the 0° angle, with higher velocities between the shaft and the blades. The results emphasize that greater chamfer angles lead to improved fluid engagement and momentum transfer, resulting in more effective mixing under higher operational conditions.

In Figure 10, the velocity profiles further elucidate the effects of chamfered angles and Reynolds numbers. At Re = 13.8, the plots reveal a lack of substantial fluid motion, with velocities remaining low. As the Reynolds number increased to 415, the median plane profiles reflect a significant improvement in fluid dynamics. The higher velocities are observed throughout the plane, indicating robust mixing across all impeller configurations. The 0° impeller configuration shows the lowest velocity profile, characterized by minimal fluid circulation. In contrast, increasing the chamfered angle correlates with an increase in velocity, with the 67.5° chamfered impeller displaying the highest velocity values. This configuration demonstrates a substantial increase in active fluid motion and circulation efficiency compared to the other designs. The velocity distributions indicate a more pronounced flow pattern, suggesting that the chamfered design effectively overcomes the challenges posed by yield stress fluids.

Figures 11 and 12 illustrate the effects of blade chamfered angles on tangential velocity profiles, along the impeller line and the median line, respectively, at a constant Reynolds number of Re = 13.8.

In Figure 11, the tangential velocity profiles measured at the impeller line illustrate distinct variations as yield stress increases from τ_0 = 0 Pa to τ_0 = 30 Pa. At zero yield stress, the tangential velocity profiles for all chamfered angles exhibit robust fluid movement, peaking near the impeller tips and decreasing towards the tank walls. Notably, at τ_0 = 0 Pa, the chamfered designs demonstrate markedly improved fluid movement compared to the 0° configuration, indicating superior mixing efficiency facilitated by the chamfered profiles.

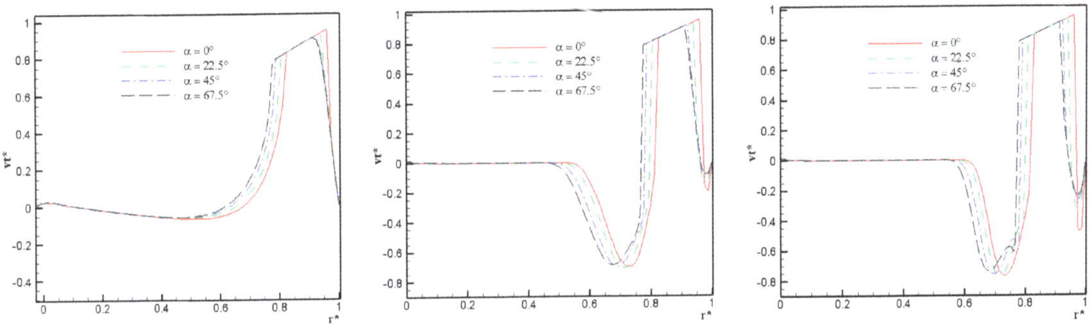

(**a**) (**b**) (**c**)

Figure 11. Effect of blade chamfered angles on tangential velocity profiles along the impeller line for different yield stress: (**a**) $\tau_0 = 0$ Pa; (**b**) $\tau_0 = 1$ Pa; (**c**) $\tau_0 = 30$ Pa for Re = 13.8.

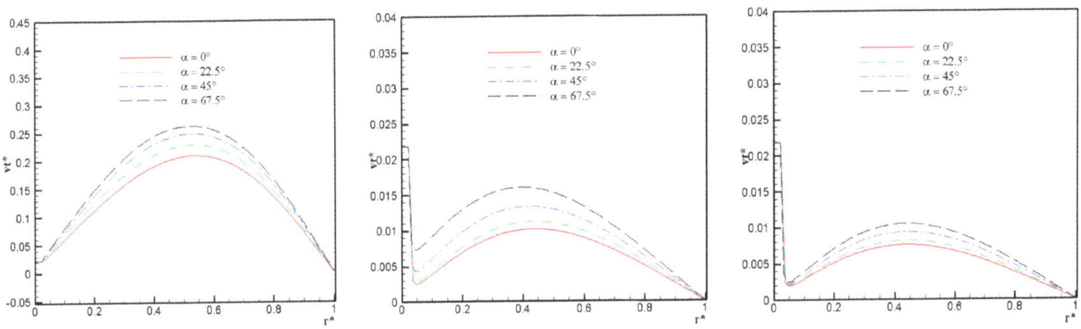

(**a**) (**b**) (**c**)

Figure 12. Effect of blade chamfered angles on tangential velocity profiles along the median line for different yield stress: (**a**) $\tau_0 = 0$ Pa; (**b**) $\tau_0 = 1$ Pa; (**c**) $\tau_0 = 30$ Pa for Re = 13.8.

However, as yield stress is introduced, particularly at τ_0 = 1 Pa and τ_0 = 30 Pa, the impact on fluid dynamics becomes pronounced. For τ_0 = 1 Pa, the tangential velocities show a marked decrease and become negative in proximity to the blades, particularly for the 0° angle, indicating a significant reduction in mixing efficiency. While the chamfered angles again exhibit better performance, the advantages are less pronounced than at zero yield stress.

At τ_0 = 30 Pa, the overall reduction in tangential velocities across all configurations highlights the challenges posed by increased yield stress. The 67.5° chamfered impeller shows the highest velocities despite the elevated yield stress but still exhibits weak fluid movement, with a significant stagnant region evident inside the vessel.

In Figure 12, the tangential velocity profiles measured at the median line complement the findings from the impeller line, further elucidating the influence of chamfered angles on fluid dynamics under varying yield stress conditions. For τ_0 = 0 Pa, the velocity profiles mirror those observed in the impeller plane, with all configurations exhibiting significant fluid movement. However, as yield stress is introduced, the profiles demonstrate notable shifts.

At τ_0 = 1 Pa, the profiles illustrate reduced velocities across the board, with the 0° angle revealing the least fluid movement. The chamfered designs show an advantage, with increasing angles correlating with enhanced velocity distributions, reaffirming that chamfered anchor impellers are more effective in facilitating fluid motion, particularly at lower yield stress conditions.

For the highest yield stress condition (τ_0 = 30 Pa), the median plane profiles reflect a substantial decrease in overall velocities, underscoring the considerable impact of yield stress on flow dynamics. The 67.5° chamfered impeller emerges as the most effective configuration, achieving the highest tangential velocities compared to other configurations. The pronounced flow patterns indicate that the chamfered design effectively counters the resistance posed by yield stress; however, as yield stress increases, the benefits become less pronounced compared to lower yield stress conditions.

The analysis of both the impeller and median lines reveals that chamfered anchor impellers significantly enhance fluid dynamics across varying yield stress conditions. Notably, at zero yield stress, the chamfered designs exhibit superior mixing efficiency, reinforcing the importance of impeller design in optimizing fluid motion. As yield stress increases, while the effectiveness of the chamfered angles remains advantageous, the benefits are less pronounced.

Figure 13 compares the temperature field distribution in a stirred system using the classical anchor (α = 0°) and the chamfered anchor (α = 67°). In the classical anchor configuration, the thermal profile exhibits a more layered structure, with higher temperatures concentrated on the heated wall and cooler regions towards the center. This indicates that the classical design is less effective at disrupting thermal stratification, resulting in insufficient heat transfer throughout the fluid volume. The central and interior areas show lower temperatures, suggesting limited mixing capability, which is particularly problematic for viscoplastic fluids due to their yield stress behavior, which can inhibit fluid motion.

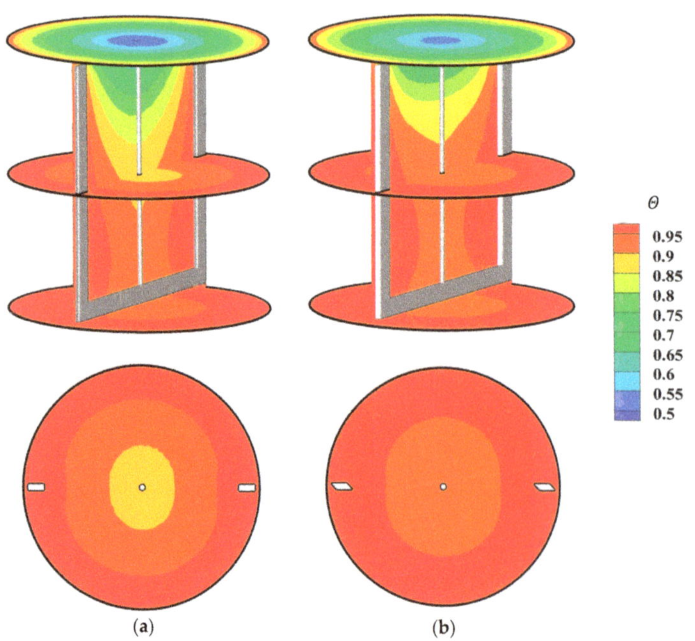

Figure 13. Temperature field distribution for different impeller designs: (**a**) α = 0°; (**b**) α = 67° (Re = 13.8, τ_0 = 1 Pa and Pr = 7).

In contrast, the chamfered anchor configuration ($\alpha = 67^{\circ}$) demonstrates a more uniform temperature distribution, characterized by more evenly distributed isotherms that extend radially towards the center. This configuration effectively facilitates heat distribution throughout the stirred volume. The more uniform temperature field, coupled with reduced temperature gradients, indicates that the chamfered design mitigates thermal stratification by enhancing the agitation and promoting more efficient mixing. As a result, the system exhibits more homogeneous thermal behavior compared to the classical design ($\alpha = 0^{\circ}$), with a more consistent temperature profile throughout the vessel.

Figure 14 presents the shear rate contours for the classical anchor ($\alpha = 0^{\circ}$) and chamfered anchor ($\alpha = 67.5^{\circ}$), illustrating distinct differences in shear distribution and fluid dynamics within the vessel at Re = 13.8 and $\tau_0 = 1$. The chamfered anchor demonstrates a more uniform shear distribution throughout the fluid domain, while the classical design tends to localize shear near the impeller blades. This uniformity in shear is advantageous for minimizing dead zones and ensuring consistent fluid movement, which is particularly important for Bingham fluids that require adequate shear to overcome yield stress and initiate flow.

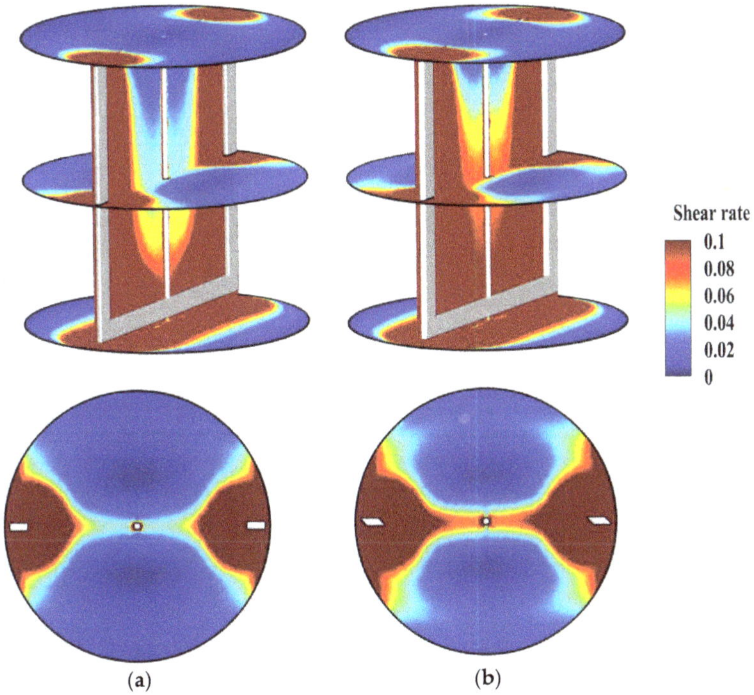

Figure 14. Shear rate distribution for different impeller designs: (**a**) $\alpha = 0^{\circ}$; (**b**) $\alpha = 67^{\circ}$ (Re = 13.8, $\tau_0 = 1$ Pa and Pr = 7).

In the vertical plane, the chamfered anchor shows a broader spread of shear along the vessel height, indicating enhanced vertical fluid motion and potentially more effective mixing from top to bottom. In contrast, the classical anchor generates higher shear near the mid-height and bottom regions but has a limited influence on the upper portion of the vessel, resulting in less uniform mixing.

Radial mixing patterns further emphasize the benefits of the chamfered design, as the high-shear zones extend further from the impeller, suggesting that these configurations promote better lateral mixing. The classical anchor, with its more confined shear regions, may be effective for localized mixing near the blades but could leave the outer regions under-agitated.

Figure 15 shows the variations in power consumption for different chamfered anchor angles as a function of the Reynolds number. It can be observed that the power number decreases with an increasing Reynolds number, which is attributed to the reduction in fluid viscosity at a constant impeller speed. This indicates that lower viscosity results in less power being required for agitation. A closer examination of the initial region reveals a slight increase in power at lower Reynolds numbers as the chamfered angle increases. However, at higher Reynolds numbers, power consumption becomes nearly identical across the different angles. The modifications in the anchor designs effectively maintain low power requirements, making these configurations advantageous for achieving the desired agitation performance without significant increases in energy consumption.

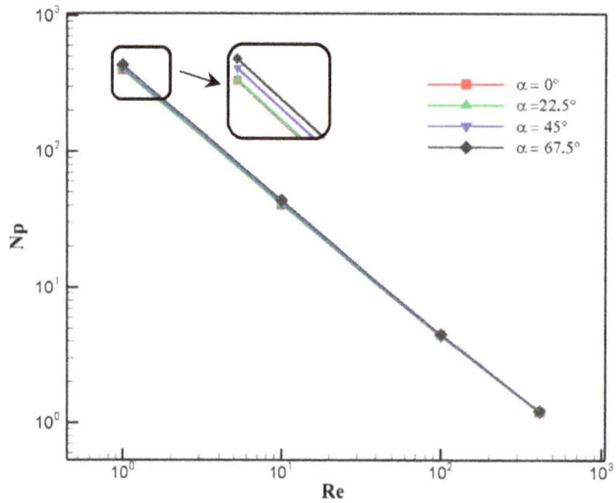

Figure 15. Power number variation for different Reynolds number and different impeller designs ($\tau_0 = 1$ Pa).

4. Conclusions

This study evaluates the hydrodynamic and thermal performance of stirring systems for viscoplastic fluids using close-clearance anchor impellers with chamfered angles of 0°, 22.5°, 45°, and 67.5°. The numerical simulations were conducted using ANSYS Fluent 2023, with the finite-volume method to solve the Navier–Stokes equations. The regularized version of the Bingham model, proposed by Bercovier and Engelman, was integrated into the solver using the User-Defined Function (UDF) framework to model the flow characteristics.

The results indicate that among the tested configurations, the 67.5° chamfered angle demonstrated the best performance, showing the most extensive fluid motion and robust circulation. It significantly improved flow dynamics by enhancing circulation inside the vessel and near the blades and reducing stagnant zones.

The analysis of shear rate contours indicates that the chamfered anchor also enhances vertical mixing by promoting a broader spread of shear along the vessel height, improving top-to-bottom mixing. Radially, the high-shear zones extend further from the impeller in chamfered designs, facilitating better lateral mixing than the classical anchor.

Thermal analysis reveals that chamfered impellers contribute to a more homogeneous temperature distribution, effectively mitigating thermal stratification and enhancing heat transfer throughout the fluid volume.

The study demonstrates that the energy requirements for chamfered configurations remain manageable, making them a practical choice for industrial applications. Optimizing the chamfered angle thus emerges as a crucial factor in advancing mixing technologies, providing valuable insights for designing more efficient agitation systems in industries handling complex fluids.

The study incorporates yield-stress behavior; however, temperature-dependent viscosity was not explicitly considered. Future work could incorporate temperature-dependent viscosity to more accurately represent real-world conditions, especially for high-temperature processes.

Author Contributions: Conceptualization, A.B.; methodology, A.B.; software, A.B.; validation, A.B.; formal analysis, A.B. and J.C.P.; investigation, A.B.; resources, J.C.P.; writing—original draft preparation, A.B.; writing—review and editing, A.B. and J.C.P.; supervision, J.C.P.; project administration, J.C.P.; funding acquisition, J.C.P. All authors have read and agreed to the published version of the manuscript.

Funding: This research was funded by the GreenAuto Project: Green Innovation for the Automotive Industry, n° 02/C05- i01.02/2022.PC644867037-00000013 from the Incentive System to Mobilizing Agendas for Business Innovation, funded by the Recovery and Resilience Plan. This research was also supported by the Portuguese Foundation for Science and Technology, I.P. (FCT, I.P.) FCT/MCTES through national funds (PIDDAC), under the R&D Unit C-MAST, Center for Mechanical and Aerospace Science and Technology, Research Unit No. 151, reference: Projects UIDB/00151/2020 (https://doi.org/10.54499/UIDB/00151/2020) and UIDP/00151/2020 (https://doi.org/10.54499/UIDP/00151/2020).

Data Availability Statement: The original contributions presented in the study are included in the article, further inquiries can be directed to the corresponding author.

Conflicts of Interest: The authors declare no conflicts of interest.

References

1. Prajapati, P.; Ein-Mozaffari, F. CFD Investigation of the Mixing of Yield-Pseudoplastic Fluids with Anchor Impellers. *Chem. Eng. Technol. Ind. Chem. Plant Equip. Process Eng. Biotechnol.* **2009**, *32*, 1211–1218. [CrossRef]
2. Driss, Z.; Salah, A.; Driss, D.; Necib, B.; Kchaou, H.; Abid, M.S. CFD Investigation of the hydrodynamic structure around a modified anchor system. In *CFD Techniques and Energy Applications*; Springer: Cham, Switzerland, 2018; pp. 129–150.
3. Lins Barros, P.; Ein-Mozaffari, F.; Lohi, A. Gas Dispersion in Non-Newtonian Fluids with Mechanically Agitated Systems: A Review. *Processes* **2022**, *10*, 275. [CrossRef]
4. Casarez-Duran, A.A.; Paredes-Rojas, J.C.; Torres-San Miguel, C.R.; Méndez-García, S.R.; Ortiz-Hernández, F.E.; Urriolagoitia Calderón, G.M. CFD Simulation of Mixing Forest Biomass to Obtain Cellulose. *Processes* **2024**, *12*, 2250. [CrossRef]
5. Benmoussa, A.; Rahmani, L.; Rebhi, M. Regularization model effect on yield stress fluids behavior in a rotating vessel. *Adv. Appl. Fluid Mech.* **2016**, *19*, 507–515. [CrossRef]
6. Ebrahimi, M.; Tamer, M.; Villegas, R.M.; Chiappetta, A.; Ein-Mozaffari, F. Application of CFD to Analyze the Hydrodynamic Behaviour of a Bioreactor with a Double Impeller. *Processes* **2019**, *7*, 694. [CrossRef]
7. Benmoussa, A. Agitation of Viscoplastic Fluid in a Rotating Vessel Using Close Clearance Agitators. *Eng* **2023**, *4*, 2525–2541. [CrossRef]
8. Anne-Archard, D.; Marouche, M.; Boisson, H.-C. Hydrodynamics and Metzner–Otto correlation in stirred vessels for yield stress fluids. *Chem. Eng. J.* **2006**, *125*, 15–24. [CrossRef]
9. Kamla, Y.; Ameur, H.; Arab, M.I. Data on the hydrodynamics and power consumption induced by modified anchor impellers in cylindrical tanks. *Data Brief* **2021**, *39*, 107669. [CrossRef]
10. Hami, O.; Draoui, B.; Mebarki, B.; Rahmani, L.; Bouanini, M. Numerical model for laminar flow and heat transfer in an agitated vessel by inclined blades anchor. In *Proceedings of CHT-08 ICHMT International Symposium on Advances in Computational Heat Transfer*; Begel House Inc.: Danbury, CT, USA, 2008.
11. Komoda, Y.; Date, T. Enhancement of laminar mixing by an anchor impeller with rotationally reciprocating motion. *AIP Adv.* **2022**, *12*, 015013. [CrossRef]
12. Kada, B.; Lakhdar, R.; Brahim, M.; Ameur, H. Agitation of Complex Fluids in Cylindrical Vessels by Newly Designed Anchor Impellers. *Period. Polytech. Mech. Eng.* **2022**, *66*, 109–119. [CrossRef]
13. Mihailova, O.; Mothersdale, T.; Rodgers, T.; Ren, Z.; Watson, S.; Lister, V.; Kowalski, A. Optimization of mixing performance of helical ribbon mixers for high throughput applications using computational fluid dynamics. *Chem. Eng. Res. Des.* **2018**, *132*, 942–953. [CrossRef]
14. Ameur, H.; Kamla, Y.; Sahel, D. Performance of helical ribbon and screw impellers for mixing viscous fluids in cylindrical reactors. *ChemEngineering* **2018**, *2*, 26. [CrossRef]
15. Ameur, H.; Kamla, Y. Geometrical modifications of the anchor impeller to enhance the overall performances in stirred tanks. *Instal* **2020**, *6*, 42–45.
16. Ameur, H. Energy efficiency of different impellers in stirred tank reactors. *Energy* **2015**, *93*, 1980–1988. [CrossRef]

17. Mebarki, B.; Draoui, B.; Rahmani, L.; Bouanini, M.; Rebhi, M.; Benachour, E.H. Numerical study of mechanical stirring in case of yield stress fluid with circular anchor impeller. *Sens. Transducers* **2011**, *132*, 108–121.
18. Sossa-Echeverria, J.; Taghipour, F. Effect of mixer geometry and operating conditions on flow mixing of shear thinning fluids with yield stress. *AIChE J.* **2014**, *60*, 1156–1167. [CrossRef]
19. Ein-Mozaffari, F.; Upreti, S.R. Using ultrasonic doppler velocimetry and cfd modeling to investigate the mixing of non-newtonian fluids possessing yield stress. *Chem. Eng. Res. Des.* **2009**, *87*, 515–523. [CrossRef]
20. Savreux, F.; Jay, P.; Magnin, A. Viscoplastic fluid mixing in a rotating tank. *Chem. Eng. Sci.* **2007**, *62*, 2290–2301. [CrossRef]
21. Kazemzadeh, A.; Ein-Mozaffari, F.; Lohi, A.; Pakzad, L. Effect of the rheological properties on the mixing of herschel-bulkley fluids with coaxial mixers: Applications of tomography, cfd, and response surface methodology. *Can. J. Chem. Eng.* **2016**, *94*, 2394–2406. [CrossRef]
22. Baccar, M.; Mseddi, M.; Abid, M.S. Contribution numérique à l'étude hydrodynamique et thermique des écoulements turbulents induits par une turbine radiale en cuve agitée. *Int. J. Therm. Sci.* **2011**, *40*, 753–772. [CrossRef]
23. Gammoudi, A.; Ayadi, A.; Baccar, M. The hydrodynamic and thermal characterization of a yield stress fluid in stirred tanks equipped with simple helical ribbons with two stages. *Meccanica* **2017**, *52*, 1743–1766. [CrossRef]
24. Benmoussa, A.; Rahmani, L.; Draoui, B. Simulation of viscoplastic flows in a rotating vessel using a regularized model. *Int. J. Multiphysics* **2017**, *11*, 349–358.
25. Benmoussa, A.; Rahmani, L.; Rebhi, M. Effect of regularization parameter on flow field and power consumption in agitated vessel by two blade impeller. *World Appl. Sci. J.* **2015**, *3*, 1025–1031.
26. Korib, K.; Mohamed, R.; Yacine, K. Numerical study on characteristics of flow and thermal fields around rotating cylinder. *Metall. Mater. Eng.* **2020**, *26*, 71–86. [CrossRef] [PubMed]
27. Brahim, M.; Benhanifia, K.; Jamshed, W.; Al-Farhany, K.; Redouane, F.; Eid, M.R.; Hussain, S.M.; Akram, M.; Kamel, A. Computational Analysis of Viscoplastic Nanofluid Blending by a Newly Modified Anchorage Impeller within a Stirred Container. *Symmetry* **2022**, *14*, 2279. [CrossRef]
28. Youcefi, S.; Mokhefi, A.; Bouzit, M. Theoretical study of thermo-hydrodynamic performances inside an agitated tank equipped with modified two-blade design. *Therm. Sci. Eng. Prog.* **2024**, *55*, 102866. [CrossRef]
29. Mokhefi, A. Hydrodynamic and thermal performance analysis of an inclined anchor impeller designed for yield stress food mixing applications. *Food Bioprod. Process.* **2024**, *143*, 255–270. [CrossRef]
30. Bertrand, F.; Tanguy, P.A.; Brito-De La Fuente, E. A new perspective for the mixing of yield stress fluids with anchor impellers. *J. Chem. Eng. Jpn.* **1996**, *29*, 51–58. [CrossRef]
31. Bercovier, M.; Engelman, M. A finite element method for incompressible non-Newtonian flows. *J. Comput. Phys.* **1980**, *36*, 313–326. [CrossRef]
32. Rahmani, L.; Seghier, O.; Benmoussa, A.; Draoui, B. CFD study of the thermal transfer of a non-Newtonian fluid within a tank mechanically stirred by an anchor-shaped impeller. In *EPJ Web of Conferences*; EDP Sciences: Les Ulis, France, 2018; Volume 180, p. 02089.
33. Youcefi, S.; Youcefi, A. Power consumption and mixing time in rheologically complex fluids by a two-bladed impeller. *J. Mech. Sci. Technol.* **2015**, *29*, 543–548. [CrossRef]
34. Marouche, M.; Anne-Archard, D.; Boisson, H.C. A numerical model of yield stress fluid dynamics in a mixing vessel. *Appl. Rheol.* **2002**, *12*, 182–191. [CrossRef]
35. Anne-Archard, D.; Boisson, H.C.; Marouche, M. Etude expérimentale d'une cuve agitée en fluide viscoplastique. Experimental study of a stirred tank in viscoplastic fluid. In Proceedings of the 18éme Congrès Français de Mécanique, Grenoble, France, 27–31 August 2007.

fluids

MDPI

Article

Non-Uniform Turbulence Modeling in Isolated Unsteady Diffuser Computational Models for a Vaned Centrifugal Compressor

Benjamin L. Holtmann and Nicole L. Key *

Department of Mechanical Engineering, Purdue University, West Lafayette, IN 47907, USA; bholtma@purdue.edu
* Correspondence: nkey@purdue.edu; Tel.: +1-765-494-2333

Abstract: Recent advancements in computational fluid dynamics (CFD) enable new and more complex analysis methods to be developed for early design stages. One such method is the isolated unsteady diffuser model, which seeks to reduce the computational cost of unsteady CFD when modeling diffusion systems in centrifugal compressors with vaned diffusers by isolating the diffuser from the computational domain and prescribing an unsteady and periodic inlet boundary condition. An initial iteration of this computational methodology was developed and validated for the Centrifugal Stage for Aerodynamic Research (CSTAR) at the High-Speed Compressor Laboratory at Purdue University. However, that work showed discrepancies in flow structure predictions between full-stage and isolated unsteady CFD models, and it also presented a narrow scope of only a single loading condition. Thus, this work addresses the need for improvement in the modeling fidelity. The original methodology was expanded by including a more accurate, non-uniform definition of turbulence at the diffuser inlet and modeling several loading conditions ranging from choke to surge. Results from isolated unsteady diffuser models with non-uniform turbulence modeling were compared with uniform turbulence isolated unsteady diffuser models and full-stage unsteady models at four loading conditions along a speedline. Flow structure predictions by the three methodologies were compared using 1D parameters and outlet total pressure and midspan velocity contours. The comparisons indicate a significant improvement in 1D parameter and flow structure predictions by the isolated unsteady diffuser models at all four loading conditions when including more accurate non-uniform turbulence, without a corresponding increase in computational cost. Additionally, both isolated diffuser methodologies accurately track trends in 3D flow structures along the speedline.

Keywords: centrifugal compressor; diffuser; CFD; turbomachinery

1. Introduction

Recent improvements in the robustness and fidelity of computational models in turbomachinery have allowed engineers to increase the scope of their early design assessments. This is often implemented in two ways. First, increasingly complex and large models are made to simulate entire turbomachinery stages. Second, high fidelity models of individual components can be simulated with high accuracy, allowing for detailed design changes to be made early in the design process. The design of centrifugal compressors, which feature extremely complicated and unsteady flows in the impeller and vaned diffuser, has especially benefited from these improvements [1].

The flow at the outlet of an impeller in a centrifugal compressor is characterized by the jet-wake model. First hypothesized by Dean and Senoo [2], the model categorizes the impeller outlet flow into two distinct regions, the first being the jet—a region of high momentum flow adjacent to the impeller blade pressure side—and the second being the wake—a region of low momentum flow adjacent to the impeller blade suction side. The model was initially supported by circumstantial evidence from flow visualizations in an impeller using water as the working fluid [3] and simplified component studies [4,5].

Eckardt made the first direct measurements of the jet-wake flow structure [6,7]. An example of a computed jet-wake flow structure at the outlet of a centrifugal compressor is shown in Figure 1.

When creating computational models for centrifugal compressors, considerations are made to ensure correct modelling of the impeller–diffuser interaction. This is due to the close coupling of the impeller and diffuser performance. The jet-wake flow structure exiting the impeller is the predominant driver in this interaction [8]. The impeller outlet flow, and specifically the average exit flow angle, is closely correlated with the diffuser performance [9]. Conversely, the potential field of diffuser vane leading edges imposes an unsteady condition on the impeller outlet flow, which contributes to tip leakage and potentially distinct regions of reversed flow [10]. Two different types of computational model can be used to accurately capture the impeller-diffuser interaction and pitch change. Reynolds Averaged Navier–Stokes (RANS) simulations solve for time-averaged solutions of the Navier–Stokes equations. When implementing this methodology, the interface between the rotating (impeller) and stationary (diffuser) domains is most frequently modeled using a mixing plane. This interface option for the RANS equations accounts for changes in reference frame and/or pitch [11–13]. While the performance results from this computational methodology and interface are frequently deemed "close enough" approximations, they do not accurately capture the unsteady and periodic jet-wake flow at the impeller outlet. Therefore, designers attempting to influence specific flow structures with design changes might be misled by RANS results. Unsteady RANS (URANS) simulations alleviate this issue by capturing time-varying properties in the impeller and diffuser domains. This leads to flow structure predictions that more closely resemble experimental results. However, compared to RANS, URANS simulations suffer from significantly increased computational costs. When only small design iterations are considered, these increased computational costs can be prohibitively expensive.

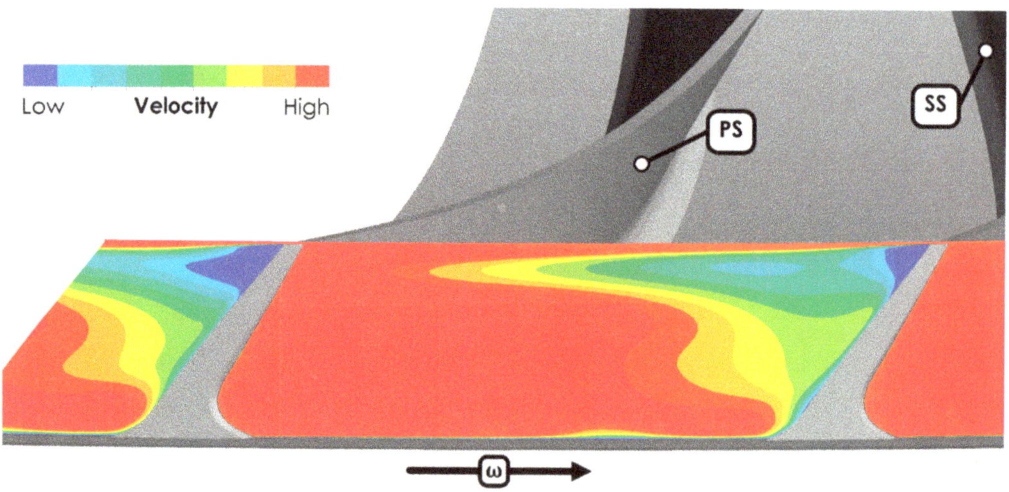

Figure 1. Jet-wake interaction at impeller outlet [1].

A method developed by Giles reduces the computational cost of URANS. The time-inclination method proposes a technique that transforms the time coordinates of the rotor and stator in the circumferential direction [14]. This allows for URANS simulations to use single passage computational domains with unequal pitches at interfaces instead of more costly full annulus simulations. Furthermore, several computational methods exist that are suited to simulating isolated blade rows. The harmonic balance method, which discretizes the RANS equations into mathematically steady ones in which excitation frequencies appear as parameters, is suited for calculation of aerodynamic forces used in flutter and forced response simulations [15]. Isolated steady diffuser models for the vaned diffuser of a centrifugal compressor were developed by Everitt [16]. The diffuser inlet profile was taken from the impeller outlet of a single passage steady simulation using a mixing plane. The isolated steady diffuser model benefited from a reduced computational cost and was successfully applied to stall inception studies. Isolated blade row models are sensitive to the fidelity of the applied boundary condition at the inlet. An approach to modeling realistic and coherent boundary conditions at the inlet of the domain was developed by Semlitsch. This method, which is based on the method of characteristics, was successfully applied and validated at a turbine inlet [17].

These methods reduce the computational cost of URANS and provide alternatives for simulating isolated blade rows. However, in the case of the harmonic balance and isolated steady diffuser method, they neglect to transfer the periodic jet-wake flow from the impeller outlet to the diffuser inlet. Therefore, isolated unsteady diffuser models were developed for the CSTAR Gen. 1 geometry with the goal of further decreasing the computational cost of modeling time-varying flow properties in the vaned diffuser of a centrifugal compressor while also maintaining an accurate jet-wake flow structure at the diffuser inlet [1]. The challenge of applying a complex, unsteady, and periodic diffuser inlet boundary condition is further compounded by the close vicinity of the diffuser vanes to the domain inlet. An analysis which included comparing the computational results from isolated unsteady diffuser models with full-stage unsteady models and Laser Doppler Velocimetry (LDV) experimental data indicated that the isolated unsteady diffuser methodology was successful in closely matching both higher-fidelity CFD predictions and experimentally observed flow structures. However, this initial study only modeled a single geometry and loading condition (design point) [18]. The application of the methodology to a new geometry, CSTAR Gen. 2.5, is detailed in [19]. This geometry produced difficulties for the modeling approach due to the reduced vaneless space size when compared to the Gen. 1 geometry. Additionally, a detailed analysis of the flow development predicted by the full-stage and isolated models through the diffuser passage indicated a significant difference in the size and propagation of a separation region on the diffuser vane pressure side. In both models, this region originates near the shroud and diffuser vane pressure side and propagates towards the hub and passage suction side. The isolated diffuser models, described in [1,18,19], suffered from an overprediction of the size and speed of propagation of this region.

This study focuses on the improvement of the isolated unsteady diffuser computational methodology by defining a more accurate turbulence profile at the diffuser inlet. The model is then applied to four loading conditions: high loading (HL), design point (DP), low loading (LL), and near choke (NC). A comparison between the results from isolated unsteady diffuser models with and without improved turbulence modeling and full-stage unsteady models is made. A focus is placed on the modeling of the separation region in the diffuser passage by the two isolated diffuser models. However, additional 1D performance parameters are compared as well as 3D flow structures at various locations throughout the entire diffuser passage. Thereby a thorough analysis of the performance and flow structure predictions is made.

2. Experimental Facility and Methodology

The Centrifugal Stage for Aerodynamic Research (CSTAR) at the Purdue University High-Speed Compressor Laboratory was used for this work. The research facility consists of the infrastructure needed to conduct high-speed tests of a low-specific-speed centrifugal compressor and stationary diffusion system at design scale. The compression system is intended for an axi-centrifugal engine architecture [20]. As mentioned in Section 1, this study focuses on the CSTAR Gen 2.5 geometry, Figure 2. Compared to Gen. 1, this geometry included not only an additively manufactured diffuser but also a deswirl to turn the flow back to axial. Additionally, the diffuser leading edge radius ratio was reduced from 1.08 in Gen. 1 to 1.05 in Gen. 2.5 [21].

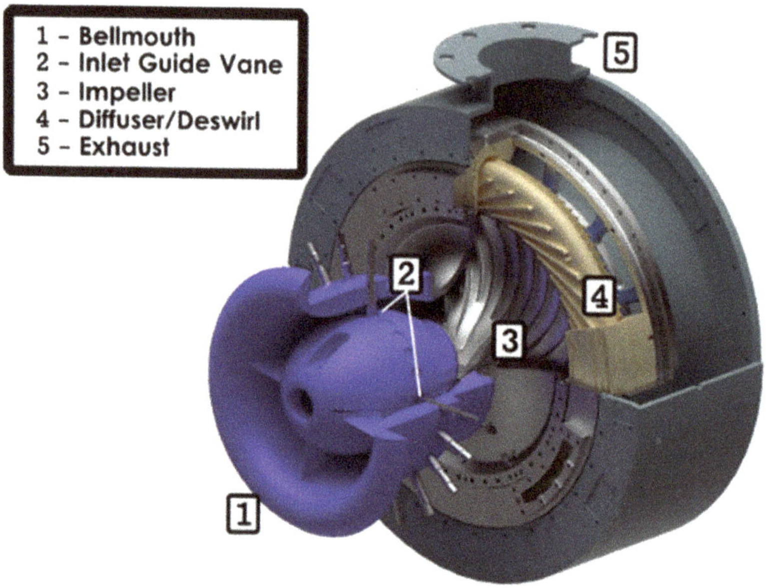

Figure 2. CSTAR Gen 2.5 compressor cutaway [22].

At the design point, which is also the best efficiency point, CSTAR Gen. 2.5 operates at about 20,800 rpm, and the total pressure ratio (TPR) is approximately 3. Inlet guide vanes are used to simulate the swirl from the latter stages of an axial compressor. The impeller consists of 15 main and 15 splitter blades, which start at 34% of the passage. Following the impeller is an additively manufactured diffusion system, with a diffuser consisting of 37 wedged vanes and a deswirl consisting of 99 vanes [23]. The facility is highly instrumented, including a significant amount of instrumentation in the diffusion system. The diffuser and deswirl include built-in static pressure taps along the vane surfaces and total pressure rakes at the vane leading edges and passage outlets. Scanivalve pressure scanning units combined with two Scanivalve rack-mounted enclosures containing DSA 3016 digital sensor arrays provide the pressure measurements. Total temperature rakes are included at the deswirl outlet. An Agilent 34980A Multifunction Switch/Measure mainframe is used to acquire temperature measurements. Combined with total pressure measurements in the facility inlet, total pressure and temperature ratio can be calculated to evaluate the performance and efficiency of the system.

The four chosen loading conditions (high loading, design point, low loading, and near choke) were defined by experimental data at 100% N_c, Figure 3. The design point loading condition was set at the best efficiency point (Figure 3b). The high loading, low loading, and near choke conditions were chosen such that trends in the flow structures could be assessed along the speedline. An additional near-surge loading condition was attempted; however, difficulties arose with the convergence of the full-stage unsteady model. Therefore, the results from that loading condition are omitted from this research.

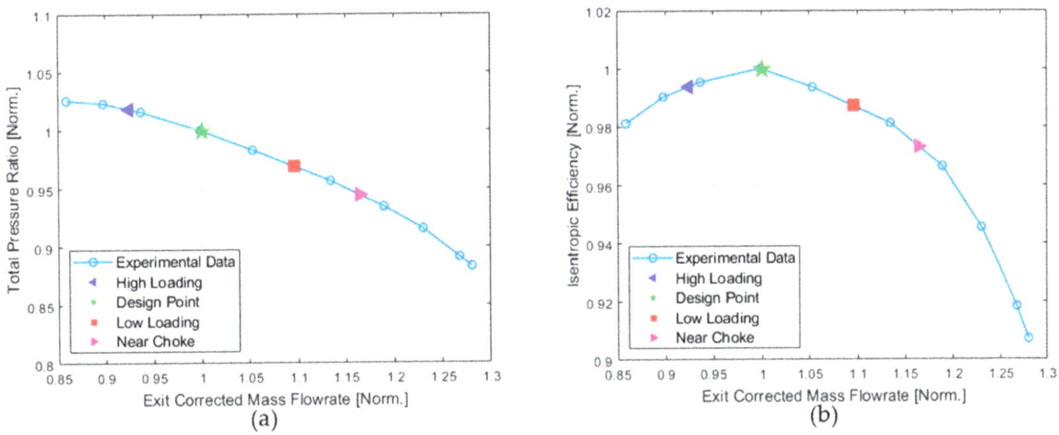

Figure 3. Compressor performance map (**a**) and efficiency map (**b**) indicating relative positions of high loading, design point, low loading, and near choke conditions.

3. Computational Methodology

The isolated unsteady diffuser computational methodology, as described in [1,18,19], is intended for diffuser design studies where only small changes in geometry are made between design iterations. These changes are unlikely to have a strong impact on the diffuser potential field, which is the primary means through which the diffuser interacts with the impeller outflow [8]. Therefore, a constant inlet profile can be used. The methodology relies on the diffuser inlet profile from a converged, full-stage, unsteady simulation. This inlet profile can then be applied, using the isolated unsteady methodology, to slightly different diffuser geometries. The best-performing diffuser design can then be assessed using higher fidelity CFD, if required. In short, the methodology makes use of the significantly reduced computational cost of isolated unsteady diffuser models to quickly iterate diffuser design in early design stages. Computational models for this study were generated in and simulated using Ansys CFX 2024 R1. Structured meshes were made for each component using Ansys TurboGrid 2024 R1, with the exception being an outlet region added to the isolated diffuser model. The outlet region, which models the curvature of the turn-to-axial and extends the computational domain to avoid reflections, is an unstructured mesh created using Ansys Meshing 2022. Outlet regions in both the full-stage and isolated models use expansion ratios such that the mesh becomes increasingly coarse towards the boundaries. The approximate element counts for both models are shown in Table 1. For the isolated diffuser model, the same diffuser domain from the full-stage model is used. The additional outlet region has an element count of approximately 3,000,000. Therefore, the isolated diffuser model has about half the element count of the full-stage model.

Table 1. Approximate element count of single passage computational domains.

Domain (Single Passage)	Approximate Element Count
Inlet Strut	1,000,000
IGV	4,000,000
Impeller	12,000,000
Diffuser (x2)	16,000,000
Deswirl	4,000,000
Total	37,000,000

The single-passage, full-stage unsteady computational domain, displayed in Figure 4a, includes the inlet, IGVs, impeller, diffuser, and deswirl. The impeller domain includes one main blade and one splitter blade. At the inlet, ambient total pressure and total temperature at sea-level are defined. Additionally, medium intensity turbulence (5%) and an eddy viscosity ratio of 10 is assumed. An exit-corrected mass flowrate boundary condition is used at the outlet. The specific values are determined from experimental results and chosen to match the corresponding loading conditions in Figure 3. A study of computational modeling decisions applied to the CSTAR geometry by Gooding indicated the importance of modeling surface roughness [21]. Therefore, the surface roughness was defined for each computational domain based off measurements from a profilometer. A temperature profile, taken from experimental data, was defined for the impeller shroud. This temperature profile was adjusted for each loading condition. The remaining walls in the domain were modeled as adiabatic.

The Baseline Explicit Algebraic Reynolds Stress Model (BSL-EARSM) was used as the turbulence model for the full-stage and isolated diffuser models. This model was used due to a previous study in CSTAR finding that it produced the best combination of results, accuracy, and computational cost [21]. BSL-EARSM combines the baseline k-ω (turbulence kinetic energy—turbulence eddy frequency) model with the Reynolds Stress Model. This model is higher fidelity than basic, two-term models but has reduced computational costs compared to full Reynolds Stress Models. In the full-stage model, the impeller–diffuser interface is modeled using the Ansys Transient Blade Row Time Transformation (TT) model, which was originally proposed by Giles [14]. Time Transformation handles unsteady, unequal pitch problems "by transforming the time coordinates of the rotor and stator in the circumferential direction in order to make the models fully periodic in 'transformed' time" [24]. When using TT, Ansys recommends the pitch ratio between the domains be no more than 20%. This limit is the result of an inequality dependent on the Mach number associated with the rotor rotational speed and the Mach number associated with the circumferential velocity component [11]. Therefore, two diffuser passages were modeled. The remaining interfaces in the full-stage model feature mixing planes. A frozen rotor was used for the interface between the diffuser and the outlet in the isolated diffuser models. This method was chosen due to the pitch ratio between domains being unity. Additionally, the frozen rotor converged more quickly than a corresponding method using a mixing plane and including the deswirl [19]. Transient convergence for the full-stage and isolated models was monitored using the Clark and Grover method [25].

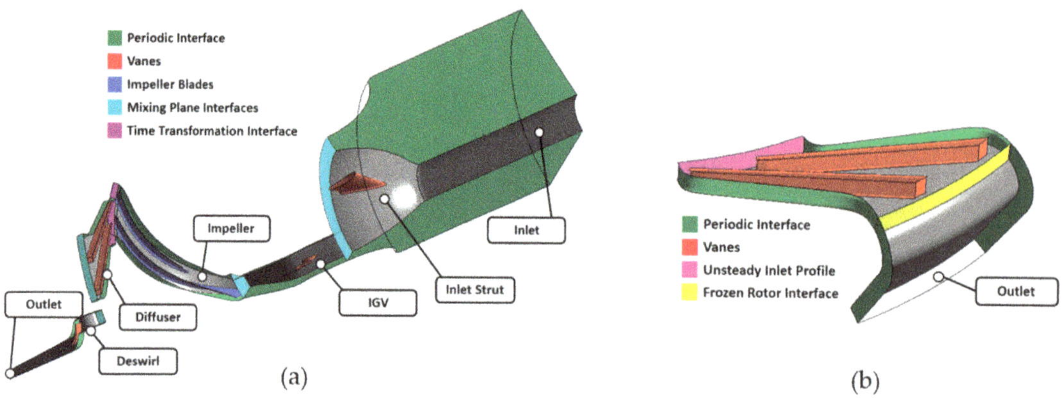

Figure 4. Definition of domains and interfaces for full-stage unsteady model (**a**) and isolated unsteady diffuser model (**b**).

Detailed descriptions of the setup of the isolated unsteady diffuser methodology are found in [1,18,19]. A short description will be provided here, with a focus placed on the differences in the methodology used for this study compared to the previous ones. Once a full-stage unsteady model is converged, the diffuser inlet profile can be extracted at each time step across one blade passing period. For the full-stage models used in this study, one blade passing period was defined with 80 time steps, or roughly 0.3 μs per time step. Then, using a MATLAB (R2023b) script, the individual profiles were compiled into a single unsteady and periodic profile. A detailed description of how this is achieved is found in [18]. In previous iterations of the isolated diffuser methodology, this profile consisted of five variables: total pressure, total temperature, and normalized velocity components (x, y, and z, or radial, circumferential, and axial). These variables defined the momentum, heat transfer, and velocity direction at the diffuser inlet for the isolated models. Ansys allows for Cartesian and cylindrical velocity components and spatial coordinates to be used interchangeably. For ease of visualization, cylindrical coordinates were used in this study.

To improve the modeling fidelity of isolated diffuser models, a more accurate, non-uniform, turbulence definition was given at the diffuser inlet. This was motivated by a comparison of full-stage unsteady and isolated unsteady diffuser results indicating less mixing of high momentum and low momentum flows in the isolated models. Instead of assuming medium-intensity turbulence (5%) at the diffuser inlet, turbulence kinetic energy (k) and turbulence eddy frequency (ω) were included in the unsteady periodic diffuser inlet profile. These two variables were also extracted from a full-stage, unsteady converged solution and compiled using the same methodology described in [19]. Turbulence kinetic energy and eddy frequency were chosen due to the specified turbulence model (BSL-EARSM) using them more predominantly in its calculations. It was assumed that using these variables would cause fewer issues with convergence and overconstraining of the computational domain. A summary of the components defining the isolated diffuser inlet profile, and their corresponding defining variables, is provided in Table 2.

This study includes a comparison of results from full-stage unsteady, uniform turbulence (UT) isolated unsteady diffuser, and non-uniform turbulence (NUT) isolated unsteady diffuser. Each computational methodology was evaluated at the four specified loading conditions (Figure 3). This means that this study necessitated the development and analysis of 12 computational models.

Multiple existing studies comparing computational results to experimental data in CSTAR have shown that the computational methodology previously employed produced good results and matched experimental data closely [1,22,26]. The full-stage unsteady computational methodology developed herein is heavily based on the previous existing ones. Therefore, the computational results from the isolated diffuser model presented in this paper were only compared to full-stage unsteady results, with a lesser focus placed on matching experimental data results. This is also in-line with the goal of the isolated unsteady diffuser methodology to reduce computational costs in early computational diffuser design studies.

Table 2. Summary of isolated unsteady diffuser model inlet profile definition.

Inlet Profile Component	Defined by
Momentum Heat Transfer	Total Pressure Total Temperature
Velocity Direction	Normalized Radial Velocity Normalized Circumferential Velocity Normalized Axial Velocity
Additionally Included for Models with Non-Uniform Turbulence	
Turbulence	Turbulence Kinetic Energy (k) Turbulence Eddy Frequency (ω)

4. Computational Results and Discussion

While matching experimental data was given less emphasis, it was still important that the full-stage unsteady results were physically representative. Figure 5 shows a comparison of the TPR and isentropic efficiency predictions by the full-stage unsteady computational models at all four loading conditions compared to the experimental data. The computational results discussed for unsteady models presented in this section are all time-averaged. Full-stage steady state (RANS) models are included in the data set as an additional reference. These models were generated using the same computational domains as described in Section 3 for the full-stage unsteady models, with the exception being that the impeller–diffuser interface was modeled using a mixing plane instead of TT. The total pressure and total temperature ratios for the computational models were calculated using time-averaged values of total pressure and total temperature at nodes in the meshes that corresponded to the physical locations of total pressure and total temperature rake elements in the physical compressor. Both full-stage computational methodologies overpredict TPR at all four loading conditions (Figure 5a). The results from the steady and unsteady models are very similar, with the steady model having slightly more accurate TPR predictions near choke and at high loading, but the unsteady model has a more consistent TPR overprediction along the speedline. These results are reiterated in the comparison of predicted isentropic efficiencies (Figure 5b). In this case, the unsteady model produced slightly more accurate isentropic efficiency values at the design point and high loading. The overprediction of TPR and corresponding underprediction of isentropic efficiency is most likely due to an overprediction of the total temperature ratio (TTR) in the computational models. This is due to incorrect modeling of the heat transfers in the domains, as most walls are assumed to be adiabatic when, in reality, some losses due to heat transfer are expected and observed in the experimental results. A similar trend in the computational results for CSTAR Gen. 2.5 was determined by Clement [22].

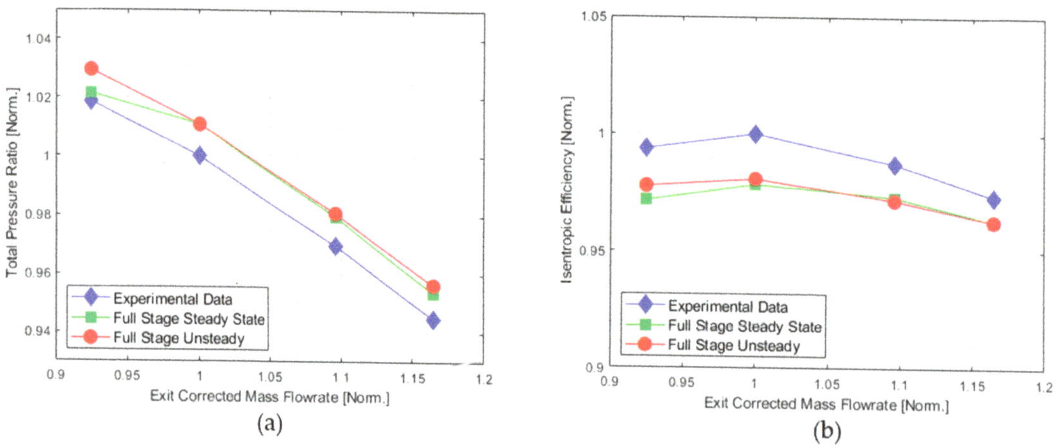

Figure 5. Comparison of total pressure ratio (**a**) and isentropic efficiency (**b**) from experimental data, full-stage steady state computational models and full-stage unsteady computational models at all four loading conditions.

The results from a comparison of 1D performance parameters typically used to judge diffuser performance are shown in Figure 6. The static pressure recovery coefficient, C_p (Figure 6a), measures the static pressure difference between the diffuser outlet and inlet, normalized by the difference in total and static pressure at the diffuser inlet. The UT and NUT isolated diffuser models underpredict static pressure recovery at all loading conditions along the speedline. The NUT isolated model is consistently closer to the value predicted by the full-stage unsteady model. This is especially true for the low loading and near choke loading conditions. The underprediction in static pressure recovery is also more consistent for the NUT isolated diffuser model. The comparison of total pressure loss shows similar trends (Figure 6b). The total pressure loss coefficient, K, measures the total pressure difference between the diffuser outlet and inlet, normalized by the difference in total and static pressure at the diffuser inlet. Once again, the difference in the predictions by the UT and NUT isolated diffuser models are most evident at lower loading conditions. The predicted total pressure losses by the NUT isolated diffuser model are more consistent along the speedline. Both models successfully track the trend of increasing C_p and decreasing K with increasing loading.

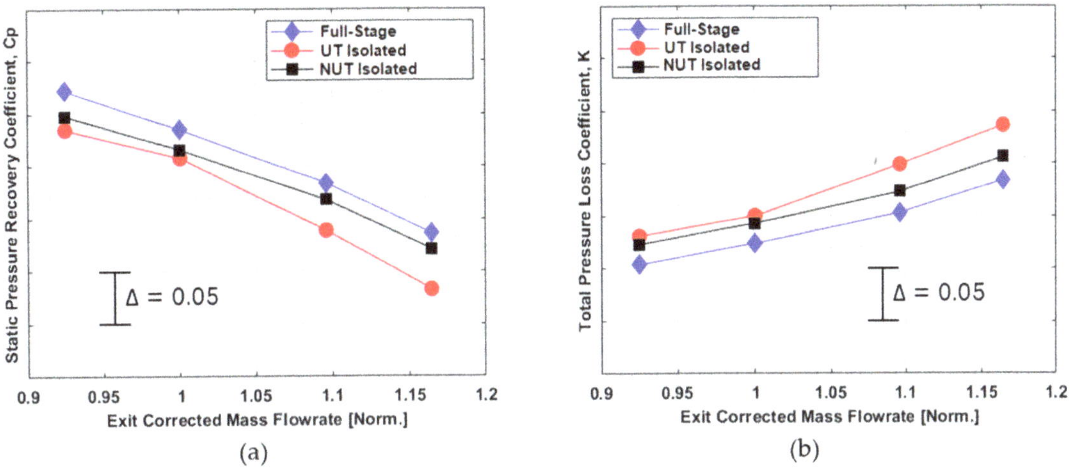

Figure 6. Comparison of static pressure recovery coefficient, C_p (**a**), and total pressure loss coefficient, K (**b**), for full-stage unsteady, UT isolated unsteady diffuser, and NUT isolated unsteady diffuser models.

The comparison of 1D performance parameters indicates that the lack of mixing of high and low momentum in the isolated unsteady diffuser models leads to lower predicted performance values. Modeling non-uniform turbulence at the diffuser inlet increases the mixing and, therefore, also increases the predicted performance. The consistent performance of the NUT isolated unsteady diffuser model along the speedline is especially important. The result is a significant improvement to the UT isolated unsteady diffuser model.

Despite the positive performance regarding predicting 1D performance parameters, it is also important that the 3D flow structure predictions by all three methodologies closely match. Therefore, several contours were generated to compare the flow structures observed at specific locations in the diffuser passage. Diffuser outlet total pressure and midspan velocity contours are compared for full-stage unsteady, UT isolated unsteady diffuser, and NUT isolated unsteady diffuser models at all four loading conditions. In these comparisons, the isolated diffuser models were analyzed with respect to their ability to predict flow structures that were consistent along the speedline.

The first flow structure comparison is made using diffuser passage outlet total pressure contours, as shown in Figure 7. The contours are generated forward looking aft, with 0% span defined as the hub and 100% span defined as the shroud. The impeller spins from low pitch (diffuser suction side) to high pitch (diffuser pressure side). The total pressure values at each loading condition are normalized by the area-averaged total pressure at the diffuser inlet for the design point loading condition. Compared to the full-stage results, both isolated models suffer from an overprediction of the separation regions on the diffuser vane suction and pressure sides at all loading conditions. The overprediction is more pronounced at lower loading conditions. The increased separation region leads to more throughflow through the core and, therefore, increased total pressure at the core. When comparing the two isolated models, it is evident that modeling non-uniform turbulence at the inlet significantly improves the flow structures. Although the separation regions are still overpredicted, the total pressure near the endwalls is more similar to that observed in the full-stage model. Additionally, the prediction of the core location is improved. The shape of the high-momentum region in the NUT isolated diffuser model, at all loading conditions, closely resembles that of the full-stage model.

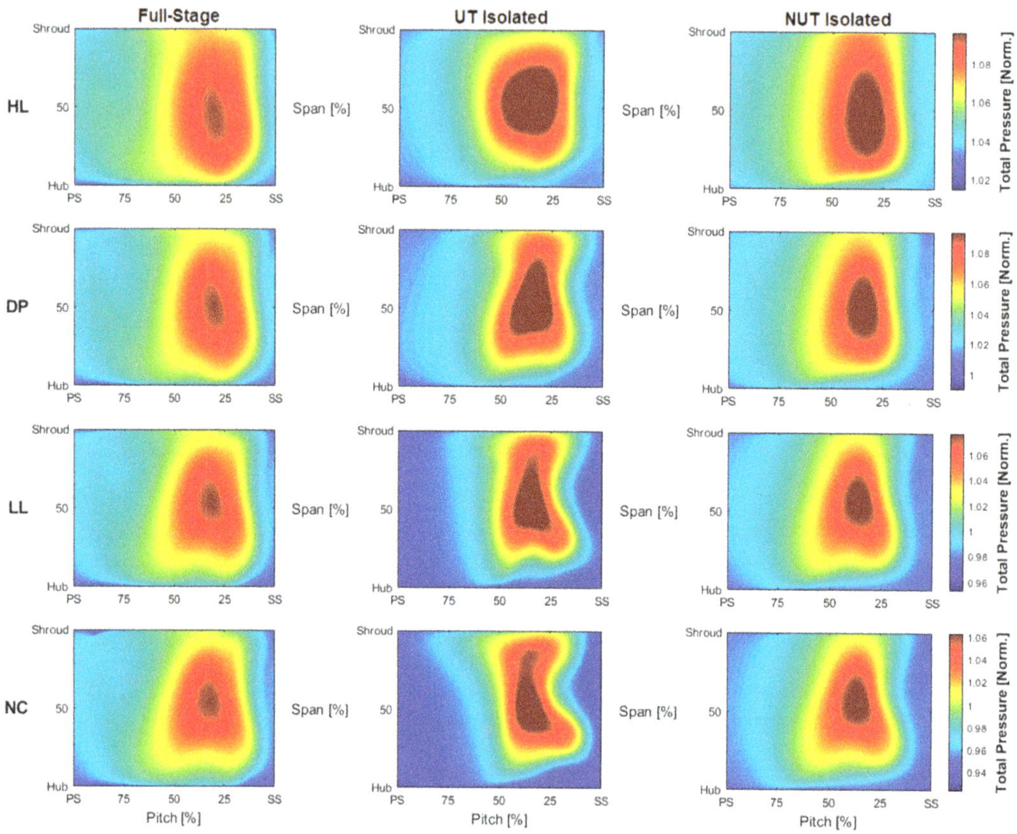

Figure 7. Diffuser passage outlet normalized total pressure contours for full-stage unsteady, UT isolated unsteady diffuser, and NUT isolated unsteady diffuser models at high loading (HL), design point (DP), low loading (LL), and near choke (NC).

To further analyze the flow structures through the diffuser passage, midspan normalized absolute velocity magnitude contours were generated for all three methodologies at all four loading conditions, Figure 8. The velocity values were normalized using the wheelspeed at 100% Nc. All contours, regardless of loading condition, predict a separation region on the diffuser vane pressure side. The size of the separation region is overpredicted by the UT isolated model. The NUT isolated model has a more accurate prediction of the size and shape of the region at all loading conditions. The contours indicated that both isolated methodologies accurately predict the velocity direction in the inlet region. This is most evident when analyzing the small shock formations and stagnation regions at the vane tip. Additionally, the isolated models indicate the same trends in separation region size and velocity magnitude as the full-stage model along the speedline.

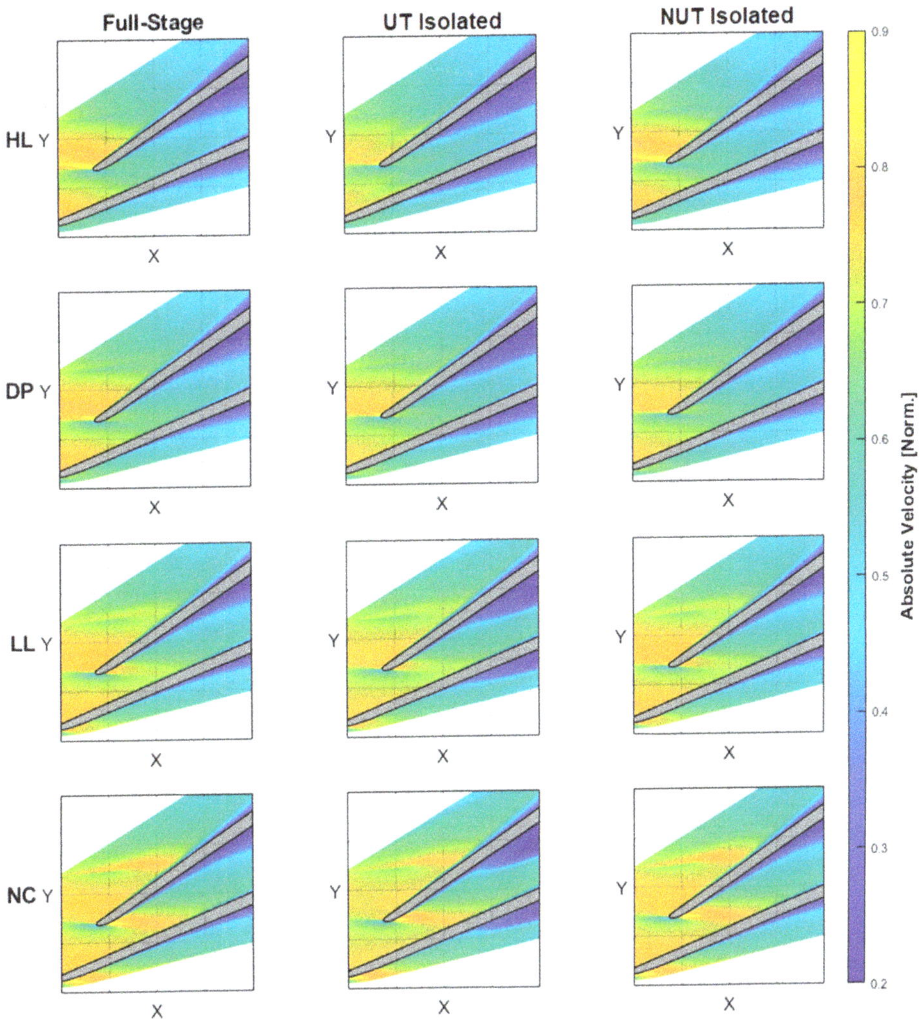

Figure 8. Normalized midspan absolute velocity magnitude contours for full-stage unsteady, UT isolated unsteady diffuser, and NUT isolated unsteady diffuser models at high loading (HL), design point (DP), low loading (LL), and near choke (NC).

The velocity predictions in the diffuser passage are further analyzed using contours indicating the difference between the full-stage and isolated models, Figure 9. In these contours, positive values indicate that the specific isolated diffuser model is overpredicting the velocity magnitude compared to the full-stage model, and negative values indicate that the specific isolated diffuser model is underpredicting the velocity magnitude compared to the full-stage model. The most obvious observation, when comparing the two isolated methodologies, is that the inclusion of non-uniform turbulence in the boundary profile significantly improved the velocity magnitude predictions throughout the entire diffuser passage. In the results from the UT isolated model, there is a small underprediction of the velocity magnitude in the inlet region. The largest differences are in the passage, where the overprediction of the size of the separation region is evident. This overprediction leads to increased velocity through the core of the passage. On the other hand, the NUT model does not suffer from these inaccuracies. The largest velocity differences are found at high loading and appear in the diffuser inlet region and along the vane suction side. At lower loading conditions, especially the low loading and near choke conditions, velocity differences are hardly distinguishable.

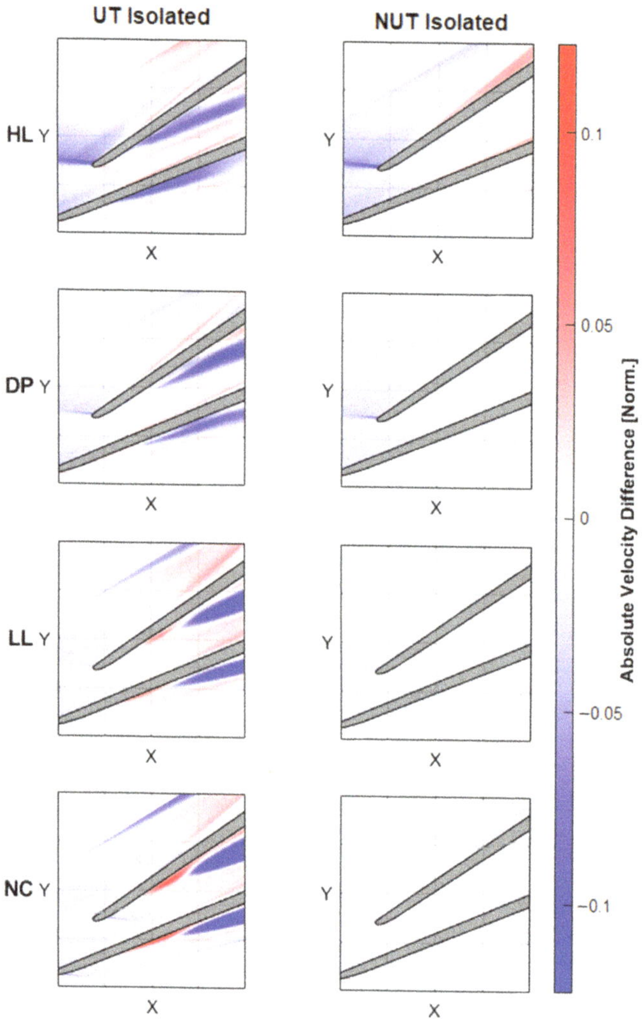

Figure 9. Difference in midspan normalized absolute velocity magnitude predictions for full-stage unsteady, UT isolated unsteady diffuser, and NUT isolated unsteady diffuser models at high loading (HL), design point (DP), low loading (LL), and near choke (NC).

When analyzing the flow structure predictions by the UT and NUT isolated unsteady diffuser models, the ability to track trends along the speedline was considered. The results from Figures 7–9 indicate that both isolated models are successfully able to do this. The trends in size of the separation region in the diffuser passage and magnitude of total pressure and velocity throughout the passage are correctly tracked along the speedline. The benefit of including non-uniform turbulence in the inlet condition comes in the accuracy of the flow structure predictions. The NUT isolated unsteady diffuser model consistently produces more accurate flow structure predictions when compared to the UT isolated unsteady diffuser model and the full-stage results. However, the UT isolated model is also able to successfully capture trends in the full-stage unsteady model along the speedline. Therefore, when using the isolated unsteady diffuser methodology to accurately model trends along a speedline, non-uniform turbulence modeling is not a must.

At the design point loading condition, total pressure differences between the full-stage model and the two isolated diffuser models at the diffuser throat, 50% chord, and passage outlet are compared, as shown in Figure 10. Again, positive values indicate an overprediction of total pressure by the isolated model compared to the full-stage results, and negative values indicate an underprediction. The modeling fidelity benefit of including non-uniform turbulence is already evident at the throat. The UT isolated unsteady diffuser model is plagued by a significant underprediction of total pressure near the endwalls, especially near the pressure-side hub. This misrepresentation of the flow at the throat is significantly reduced when including non-uniform turbulence in the inlet profile. Additionally, as the flow progresses through the diffuser passage, the magnitude differences in total pressure in the NUT isolated unsteady diffuser model decrease. At the 50% chord location, a more significant underprediction of total pressure is captured by the UT isolated diffuser model between 50% and 75% pitch and around 25% span. At this location, the separation region predicted by the model near the vane pressure side and hub at the throat migrates spanwise downwards towards the core of the flow. This then causes the increased total pressure observed near the vane suction side. This flow pattern is not observed in the NUT model. Instead, at the 50% chord location, only small differences in total pressure exist throughout the contour. A pattern in the differences was not distinguishable at this location. The results at the passage outlet reiterate those in Figure 7. Decreased total pressure is observed near the endwalls for both isolated models, and this, in turn, leads to increased total pressure through the core. The magnitude differences in the NUT isolated model are lower than in the UT model.

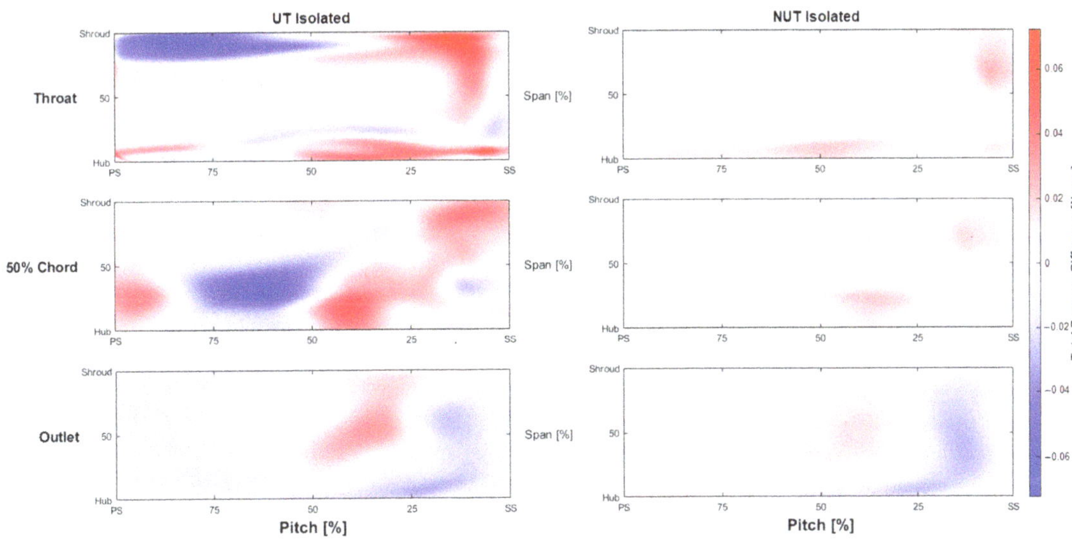

Figure 10. Normalized total pressure differences at diffuser throat, 50% chord, and diffuser passage outlet at design point loading for UT isolated unsteady diffuser and NUT isolated unsteady diffuser models.

Lastly, the computational cost to solve was quantified as the product of the time until a converged solution was reached, CPUs used, and element count in the solved mesh for each of the computational methodologies at the design point loading condition. The numerical results are summarized in Table 3. Both isolated models are computationally much less costly than the full-stage unsteady model. Including non-uniform turbulence in the isolated diffuser methodology only slightly increased the computational cost. The trends observed in Table 3 were true when conducting the same analysis for all four loading conditions along the speedline.

Table 3. Computational cost to solve for UT isolated diffuser, NUT isolated diffuser, and full-stage models at the design point loading condition.

	Cost to Solve [Time × CPUs × Element Count]	Normalized Cost to Solve
UT Isolated	8.26×10^{14}	14.2%
NUT Isolated	8.35×10^{14}	14.3%
Full Stage	5.84×10^{15}	100%

5. Conclusions

The goals of this study were twofold; first, to increase the fidelity of isolated unsteady diffuser modeling by including non-uniform turbulence kinetic energy and turbulence eddy frequency in the diffuser inlet profile. Secondly, the goal was to apply the isolated unsteady diffuser computational methodology with non-uniform inlet turbulence quantities to multiple loading conditions along the speedline. A focus was placed on the ability of the isolated unsteady diffuser models to track trends along the speedline and accurately predict 3D flow structures at each loading condition.

Both UT and NUT isolated diffuser models were successful in tracking trends of 1D performance parameters along the speedline. Including non-uniform turbulence in the diffuser inlet profile improved the predictions at all loading conditions, especially at the low loading and near choke conditions. Diffuser outlet total pressure and midspan velocity contours indicate that the UT isolated diffuser model captured the 3D flow structure trends from the full-stage unsteady model along the speedline. However, there are notable differences in the structures. These were most significant regarding the prediction the separation regions along the diffuser vane pressure and suction sides. The overprediction of these separation regions resulted in increased total pressure and velocity magnitude through the core. When non-uniform turbulence is included in the isolated diffuser model, these differences are reduced. While a small overprediction of the total pressure through the core still exists, the size of the separation regions is significantly more accurate when compared to the full-stage model. The magnitude of the velocity differences at midspan are also more accurate.

Overall, the inclusion of non-uniform turbulence modeling at the diffuser inlet significantly improved the performance and flow structure predictions. Since these improvements are only associated with a minor increase in computational cost, it is recommended that future implementations of the isolated unsteady diffuser computational methodology include higher fidelity non-uniform turbulence modeling. The chosen turbulence variables of turbulence kinetic energy and turbulence eddy frequency can be varied depending on the chosen turbulence model.

The results from this study, and previous ones [1,18,19], indicate that isolated unsteady diffuser models are able to accurately capture 3D flow structures at multiple loading conditions closer to choke and surge and track trends along speedlines. Future improvements should focus on the implementation of the methodology in diffuser design studies and determination of when the coupling between the impeller and diffuser is affected such that the constant inlet profile assumption with changing diffuser geometries cannot be utilized.

Author Contributions: Conceptualization, B.L.H.; Methodology, B.L.H.; Validation, B.L.H.; Formal analysis, B.L.H.; Writing—original draft, B.L.H.; Writing—review & editing, N.L.K.; Supervision, N.L.K.; Project administration, N.L.K. All authors have read and agreed to the published version of the manuscript.

Funding: This research received no external funding.

Data Availability Statement: The datasets presented in this article are not readily available because the data are part of an ongoing study. Requests to access the datasets should be directed to Nicole Key (nkey@purdue.edu).

Acknowledgments: The authors are grateful to Rolls-Royce for allowing the use of the rig geometry and the experimental data for validating this computational study.

Conflicts of Interest: The authors declare no conflicts of interest.

References

1. Shapochka, M.Y. Stability Enhancement in Aeroengine Centrifugal Compressors Using Diffuser Recirculation Channels. Master's Thesis, Purdue University, West Lafayette, IN, USA, 2022.
2. Dean, R.C.; Senoo, Y. Rotating Wakes in Vaneless Diffusers. *J. Basic Eng.* **1960**, *82*, 563–570. [CrossRef]
3. Fischer, K.; Thoma, D. Investigation of the Flow Conditions in a Centrifugal Pump. *J. Fluids Eng.* **1932**, *54*, 141–154. [CrossRef]
4. Moore, J. A Wake and an Eddy in a Rotating, Radial-Flow Passage—Part 1: Experimental Observations. *J. Eng. Power* **1973**, *95*, 205–212. [CrossRef]
5. Moore, J. A Wake and an Eddy in a Rotating, Radial-Flow Passage—Part 2: Flow Model. *J. Eng. Power* **1973**, *95*, 213–219. [CrossRef]
6. Eckardt, D. Detailed Flow Investigations Within a High-Speed Centrifugal Compressor Impeller. *J. Fluids Eng.* **1976**, *98*, 390–399. [CrossRef]
7. Eckardt, D. Investigation of the Jet-Wake Flow of a Highly-Loaded Centrifugal Compressor Impeller; NASA, TM-75232. 1978. Available online: https://ntrs.nasa.gov/citations/19780008108 (accessed on 24 July 2024).
8. Ziegler, K.U.; Gallus, H.E.; Niehuis, R. A Study on Impeller-Diffuser Interaction—Part I: Influence on the Performance. *J. Turbomach.* **2003**, *125*, 173–182. [CrossRef]
9. Everitt, J.N.; Spakovszky, Z.S.; Rusch, D.; Schiffmann, J. The Role of Impeller Outflow Conditions on the Performance of Vaned Diffusers. *J. Turbomach.* **2017**, *139*, 041004. [CrossRef]
10. Shum, Y.K.P.; Tan, C.S.; Cumpsty, N.A. Impeller–Diffuser Interaction in a Centrifugal Compressor. *J. Turbomach.* **2000**, *122*, 777–786. [CrossRef]
11. ANSYS. *CFX-Solver Modeling Guide*, 1st ed.; ANSYS: Canonsburg, PA, USA, 2017; Volume 18.
12. Dawes, W.N. A Simulation of the Unsteady Interaction of a Centrifugal Impeller With Its Vaned Diffuser: Flow Analysis. *J. Turbomach.* **1995**, *117*, 213–222. [CrossRef]
13. Denton, J.D. The Calculation of Three-Dimensional Viscous Flow Through Multistage Turbomachines. *J. Turbomach.* **1992**, *114*, 18–26. [CrossRef]
14. Giles, M.B. Calculation of Unsteady Wake/Rotor Interaction. *J. Propuls.* **1987**, *4*, 356–362. [CrossRef]
15. Ekici, K.; Hall, C. Nonlinear Analysis of Unsteady Flows in Multistage Turbomachines Using Harmonic Balance. *AIAA J.* **2007**, *45*, 1047–1057. [CrossRef]
16. Everitt, J.N.; Spakovszky, Z.S. An Investigation of Stall Inception in Centrifugal Compressor Vaned Diffuser. *J. Turbomach.* **2013**, *135*, 011025. [CrossRef]
17. Semlitsch, B. Imposing Realistic Unsteady Flow Structures Through Inflow Boundary Conditions. In Proceedings of the 30th AIAA/CEAS Aeroacoustics Conference, Rome, Italy, 4–7 June 2024. [CrossRef]
18. Holtmann, B.; Shapochka, M.; Key, N. Development and Validation of an Isolated Unsteady Diffuser Computational Model for a Vaned Centrifugal Compressor. In Proceedings of the ASME Turbo Expo, London, UK, 24–28 June 2024. [CrossRef]
19. Holtmann, B. Expanding the Scope of Isolated Unsteady Diffuser Computational Modeling. Master's Thesis, Purdue University, West Lafayette, IN, USA, 2024.
20. Methel, J.; Gooding, W.J.; Fabian, J.C.; Key, N.L.; Whitlock, M. The Development of a Low Specific Speed Centrifugal Compressor Research Facility. In *ASME Turbo Expo 2016: Turbomachinery Technical Conference and Exposition, Volume 2D: Turbomachinery, Seoul, Republic of Korea, 13–17 June 2016*; American Society of Mechanical Engineers: New York, NY, USA, 2016. [CrossRef]
21. Gooding, W.J.; Meier, M.A.; Key, N.L. The Impact of Various Modeling Decisions on Flow Field Predictions in a Centrifugal Compressor. *J. Turbomach.* **2021**, *143*, 101006. [CrossRef]
22. Clement, J.T. Experimental and Numerical Evaluation of Stationary Diffusion System Aerodynamics in Aeroengine Centrifugal Compressors. Master's Thesis, Purdue University, West Lafayette, IN, USA, 2024.
23. Clement, J.T.; Coon, A.T.; Key, N.L. Development of an Additively Manufactured Stationary Diffusion System for a Research Aeroengine Centrifugal Compressor. *J. Turbomach.* **2025**, *147*, 031004. [CrossRef]
24. ANSYS. *CFX-Solver Theory Guide*, 1st ed.; ANSYS: Canonsburg, PA, USA, 2016; Volume 17.

25. Clark, J.P.; Grover, E.A. Assessing Convergence in Predictions of Periodic-Unsteady Flowfields. *J. Turbomach.* **2006**, *129*, 740–749. [CrossRef]
26. Coon, A.T. Utilization of Additive Manufacturing in the Development of Stationary Diffusion Systems for Aeroengine Centrifugal Compressors. Master's Thesis, Purdue University, West Lafayette, IN, USA, 2023.

fluids

MDPI

Article

An Analysis and Comparison of the Hydrodynamic Behavior of Ships Using Mesh-Based and Meshless Computational Fluid Dynamics Simulations

Davide Caccavaro [1,2,*], Bonaventura Tagliafierro [3,4], Gianluca Bilotta [5], José M. Domínguez [6], Alessio Caravella [1,2], Roberto Gaudio [5], Alfredo Cassano [2], Corrado Altomare [4] and Agostino Lauria [7]

1 Department of Computer Engineering, Modelling, Electronics and Systems Engineering (DIMES), University of Calabria (UNICAL), Via P. Bucci, Cubo 42C, 87036 Rende, Italy
2 Institute on Membrane Technology, National Research Council, CNR-ITM, Via P. Bucci, Cubo 17C, 87036 Rende, Italy; a.cassano@itm.cnr.it
3 Department of Electrical Engineering, Uppsala University, 753 10 Uppsala, Sweden
4 Laboratori d'Enginyeria Marítima, Universitat Politècnica de Catalunya–BarcelonaTech (UPC), 08034 Barcelona, Spain
5 Department of Civil Engineering, University of Calabria, 87036 Rende, Italy; roberto.gaudio@unical.it (R.G.)
6 Environmental Physics Laboratory, CIM-UVIGO, Universidade de Vigo, 34002 Ourense, Spain; jmdominguez@uvigo.es
7 Department of Engineering for Innovation, University of Salento, 73100 Lecce, Italy; agostino.lauria@unisalento.it
* Correspondence: davide.caccavaro@dimes.unical.it

Abstract: This paper presents a comparison of two turbulence models implemented in two different frameworks (Eulerian and Lagrangian) in order to simulate the motion in calm water of a displacement hull. The hydrodynamic resistance is calculated using two open-source Computational Fluid Dynamics (CFD) software packages: OpenFOAM and DualSPHysics. These two packages are employed with two different numerical treatments to introduce turbulence closure effects. The methodology includes rigorous validation using a Wigley hull with experimental data taken from the literature. Then, the validated frameworks are applied to model a ship hull with a 30 m length overall (LOA), and their results discussed, outlining the advantages and disadvantages of the two turbulence treatments. In conclusion, the resistance calculated with OpenFOAM offers the best compactness of results and a shorter simulation time, whereas DualSPHysics can better capture the free-surface deformations, preserving similar accuracy.

Keywords: ship; Wigley; OpenFOAM; DualSPHysics; free-surface flows

1. Introduction

The study of the performance and dynamics of ship hulls is a subject of interest to all sectors in the naval world, from commercial ships to high-performance sailing ships, and even from displacement to planing hulls [1]. In the past, this type of analysis was exclusively performed by means of empirical models, simplified formulas, and tank tests. Nowadays, thanks to the development of computer technology, Computational Fluid Dynamics (CFD) approaches, based on the numerical resolution of viscous flow differential Navier–Stokes (NS) equations, have become largely available. CFD is becoming cheaper and faster and has provided increasingly accurate results, using highly complex numerical tools to study the performance and dynamics of ship hulls [2–4]. Fluid dynamics investigations can be carried out using the mesh-based CFD approach, which has the ability to calculate any complicated flow, but has the difficulty of deforming meshes (and also their possible distortion), or the particle method, which is based on the Lagrangian description of flow and is applied where the mesh-based method has difficulty [5].

CFD makes it possible to describe the turbulent motion that is generated around solid–fluid interfaces, such as hulls or aircraft [6]. Since exact solutions for turbulent flows are very complex and difficult to obtain computationally owing to heavy resolution requirements [7], turbulence closure models have been used to achieve practical calculation times [8]. Naturally, the model approximates the exact equations of motion. By their nature, turbulence models are primarily based on dimensional analysis combined with intuition and experimental observations [9].

Eulerian-based descriptions have been used to investigate the hydrodynamic performance of hulls. Wu et al. [10] studied a trimaran using CFD via Reynolds-averaged Navier–Stokes (RANS) equations. The results of a seakeeping experiment for the high-speed trimaran carried out at the China Ship Scientific Research Center (CSSRC) towing tank were also studied. Rather good agreement was shown between the computational and experimental results. Zha et al. [11] applied a RANS solver and a shear-stress transport (SST) k-ω turbulence model [12] to study hydrodynamic resistance at different Froude numbers (Fr) by comparing different types of hulls. Sulisetyono and Alifrananda [13] used RANS equations and an SST k-ω model to compare the experimental data of the DTMB 5415 hull with the NUMECA numerical code [14] and observed that the wake profile generated by the ship model compares well with the real case and the resistance results are very close to it. Desouky and Elhenawy [15] utilized a RANS solver and a k-ϵ turbulence model to evaluate the free-surface flow around the KSC hull using the STAR-CCM+ (software version 14.02.010-R8) [16] and determined the hydrodynamics of the ship in the Suez Canal with good accuracy to understand the effects of squat in order to allow them to prevent ships in transit from having accidents. Islam and Guedes Soares [17] applied RANS equations and an SST k-ω model to estimate resistance in calm water and in the presence of waves for a ferry model using the open-source OpenFOAM toolkit. The paper showed that the solver is good for estimating resistance in calm water, but still has some shortcomings when it simulates waves with free motion. Furthermore, in another piece of work [18], for uncertainty in ship resistance prediction with OpenFOAM, both the factor of safety (F_s) and the correction factor (C_i) were estimated. The study was conducted on four different hull shapes, two container ship models (KCS, DTC), a crude oil carrier (KVLCC2), and a bulk carrier (JBC), and they were simulated in calm sea conditions; from the results the total drag coefficient, sinking, and trim were predicted. Kim et al. [19] studied the KSC hull model via STAR CCM+. The resistance performance and flow structure around the ship were analyzed. This study confirms and shows that full-scale results can be obtained at model scale by applying virtual fluid instead of full-scale numerical simulations, which require more computational resources.

Likewise, Lagrangian-based numerical approaches have been used to study hull dynamics. Kawamura et al. [20] validated the accuracy of the Smoothed Hydrodynamic Particle (SPH) method by comparing it with free and constrained motion tests in storm conditions, using a simplified ship model. The results showed that the SPH simulation has good potential for the quantitative assessment of vessel safety in seagoing situations. Priyambada and Tarwidi [21] applied artificial viscosity to study the vertical displacement of a floating object on free-surface flow, which was evaluated by varying the frequency of the piston wavemaker, and Eriksson [22] chose the KSC hull as a case study and evaluated the change in total drag coefficient using the SPH and laminar-and-sub-particle scale (SPS) models. Cheng et al. [23] used a weakly compressible SPH implementation to investigate the water entry and the ensuing slamming loads on a ship cross-section, and then extended the validity of their analysis to a three-dimensional (3D) framework. The emphasis of this work was on the stability and robustness of their implementation, which substantiated into capturing the spray features and the hull pressure field. Tagliafierro et al. [24] studied the spray phenomena around a planing hull by applying DualSPHysics code [25]. They calculated the hydrodynamic drag and running attitudes for a wide Fr range by comparing them with the experimental results and reproduced the spray around the planing hull. Capasso et al. (2023) [26,27] studied the modeling of free-surface flows interacting with a planing hull under regular wave conditions using DualSPHysics. Overall, the results are in good agreement with the experiments.

Throughout the reviewed literature, mesh-based and meshless methods have been shown to possess strengths and weaknesses for the hydrodynamic simulation of hulls in calm or rough water conditions [6,28,29]. With respect to the usability of each approach, code-to-code comparisons become very important to optimize their usage for engineering applications. In this regard, Gruwez et al. [30] compared the capability of OpenFOAM and DualSPHysics in reproducing a large-scale wave channel experiment of bichromatic wave transformations on a steeply sloping, very shallow shoreline, showing similar outcomes. The results showed that overall, OpenFOAM provides the highest model skill, but has the highest computational cost. DualSPHysics is shown to have reduced model performance, but still comparable to OpenFOAM and at a lower computational cost. Meringolo et al. [31] performed a first comparison between the SPH method and Finite Volume Method (FVM) for multiphase flows. The numerical results were compared with those obtained experimentally, showing good agreement. The results showed very good agreement in terms of potential and kinetic energy component evolution among the proposed model, the Large Eddy Simulation (LES)–FVM solver, and the single-phase δ-SPH model.

By recognizing the high value of turbulence models for engineering applications, in this work, two open-source tools, OpenFOAM (v2012) (mesh-based) and DualSPHysics (v5.2) (mesh-free), largely used for most applications reported here, are used to study the fluid flow velocity, vorticity, and forces acting on a ship surface in calm water conditions.

The main objective of this work is to compare the two open-source models in studying ship hydrodynamics in terms of total resistance, free-surface prediction, and wake evolution, as well as perform an analysis of the simulation runtimes. In order to validate the adopted numerical models, the experimental results obtained for Wigley hull [32,33] were used for comparison. Validation was carried out at three different Fr values: 0.25, 0.316, and 0.408, respectively. Furthermore, the validated frameworks were applied to model newly designed ship hulls with a 30 m length overall (LOA), and the results obtained with the SPH and FVM approaches at different Fr values were compared and analyzed.

The rest of this paper is organized as follows. In Section 2, the numerical model and numerical methods are introduced; in Section 3, the numerical setup is described; in Section 4, the case studies are illustrated; in Section 5, the results are presented and discussed; Section 6 provides the concluding remarks.

2. Numerical Models

This section describes the numerical models used by OpenFOAM and DualSPHysics for our case study.

2.1. Eulerian Approach in OpenFOAM

We consider the Navier–Stokes equations in a domain characterized by the presence of different fluids. The continuity and momentum equations are written for the Eulerian solver, considering the incompressible fluid approximation. This system of equations can be expressed as

$$\frac{\partial u_i}{\partial x_i} = 0 \tag{1}$$

$$\frac{\partial(\rho u_i)}{\partial t} + u_j \frac{\partial(\rho u_i)}{\partial x_j} = -\frac{\partial p}{\partial x_i} + \mu \frac{\partial^2 u_i}{\partial x_i x_j} + g_i \tag{2}$$

where u_i (i = 1, 2, 3) denotes flow velocity on the x-, y-, and z-axes, respectively, ρ represents the mixture density of the two phases separately considered, p denotes pressure, μ is the dynamic viscosity of the fluid, and g_i represents the components of the gravitational acceleration.

The Reynolds averaging approach was adopted to eliminate the fluctuating components of the solution variables in the instantaneous NS equations to reduce the computational cost. The average of the continuity and NS equations yields

$$\frac{\partial \bar{u}_i}{\partial x_i} = 0 \tag{3}$$

$$\frac{\partial(\rho\bar{u}_i)}{\partial t} + \bar{u}_j\frac{\partial(\rho\bar{u}_i)}{\partial x_j} = -\frac{\partial \bar{p}}{\partial x_i} + \mu\frac{\partial^2 \bar{u}_i}{\partial x_i x_j} - \frac{\partial(\rho\overline{u_i'u_j'})}{\partial x_j} + g_i \tag{4}$$

where $\bar{u}_i$ and $\bar{p}$ represent the ensemble-averaged velocity and pressure, respectively; the product $\rho\overline{u_i'u_j'}$ denotes the Reynolds stress component. Thereafter, the RANS equations govern the transport of flow. The Reynolds stress in Equation (4) is modeled as a function of turbulent viscosity μ_T and the mean velocity gradients according to the Boussinesq hypothesis:

$$-\rho\overline{u_i'u_j'} = \mu_T\left(\frac{\partial \bar{u}_i}{\partial x_j} + \frac{\partial \bar{u}_j}{\partial x_i}\right) - \frac{2}{3}\rho k\delta_{ij}, \tag{5}$$

where k is the turbulent kinetic energy (TKE) and δ_{ij} is the Kronecker delta. The latter is a function of two discrete variables, in particular, of two variables on the integers or naturals, which is worth 1 if their values coincide, and worth 0 otherwise.

To obtain the turbulent viscosity, we use the SST k-ω turbulence model, which is a two-equation eddy-viscosity model. It is a hybrid model combining the Wilcox k-ω and the k-ϵ models. The use of a k-ω formulation in the inner parts of the boundary layer makes the model directly usable all the way down to the wall through the viscous sub-layer; hence, the SST k-ω model can be used as a Low-Re turbulence model without any extra damping functions. The SST formulation also switches to a k-ϵ behavior in the free stream and thereby avoids the common k-ω problem whereby the model is too sensitive to the inlet free-stream turbulence properties. A blending function, F_1, activates the Wilcox model near the wall and the k-ϵ model in the free stream. The model is proposed by Menter [34]. In the SST k-ω model, the viscosity term μ_T is defined as:

$$\mu_T = \rho\frac{k}{\omega} \tag{6}$$

while the TKE per unit mass k and the dissipation rate ω obey the following equations:

$$\rho\frac{\partial k}{\partial t} + \rho\bar{u}_j\frac{\partial k}{\partial x_j} = \tau_{ij}\frac{\partial \bar{u}_i}{\partial x_j} - \beta^*\rho k\omega + \frac{\partial}{\partial x_j}\left[(\mu + \sigma_k\mu_T)\frac{\partial k}{\partial x_j}\right] \tag{7}$$

$$\rho\frac{\partial \omega}{\partial t} + \rho\bar{u}_j\frac{\partial \omega}{\partial x_j} = \alpha\frac{\omega}{k}\tau_{ij}\frac{\partial \bar{u}_i}{\partial x_j} - \beta\rho\omega^2 + \frac{\partial}{\partial x_j}\left[(\mu + \sigma_\omega\mu_T)\frac{\partial \omega}{\partial x_j}\right] \tag{8}$$

In the above equations, $\alpha = 0.52$; $\sigma_k = 0.5$; $\sigma_\omega = 0.5$, whereas β and β^* are computed through auxiliary functions, as presented in Wilcox [35]. The second selected turbulence model, the SST k-ω [12], has several relatively minor variations with respect to the original SST version [34,36]. The TKE and the dissipation rate are computed using the following equations:

$$\rho\frac{\partial k}{\partial t} + \rho\bar{u}_j\frac{\partial k}{\partial x_j} = \widetilde{P}_k - \beta^*\rho k\omega + \frac{\partial}{\partial x_j}\left[(\mu + \sigma_{k1}\mu_T)\frac{\partial k}{\partial x_j}\right] \tag{9}$$

$$\rho\frac{\partial(\omega)}{\partial t} + \rho\bar{u}_j\frac{\partial(\omega)}{\partial x_i} = \alpha\rho S^2 - \beta\rho\omega^2 + \frac{\partial}{\partial x_j}\left[(\mu + \sigma_\omega\mu_T)\frac{\partial \omega}{\partial x_j}\right] + 2(1 - F_1)\rho\sigma_{\omega,2}\frac{1}{\omega}\frac{\partial k}{\partial x_j}\frac{\partial \omega}{\partial x_j} \tag{10}$$

where $\widetilde{P}_k$ represents a production limiter used in the model to prevent the build-up of turbulence in stagnation regions [12], and F_1 represents the blending function [12,34,36,37]. All the constants are predicted through a blend from the corresponding constants [e.g., $\alpha = \alpha_1 F_1 + \alpha_2(1 - F_1)$]. This model contains the following closure coefficients: $\beta^* = 9/100$, $\sigma_{k1} = 0.85$, $\sigma_{\omega 1} = 0.5$, $\sigma_{\omega 2} = 0.856$, $\alpha_1 = 5/9$, $\alpha_2 = 0.44$, and $\beta_1 = 3/40$, as suggested in [37]. The choice of the SST k-ω model is due to its generally superior performance compared to the classical two-equation model in a variety of complex flow cases [12,37,38].

The volume of fluid (VOF) method with the artificial bounded compression technique proposed by Hirt and Nichols (1981) [39] was used to capture the free surface. The free surface was considered to be a mixture of water and air, and, accordingly, the VOF transport equation was expressed as

$$\frac{\partial \alpha}{\partial t} + \nabla \cdot \boldsymbol{U_r}(1-\alpha)\alpha = 0, \tag{11}$$

where $\boldsymbol{U_r} = \boldsymbol{U_{water}} - \boldsymbol{U_{air}}$ is the relative velocity between the water and air, and is termed "compression velocity"; α represents the volume fraction, defined as the relative volume proportion of water in a cell; and ∇ represents the nabla operator. If the volume is completely filled with air, α is 0, but if the volume is filled with water, α is 1; α between 0 and 1 indicates the multiphase fluid interface. In our case, α equal to 0.5 indicates the free water surface.

The spatial variation in fluid density and dynamic viscosity can be then expressed as

$$\rho = \alpha\rho_{water} + (1-\alpha)\rho_{air} \tag{12}$$

$$\mu = \alpha\mu_{water} + (1-\alpha)\mu_{air} \tag{13}$$

The sequential solution implies that the velocity components are solved independently and only coupled through explicit terms. It follows a stability criterion that limits the duration of the time step, known as the Courant–Friedrichs–Lewy (CFL) condition [40]. The CFL number is given by

$$CFL = \frac{|u|\,\Delta t}{\Delta x} \tag{14}$$

2.2. Lagrangian Approach in DualSPHysics

The DualSPHysics model uses the meshless SPH method to describe fluid mechanics and is used to discretize a fluid volume as a collection of particles, whose motion is dictated by the NS equations. Particles represent nodal points at which physical quantities (e.g., position, velocity, density, pressure) are approximated by interpolating the values of the NS equations. The interpolation of the values of neighboring particles takes place on a compact support at a short distance. The technique is ideal for studying violent flows owing to its inherent lack of mesh distortion and can easily handle multi-phase simulations because each particle stores its own properties. The SPH method has been used to describe a variety of free-surface flows (wave propagation on beaches, breakers, impact on structures and dams) [25].

The mathematical foundations of the SPH method are based on the approximation of any quantity by convolution integrals. Any function F can be defined by

$$F(\boldsymbol{r}) = \int F(\boldsymbol{r}')\, W(\boldsymbol{r}-\boldsymbol{r}')d\boldsymbol{r}' \tag{15}$$

where W is the kernel function [41], $\boldsymbol{r}$ is the position of the point at which the function is calculated, and $\boldsymbol{r}'$ is the position of a generic calculation point. This function F can be approximated by the contributions of the interpolating particles; a summation is performed on all particles within the compact kernel support:

$$F(\boldsymbol{r}_a) \approx \sum_b F(\boldsymbol{r}_b)\, W(\boldsymbol{r}_a - \boldsymbol{r}_b, h)\frac{m_b}{\rho_b} \tag{16}$$

where a is the interpolated particle, b is a neighboring particle, m and ρ are the mass and density, respectively, $\frac{m_b}{\rho_b}$ is the volume associated with the neighboring particle, and h is the smoothing length. The kernel functions W satisfy several properties, such as positivity on compact support, normalization, and a monotonic decrease with distance [42]. Several methodologies are available; one option is the Quintic Wendland polynomial kernel [43]:

$$W(q) = \alpha_D \left(1 - \frac{q}{2}\right)^4 (2q + 1) \qquad 0 \leq q \leq 2 \tag{17}$$

where α_D is a real number such that the kernel ensures the normalization property, $q = \frac{r}{h}$ is the non-dimensional distance between the particles, and r is the distance between a certain particle a and another b. In this way, the Wendland kernel is used to calculate the interactions between the particles at a distance of up to $2h$. In the Lagrangian framework, the differential form of the NS equations can be written as a discrete version using the kernel function:

$$\frac{d\boldsymbol{v}_a}{dt} = -\sum_b m_b \frac{P_a + P_b}{(\rho_a \rho_b)} \nabla_a W_{ab} + \boldsymbol{\Gamma}_a + \boldsymbol{g} \tag{18}$$

$$\frac{d\rho_a}{dt} = \rho_a \sum_b \frac{m_b}{\rho_b} \boldsymbol{v}_{ab} \nabla_a W_{ab} + 2\delta hc \sum_b (\rho_b - \rho_a) \frac{\boldsymbol{v}_{ab} \nabla_a W_{ab}}{||\mathbf{r}_{ab}||^2} \frac{m_b}{\rho_b} \tag{19}$$

where t is the time, $\boldsymbol{v}$ is the velocity, P pressure, $\boldsymbol{g}$ is the gravitational acceleration, ∇_a is the gradient operator, W_{ab} the kernel function, whose value depends on the distance between a and b, $\boldsymbol{r_{ab}} = \boldsymbol{r}_a - \boldsymbol{r}_b$ with $\boldsymbol{r}_k$ being the position of the kth particle, and c is the speed of sound. The code allows the use of an artificial viscosity model (not employed here) or a laminar viscosity model + SPS turbulence model [42]. An SPS model in combination with a laminar viscosity model resolves small unresolved motions. Vortices that can be solved by the grid are allowed to evolve according to the NS equations, and a model is used to represent turbulence at scales below the grid. The term $\boldsymbol{\Gamma}_a$ in Equation (18) introduces dissipation, as described by

$$\boldsymbol{\Gamma}_a = \sum_b m_b \frac{4\nu_0 \boldsymbol{r}_{ab} \nabla_a W_{ab}}{(\rho_a + \rho_b)(||\mathbf{r}_{ab}||^2 + 0.01h^2)} v_{ab} + \sum_b m_b \left(\frac{\vec{\tau}_a^{ij} + \vec{\tau}_b^{ij}}{\rho_a \rho_b} \right) \nabla^i W_{ab} \tag{20}$$

The first term in Equation (20) introduces the so-called laminar viscosity [44], where the term ν_0 = 1.0×10^{-6} $m^2 s^{-1}$ refers to the fluid kinematic viscosity. The second term describes the SPS model [45], corresponding to the variationally consistent form of the symmetric formulation proposed in [44]. The SPS stress tensor, $\vec{\tau}^{ij}$, in Einstein notation in the coordinate directions i and j, is defined according to the SPS strain tensor:

$$\vec{S}^{ij} = -\frac{1}{2} \left(\frac{\partial \vec{v}^i}{\partial x^j} + \frac{\partial \vec{v}^j}{\partial x^i} \right) \tag{21}$$

where x is the position of the particle. A relationship between density and pressure binds the system of equations. DualSPHysics uses a weakly compressible SPH (WCSPH) formulation to model Newtonian fluids, and, for this formulation, the Tait equation of state is used to determine the fluid pressure, P, from the density of the particles.

$$P = \frac{c_0^2 \rho_0}{\gamma} \left(\left(\frac{\rho}{\rho_0} \right)^\gamma - 1 \right) \tag{22}$$

where ρ_0 is the density of the reference fluid, and γ is the polytropic constant. The compressibility of the fluid is adjusted so that c_0 can be artificially lowered to ensure reasonable time values.

The explicit second-order accurate Symplectic [46] time integration scheme takes the following form, considering the weakly compressible formulation, for solving the position of particle a at step $n+1$:

$$\boldsymbol{r}_a^{n+\frac{1}{2}} = \boldsymbol{r}_a^n + \frac{\Delta t}{2}\boldsymbol{v}_a^n \tag{23}$$

$$\boldsymbol{v}_a^{n+\frac{1}{2}} = \boldsymbol{v}_a^n + \frac{\Delta t}{2}\boldsymbol{F}_a^n \tag{24}$$

$$\boldsymbol{v}_a^{n+1} = \boldsymbol{v}_a^n + \frac{\Delta t}{2}\boldsymbol{F}_a^{n+\frac{1}{2}} \tag{25}$$

$$\boldsymbol{r}_a^{n+1} = \boldsymbol{r}_a^n + \Delta t\frac{\boldsymbol{v}_a^{n+\frac{1}{2}} + \boldsymbol{v}_a^n}{2} \tag{26}$$

where $\boldsymbol{F}_a = \frac{d\boldsymbol{v}_a}{dt}$, and $\boldsymbol{v}_a = \frac{d\boldsymbol{r}_a}{dt}$. The density is updated at the next time step by means of a similar two-step strategy, which reads

$$\rho_a^{n+\frac{1}{2}} = \rho_a^n + \frac{\Delta t}{2}R_a^n \tag{27}$$

$$\rho_a^{n+1} = \rho_a^n + \frac{2 + \frac{R_a^{n+\frac{1}{2}}}{\rho_a^{n+\frac{1}{2}}}\Delta t}{2 - \frac{R_a^{n+\frac{1}{2}}}{\rho_a^{n+\frac{1}{2}}}\Delta t} \tag{28}$$

where $R_a = \frac{d\rho_a}{dt}$.

As illustrated in Monaghan et al. (1999) [47], variable time stepping can be used and, considering the CFL condition, the force terms and the viscous diffusion term can be calculated as follows:

$$\Delta t = C_{CFL}\min_a\left(\sqrt{\frac{h}{|f_a|}}, \frac{h}{c + \max\limits_b \frac{|h\boldsymbol{v}_a \cdot \boldsymbol{r}_a|}{||\mathbf{r}_{ab}||^2 + 0.01h^2}}\right) \tag{29}$$

where f_a is the force per unit mass.

For this work, the modified Dynamic Boundary Condition (mDBC) [48] is used. Its predecessor approach—Dynamic Boundary Condition (DBC) [49]—has been shown to produce some non-physical gaps when free-surface flows interact with solid particles. In [50], the authors show that this issue can be addressed through the use of "Dummy" particles, following the implementation proposed in [51]. Instead, mDBC, by sharing the same particle arrangement as DBC, locates the interacting boundary surface at the middle line between the outermost layer of body particles and the fluid domain. Normal vectors are calculated with respect to this latter boundary surface. The solid body adjusts with a number of boundary particle layers to ensure the completeness of the kernel of fluid particles approaching the boundary, avoiding inconsistencies due to truncation effects. Once the geometry has been characterized by the normal vectors, the position of the interface is used to reflect the ghost nodes in the fluid domain.

An isotropic particle spacing is an important stability problem in SPH, especially in violent flows, as the particles cannot maintain a uniform distribution. It is corrected by the shifting algorithm [52]. Finally, it is important to mention that the initial condition in DualSPHysics is generated using a pre-processing tool that creates particles with an initial inter-particle distance D_p. This value also defines the resolution used in the simulations. Fluid and solid particles, whether part of a floating object or other solids, are then created following this initial spacing at the initial time step.

models and for all the considered Froude numbers, the value of the resistance coefficient is underestimated. The Lagrangian model, in this case, demonstrates better overall behavior than the Eulerian model in calculating the resistance coefficients, with smaller errors.

4. Case Study

The validated frameworks are applied to model a ship hull with a 30 m LOA. The general view and the main characteristics of the new designed hull are reported in Figure 5 and Table 7, respectively.

Table 7. Main geometric characteristics of ship.

LOA	30.00 m
B	6.00 m
D	2.00 m
S	168.96 m^2

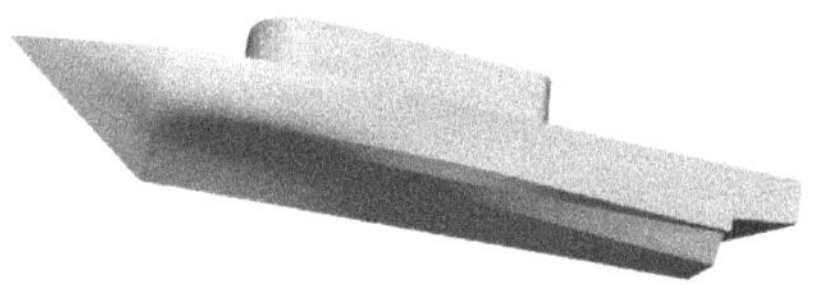

Figure 5. A general view of the new designed hull.

The two models have two different setups, as can be seen from the difference in the methods they use.

4.1. Simulations via OpenFOAM

For the construction of the case, the same design choices are adopted as in the validation for the Wigley case. The computational domain, inserted in the blockMeshdict, is proportional to that created in the validation. The number of cells is 6,245,338. The cell size ranges from a minimum size of 0.0018 m to a maximum size of 1.2 m. Also, here, the turbulence model is the SST k-ω model. It is solved by the routine interFoam.

4.2. Simulations via DualSPHysics

For the construction of the case, the same design choices are adopted as in the validation for the Wigley case, and we also use discretization proportional to the validation case. The particle size has D_p = 0.20 m. The number of particles generated is 7,580,338. This value is sufficient to accurately discretize the external characteristics of the hull, as the surface is smooth and free of sudden variations. After the static convergence, the domain is dimensioned according to the requirements of the wake development (Table 8).

Table 8. Domain simulation.

UCL	22.5 m
DCL	67.5 m
CW	50 m
CD	12 m

The boundary conditions and parameters are also the same as in the previous case. Also, here, the turbulence model is the SPS model.

5. Results and Discussion

This section summarizes the results obtained from the newly designed ship hull by elaborating data derived from the numerical simulations. Three Fr values are considered, 0.233, 0.291, and 0.408, which correspond to speeds of 4.0, 5.0, and 7.0 m/s, respectively, highlighting the resistance of the two frameworks.

Table 9 shows the resistance coefficients obtained from OpenFOAM and DualSPHysics and at different values of Fr. They give similar results. The forces obtained from the Eulerian approach are slightly lower than those obtained from the Lagrangian approach except for $Fr = 0.291$. However, the differences in values between the two models show an absolute error rate of no more than 11%. This leads to an awareness that the two models used show very similar force and drag coefficient results despite the difference in the basic application of the two methods.

Table 9. Resistance coefficient comparison.

Framework	Fr	R_T [kN]	$C_T \times 10^{-3}$	Runtime [h]
OpenFOAM	0.233	26.573	19.683	2.08
OpenFOAM	0.291	51.715	24.515	2.1
OpenFOAM	0.408	114.338	27.654	2.22
DualSPHysics	0.233	29.622	21.941	16.8
DualSPHysics	0.291	49.966	23.686	14.62
DualSPHysics	0.408	127.649	30.873	12.45

Furthermore, Table 9 shows the execution time for the two models, suggesting very big computational time differences between the models. The simulations with OpenFOAM have runtimes that are, on average, an order of magnitude shorter than those with DualSPHysics, in addition to the fact that the two numerical methods are fundamentally different. Another difference between the two methods is the hardware used. OpenFOAM simulations are run on CPUs, which are very powerful hardware for these simulations. DualSPHysics simulations are run on GPUs, and it is well known that the SPH method has a high computational cost. Simulation time is an important factor in hull optimization. In addition to optimization time, a significant reduction in CFD time improves the timing of obtaining results.

Figure 6 shows the total resistance obtained in the two approaches. The box size indicates the range of variation in resistance values, and the line in the box indicates the average value. The compactness of values is defined by the difference in the two extremes (whiskers) extending 1.5 times from the interquartile range from the top and bottom of the box. In addition, all outliers (which are the values that go outside the range) are eliminated.

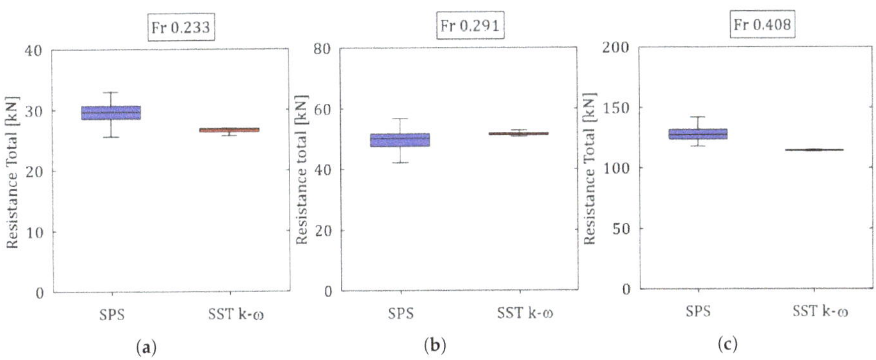

Figure 6. Box Plot of compactness of simulations: (**a**) $Fr = 0.233$; (**b**) $Fr = 0.291$; (**c**) $Fr = 0.408$.

OpenFOAM shows greater compactness of the results in the simulation. Only the *Fr* of 0.408 shows a small difference between the values, which could already be seen from the value of the resistances.

Flow Visualization

In Figures 7–9, there are six different panels representing the free-surface elevation in two different views (OpenFOAM and DualSPHysics) for three different simulation instants (0.25, 0.5, and the end of the simulation time, *T*), all of them related to the simulation with $Fr = 0.291$. In Figures 7–9, on the left (a,c,e), there are those generated by OpenFOAM and, on the right (b,d,f), those generated by DualSPHysics.

Figure 7 shows the free-surface elevation scaled with the LOA ship length. It can be seen that quantitative and qualitative agreement is reached around the ship at the time of the final simulation for both wave position and amplitude. In the first quarter of the simulation, the two simulations are similar in the bow and stern areas. In the middle of the simulations, the free surface reproduced in OpenFOAM shows more uplift. At the end of the two simulations, the two models show a good representation of the free surface with small differences. OpenFOAM simulates the free surface around the hull well, but does not propagate it beyond the end of the first refinement box owing to cell variation. DualSPHysics, on the other hand, shows better free-surface quality and excellent propagation, although it has a small reflection at the end of the domain generated by the side wall.

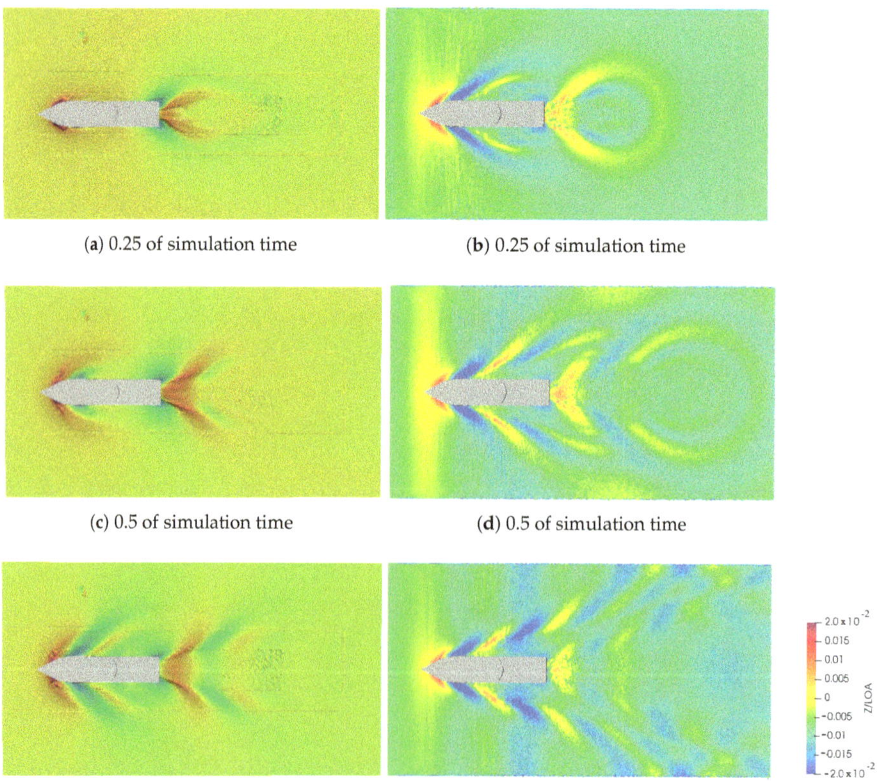

(**a**) 0.25 of simulation time (**b**) 0.25 of simulation time

(**c**) 0.5 of simulation time (**d**) 0.5 of simulation time

(**e**) Final simulation time (**f**) Final simulation time

Figure 7. Free-surface representation of the used models: (**a**) 0.25*T* in OpenFOAM; (**b**) 0.25*T* in DualSPHysics; (**c**) 0.5*T* in OpenFOAM; (**d**) 0.5*T* in DualSPHysics; (**e**) *T* in OpenFOAM; (**f**) *T* in DualSPHysics.

Figure 8 shows the velocity field at three different simulation times. The velocity field expressed by the OpenFOAM numerical model has a large area of velocity reduction around the bow and a narrow area of velocity reduction in the wake, in contrast to the velocity field generated by the DualSPHysics model, which has a much smaller area of velocity reduction around the bow and a very large area of velocity reduction in the wake. At one quarter of the simulation, the two simulations are nearly identical. At half of the simulations, the velocity field generated by OpenFOAM shows a larger area of velocity reduction around the bow than the velocity field developed in DualSPHysics. Finally, at the last instant of the simulation, similarity of the velocity field can be seen, but with a greater velocity reduction in the bow area by the simulation in OpenFOAM, and in the stern area, a narrower area of velocity reduction than in DualSPHysics. Also, from these figures, one can see the defect generated by cell variation in the OpenFOAM simulations and a small reflection in the DualSPHysics simulation.

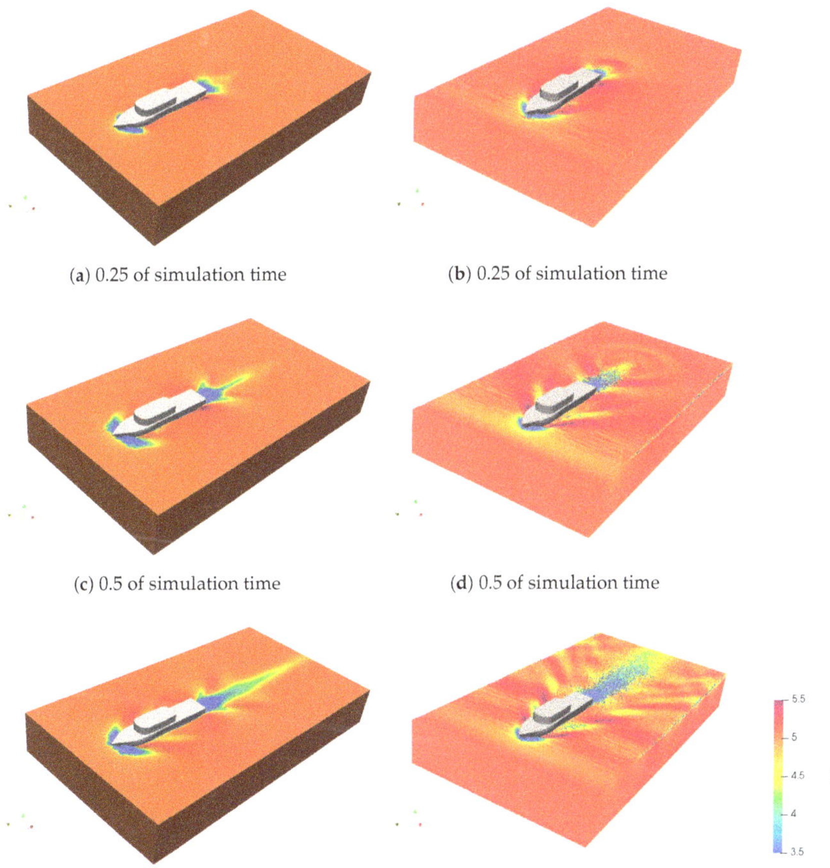

(**a**) 0.25 of simulation time (**b**) 0.25 of simulation time

(**c**) 0.5 of simulation time (**d**) 0.5 of simulation time

(**e**) Final simulation time (**f**) Final simulation time

Figure 8. Velocity representation of the used models: (**a**) 0.25T in OpenFOAM; (**b**) 0.25T in DualSPHysics; (**c**) 0.5T in OpenFOAM; (**d**) 0.5T in DualSPHysics; (**e**) T in OpenFOAM; (**f**) T in DualSPHysics.

Figure 9 shows the vorticity field at three different simulation times. During the test, both models accurately resolved the vorticity field, as evidenced by the similar vorticity fields displayed. The images represent the section along the y-direction in the middle of the domain. In the first quarter of the simulation, the two simulations are similar in the aft area, with the difference that in the DualSPHysics simulation, some particles are lost. The particles decrease owing to the anisotropic spacing of the particles, which is a stability problem of the model. In the middle of the simulations, the vorticity reproduced behind the stern in OpenFOAM shows a larger area. At the end of the two simulations, the two models show a good representation of vorticity, with minimal differences. From the vorticity field section, we can also see the difference in the accuracy of the free surface and how the simulation in DualSPHysics captures the free surface better.

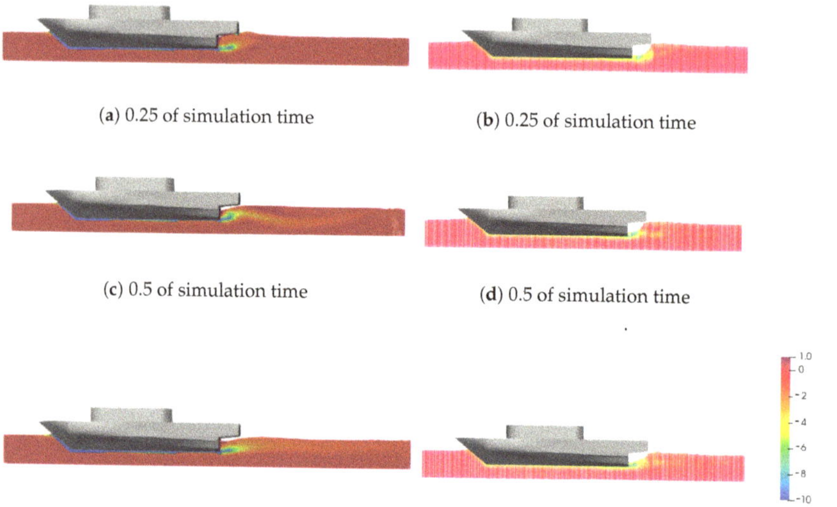

(**a**) 0.25 of simulation time (**b**) 0.25 of simulation time

(**c**) 0.5 of simulation time (**d**) 0.5 of simulation time

(**e**) Final simulation time (**f**) Final simulation time

Figure 9. Vorticity representation of the used models: (**a**) 0.25*T* in OpenFOAM; (**b**) 0.25*T* in DualSPHysics; (**c**) 0.5*T* in OpenFOAM; (**d**) 0.5*T* in DualSPHysics; (**e**) *T* in OpenFOAM; (**f**) *T* in DualSPHysics.

6. Conclusions

In this paper, two turbulence methods are presented for calculating the total resistances acting on the ship under study. The turbulence models selected for analysis are the SST k-ω model, which provides high-precision boundary layer modeling as a combination of k-ϵ and k-ω, and the Laminar-and-SPS model, which is mainly used to adequately present the viscosity and turbulent motions of the fluid flow, especially in the case of the evolution of breaking waves, but adapted to our case study. In this work, two different numerical schemes, one mesh-based and one meshless, each one employing a different turbulence method are presented for calculating the total resistances acting on the ship under study. For the mesh-based approach, the OpenFOAM model is employed. For the meshless approach, the Weaky Compressible SPH-based DualSPHysics model is used. To predict the resistance of ships with different running attitudes, numerical simulations of ships advancing at different speeds and Fr values were carried out.

A comparison of the tools may be useful for possible use in the geometry optimization field and, as a consequence, considering that the naval world is increasingly approaching an electric propulsion system, good hull optimization and knowledge of fluid dynamics would improve the range of electric ships. In the present work, the accuracy of the calculations is ensured by preventive validation of the two methods described in Section 3. The numerical results from the two methods show good agreement compared to the experimental references for all Fr values, with a maximum relative error of −15.71% at the lowest Fr for the SST k-ω turbulence model. The drag coefficients obtained by OpenFOAM and DualSPHysics and studied at the different values of Fr show very similar results, with an absolute error rate of no more than 11%. The free surface in the two simulations shows quantitative and qualitative agreement for both wave position and amplitude. Regarding the velocity field, the one expressed by the OpenFOAM numerical model has a large area of velocity reduction around the bow and a narrow area of velocity reduction in the wake, in contrast to the velocity field generated by the DualSPHysics model, which has a much smaller area of velocity reduction around the bow and a very large area of velocity reduction in the wake. As evidenced by the vorticity fields shown, both models accurately resolved the vorticity field. In general, the simulation results show that the availability of OpenFOAM and DualSPHysics as tools to predict ship resistance and their application to the development of ship design can potentially be leveraged by end-users. In this case study, the advantages of OpenFOAM are the reduced calculation time for simulations and good compactness of the results. On the other hand, DualSPHysics stands out for its simplicity of setting up the model and good reproduction of the free surface.

Author Contributions: Conceptualization, D.C., A.C. (Alessio Caravella), R.G. and A.L.; Methodology, B.T. and C.A.; Software, B.T., G.B., J.M.D. and A.C. (Alfredo Cassano); Formal analysis, B.T., G.B., J.M.D. and A.C. (Alfredo Cassano); Investigation, D.C., G.B., J.M.D. and A.C. (Alfredo Cassano); Writing—original draft, D.C.; Writing—review & editing, B.T., G.B., J.M.D. and A.C. (Alfredo Cassano); Supervision, A.C. (Alessio Caravella), R.G., C.A. and A.L.; Funding acquisition, A.C. (Alessio Caravella), R.G. and A.L. All authors have read and agreed to the published version of the manuscript.

Funding: This research was partially funded by the Italian funding program *"PON 2014–2020, Dottorati di Ricerca su tematiche Green e dell'Innovazione"*. This work was also supported by the Spanish government and the European Social Found (ESF) under the program 'Ramón y Cajal 2020' (RYC2020-030197-I/AEI/10.13039/501100011033) and the grant TED2021-129479A-I00 funded by MCIN/AEI/10.13039/501100011033 and by "European Union NextGenerationEU/PRTR", and the project ED431C 2021/44 "Programa de Consolidación e Estruturación de Unidades de Investigación Competitivas" financed by Xunta de Galicia, Consellería de Cultura, Educación e Universidade. This work was also supported by the grant CNS2022-136073 provided by MCIN/AEI/10.13039/501100011033 and by "European Union NextGenerationEU/PRTR", and the grant RYC2022-038341-I provided by MCIN/AEI/10.13039/501100011033 and by "ESF Investing in your future".

Data Availability Statement: No new data were created or analyzed in this study. Data sharing is not applicable to this article.

Conflicts of Interest: The authors declare no conflicts of interest.

References

1. Tavakoli, S.; Niazmand Bilandi, R.; Mancini, S.; De Luca, F.; Dashtimanesh, A. Dynamic of a planing hull in regular waves: Comparison of experimental, numerical and mathematical methods. *Ocean Eng.* **2020**, *217*, 107959. [CrossRef]
2. Zhang, Z.R.; Hui, L.; Zhu, S.P.; Feng, Z. Application of CFD in ship engineering design practice and ship hydrodynamics. *J. Hydrodyn. Ser. B* **2006**, *18*, 315–322. [CrossRef]
3. Wnęk, A.; Sutulo, S.; Guedes Soares, C. CFD analysis of ship-to-ship hydrodynamic interaction. *J. Mar. Sci. Appl.* **2018**, *17*, 21–37. [CrossRef]
4. De Luca, F.; Mancini, S.; Miranda, S.; Pensa, C. An Extended Verification and Validation Study of CFD Simulations for Planing Hulls. *J. Ship Res.* **2016**, *60*, 101–118. [CrossRef]
5. Wu, J.; Sun, Z.; Jiang, Y.; Zhang, G.; Sun, T. Experimental and numerical study of slamming problem for a trimaran hull. *Ships Offshore Struct.* **2021**, *16*, 46–53. [CrossRef]
6. Guan, X.S.; Sun, P.N.; Xu, Y.; Lyu, H.G.; Geng, L.M. Numerical studies of complex fluid-solid interactions with a six degrees of freedom quaternion-based solver in the SPH framework. *Ocean Eng.* **2024**, *291*, 116484. [CrossRef]
7. Fröhlich, J.; von Terzi, D. Hybrid LES/RANS methods for the simulation of turbulent flows. *Prog. Aerosp. Sci.* **2008**, *44*, 349–377. [CrossRef]
8. Argyropoulos, C.; Markatos, N. Recent advances on the numerical modelling of turbulent flows. *Appl. Math. Model.* **2015**, *39*, 693 732. [CrossRef]
9. Wilcox, D. Turbulence modeling-an overview. In Proceedings of the 39th Aerospace Sciences Meeting and Exhibit, Reno, NV, USA, 8–11 January 2001; p. 724.
10. Wu, C.S.; Zhou, D.C.; Gao, L.; Miao, Q.M. CFD computation of ship motions and added resistance for a high speed trimaran in regular head waves. *Int. J. Nav. Archit. Ocean Eng.* **2011**, *3*, 105–110. [CrossRef]
11. Zha, R.; Ye, H.; Shen, Z.; Wan, D. Numerical study of viscous wave-making resistance of ship navigation in still water. *J. Mar. Sci. Appl.* **2014**, *13*, 158–166. [CrossRef]

12. Menter, F.; Kuntz, M.; Langtry, R. Ten years of industrial experience with the SST turbulence model. *Turbul. Heat Mass Transf.* **2003**, *4*, 625–632.
13. Sulisetyono, A.; Alifrananda, M. Evaluation of the ship resistance using computational fluid dynamics at various Froude numbers. In Proceedings of the ICSEDTI 2022: Proceedings of the 1st International Conference on Sustainable Engineering Development and Technological Innovation, ICSEDTI 2022, Tanjungpinang, Indonesia, 11–13 October 2022; European Alliance for Innovation: Gent, Belgium, 2023; p. 79.
14. Vanherzeele, A. NUMECA Customer Area: Users' Guide Overview. 2015. Available online: https://www.numeca.de/ (accessed on 11 October 2024).
15. Desouky, O.; Elhenawy, Y. Prediction of a Container Ship Squat in Suez Canal Using CFD. *Port-Said Eng. Res. J.* **2022**, *26*, 32–41. [CrossRef]
16. *STAR-CCM+*, version 14.02.010-R8; CD-Adapco: New York, NY, USA, 2019.
17. Islam, H.; Guedes Soares, C. Prediction of ship resistance in head waves using OpenFOAM. In *Maritime Transportation and Harvesting of Sea Resources*; Taylor & Francis Group: Abingdon, UK, 2017; pp. 527–533.
18. Islam, H.; Guedes Soares, C. Uncertainty analysis in ship resistance prediction using OpenFOAM. *Ocean Eng.* **2019**, *191*, 105805. [CrossRef]
19. Kim, K.W.; Paik, K.J.; Lee, S.H.; Lee, J.H.; Kwon, S.Y.; Oh, D. A numerical study on the feasibility of predicting the resistance of a full-scale ship using a virtual fluid. *Int. J. Nav. Archit. Ocean Eng.* **2024**, *16*, 100560. [CrossRef]
20. Kawamura, K.; Hashimoto, H.; Matsuda, A.; Terada, D. SPH simulation of ship behaviour in severe water-shipping situations. *Ocean Eng.* **2016**, *120*, 220–229. [CrossRef]
21. Priyambada, A.R.; Tarwidi, D. 3D GPU-based SPH simulation of water waves impacting on a floating object. In Proceedings of the 2017 International Conference on Control, Electronics, Renewable Energy and Communications (ICCREC), Yogyakarta, Indonesia, 26–28 September 2017; pp. 60–65. [CrossRef]
22. Eriksson, J. Evaluation of SPH for Hydrodynamic Modeling, Using DualSPHysics. Master's Thesis, Uppsala University, Uppsala, Sweden, 2018. Available online: https://urn.kb.se/resolve?urn=urn:nbn:se:uu:diva-339557 (accessed on 10 October 2024).
23. Cheng, H.; Ming, F.; Sun, P.; Sui, Y.; Zhang, A.M. Ship hull slamming analysis with smoothed particle hydrodynamics method. *Appl. Ocean Res.* **2020**, *101*, 102268. [CrossRef]
24. Tagliafierro, B.; Mancini, S.; Ropero-Giralda, P.; Domínguez, J.; Crespo, A.; Viccione, G. Performance assessment of a planing hull using the smoothed particle hydrodynamics method. *J. Mar. Sci. Eng.* **2021**, *9*, 244. [CrossRef]
25. Domínguez, J.M.; Fourtakas, G.; Altomare, C.; Canelas, R.B.; Tafuni, A.; García-Feal, O.; Martínez-Estévez, I.; Mokos, A.; Vacondio, R.; Crespo, A.J.; et al. DualSPHysics: From fluid dynamics to multiphysics problems. *Comput. Part. Mech.* **2022**, *9*, 867–895. [CrossRef]
26. Capasso, S.; Tagliafierro, B.; Mancini, S.; De Luca, F.; Martínez-Estévez, I.; Domínguez, J.M.; Altomare, C.; Crespo, A.J.; Pensa, C.; Viccione, G. Preliminary investigation into the dynamic of planing hulls in regular waves using the smoothed particle hydrodynamics method. In Proceedings of the ASME 2023 42nd International Conference on Ocean, Offshore and Arctic Engineering, Melbourne, Australia, 11–16 June 2023; Volume 7. [CrossRef]
27. Capasso, S.; Tagliafierro, B.; Mancini, S.; Martínez-Estévez, I.; Altomare, C.; Domínguez, J.M.; Viccione, G. Regular Wave Seakeeping Analysis of a Planing Hull by Smoothed Particle Hydrodynamics: A Comprehensive Validation. *J. Mar. Sci. Eng.* **2023**, *11*, 700. [CrossRef]
28. Rakhsha, M.; Kees, C.E.; Negrut, D. Lagrangian vs. Eulerian: An Analysis of Two Solution Methods for Free-Surface Flows and Fluid Solid Interaction Problems. *Fluids* **2021**, *6*, 460. [CrossRef]
29. Tavakoli, S.; Babanin, A.V.; Hirdaris, S. The hydrodynamics of hard-chine sections entering water. In Proceedings of the ASME 2022 41st International Conference on Ocean, Offshore and Arctic Engineering, Hamburg, Germany, 5–10 June 2022; American Society of Mechanical Engineers (ASME): New York, NY, USA, 2022; Volume 5-A. [CrossRef]
30. Gruwez, V.; Altomare, C.; Suzuki, T.; Streicher, M.; Cappietti, L.; Kortenhaus, A.; Troch, P. An inter-model comparison for wave interactions with sea dikes on shallow foreshores. *J. Mar. Sci. Eng.* **2020**, *8*, 985. [CrossRef]
31. Meringolo, D.D.; Lauria, A.; Aristodemo, F.; Filianoti, P.F. Large eddy simulation within the smoothed particle hydrodynamics: Applications to multiphase flows. *Phys. Fluids* **2023**, *35*, 063312. [CrossRef]
32. Bai, K.; McCarthy, J. (Eds.) *Proceedings of the Workshop on Ship Wave-Resistance Computations: Held at Bethesda, Maryland, 13–14 November 1979*; David W. Taylor Naval Ship Research and Development Center: Washington, DC, USA, 1979; Volume 2.
33. Naz, N. Computation of Hydrodynamic Characteristics of Ships Using CFD. *Int. J. Mater. Mech. Manuf.* **2017**, *5*, 219–223. [CrossRef]
34. Menter, F.R. Two-equation eddy-viscosity turbulence models for engineering applications. *AIAA J.* **1994**, *32*, 269–289. [CrossRef]
35. Wilcox, D.C. *Turbulence Modeling for CFD*; DCW Industries: La Canada, CA, USA, 1998; Volume 2.
36. Menter, F. Zonal Two Equation k-w Turbulence Models For Aerodynamic Flows. In Proceedings of the 23rd Fluid Dynamics, Plasmadynamics, and Lasers Conference, Orlando, FL, USA, 6–9 July 1993. [CrossRef]
37. Lauria, A.; Alfonsi, G. Numerical Investigation of Ski Jump Hydraulics. *J. Hydraul. Eng.* **2020**, *146*, 04020012. [CrossRef]
38. Alfonsi, G.; Lauria, A.; Primavera, L. A study of vortical structures past the lower portion of the ahmed car model. *J. Flow Vis. Image Process.* **2012**, *19*, 81–95. [CrossRef]

39. Hirt, C.; Nichols, B. Volume of fluid (VOF) method for the dynamics of free boundaries. *J. Comput. Phys.* **1981**, *39*, 201–225. [CrossRef]
40. Courant, R.; Friedrichs, K.; Lewy, H. Über die partiellen Differenzengleichungen der mathematischen Physik. *Math. Ann.* **1928**, *100*, 32–74. [CrossRef]
41. Monaghan, J. Smoothed particle hydrodynamics. *Rep. Prog. Phys.* **2005**, *68*, 1703. [CrossRef]
42. Monaghan, J.J. Smoothed Particle Hydrodynamics. *Annu. Rev. Astron. Astrophys.* **1992**, *30*, 543–574. [CrossRef]
43. Wendland, H. Piecewise polynomial, positive definite and compactly supported radial functions of minimal degree. *Adv. Comput. Math.* **1995**, *4*, 389–396. [CrossRef]
44. Lo, E.Y.; Shao, S. Simulation of near-shore solitary wave mechanics by an incompressible SPH method. *Appl. Ocean Res.* **2002**, *24*, 275–286. [CrossRef]
45. Dalrymple, R.; Rogers, B. Numerical modeling of water waves with the SPH method. *Coast. Eng.* **2006**, *53*, 141–147. [CrossRef]
46. Leimkuhler, B.; Matthews, C. Introduction. In *Molecular Dynamics: With Deterministic and Stochastic Numerical Methods*; Springer International Publishing: Cham, Switzerland, 2015; pp. 1–51. [CrossRef]
47. Monaghan, J.J.; Cas, R.A.; Kos, A.; Hallworth, M. Gravity currents descending a ramp in a stratified tank. *J. Fluid Mech.* **1999**, *379*, 39–69. [CrossRef]
48. English, A.; Domínguez, J.; Vacondio, R.; Crespo, A.; Stansby, P.; Lind, S.; Chiapponi, L.; Gesteira, M. Modified dynamic boundary conditions (mDBC) for general-purpose smoothed particle hydrodynamics (SPH): Application to tank sloshing, dam break and fish pass problems. *Comput. Part. Mech.* **2021**, *9*, 1–15. [CrossRef]
49. Crespo, A.; Gómez-Gesteira, M.; Dalrymple, R. Boundary conditions generated by dynamic particles in SPH methods. *Comput. Mater. Contin.* **2007**, *5*, 173–184.
50. Ma, C.; Oka, M. Investigation of the influence of different boundary conditions in SPH on ship dynamics. *Ocean Eng.* **2024**, *310*, 118571. [CrossRef]
51. Adami, S.; Hu, X.; Adams, N. A generalized wall boundary condition for smoothed particle hydrodynamics. *J. Comput. Phys.* **2012**, *231*, 7057–7075. [CrossRef]
52. Lind, S.; Xu, R.; Stansby, P.; Rogers, B. Incompressible smoothed particle hydrodynamics for free-surface flows: A generalised diffusion-based algorithm for stability and validations for impulsive flows and propagating waves. *J. Comput. Phys.* **2012**, *231*, 1499–1523. [CrossRef]
53. Tafuni, A.; Domínguez, J.; Vacondio, R.; Crespo, A. A versatile algorithm for the treatment of open boundary conditions in Smoothed particle hydrodynamics GPU models. *Comput. Methods Appl. Mech. Eng.* **2018**, *342*, 604–624. [CrossRef]
54. Liu, M.; Liu, G. Restoring particle consistency in smoothed particle hydrodynamics. *Appl. Numer. Math.* **2006**, *56*, 19–36. [CrossRef]

fluids

MDPI

Article

Improved Delayed Detached-Eddy Simulation of Turbulent Vortex Shedding in Inert Flow over a Triangular Bluff Body

Matthew R. McConnell, Jason Knight and James M. Buick *

School of Electrical and Mechanical Engineering, University of Portsmouth, Portsmouth PO1 3DJ, UK; matthew.mcconnell@port.ac.uk (M.R.M.); jason.knight@port.ac.uk (J.K.)
* Correspondence: james.buick@port.ac.uk

Abstract: The Improved Delayed Detached-Eddy Simulation (IDDES) is a modification of the original Detached-Eddy Simulation (DES) design to incorporate Wall Modeled Large Eddy Simulation (WM-LES) capabilities and to extend the class of flows suitable for this methodology. For thin attached boundary layers, typically seen in external aerodynamic flows, the DES branch of the model is active, whereas with thick boundary layers, typically seen in internal flows and also wake flows, the WMLES branch is active, thus providing a numeric method suited to handling most flow cases automatically. The flow over a triangular bluff body is used to validate the suitability of the IDDES model and compare the results with experimental, DDES, and LES data. The IDDES model is found to be relatively accurate when compared with the experimental results, with recirculation length, streamwise velocity, and Reynolds stresses all showing good agreement with the experimental data. However, when compared with the DDES model, there is a ~4% overprediction of the recirculation length using the same mesh and numerical scheme. The code, with its extra complexity, is also ~3% slower to solve. The IDDES model has also been tested against different meshes, and the results show that even for a coarse mesh, there is still good agreement with the experimental data.

Keywords: hybrid RANS-LES; DDES; IDDES; turbulent flow; triangular bluff body; vortices

1. Introduction

Triangular bluff bodies are typically used as flame holders for industrial combustion systems. When a triangular bluff body is placed within a flow stream, the velocity in the wake behind the body is smaller relative to the whole flow field. A fuel can be ignited in this region, and the resultant flame will propagate within the combustion chamber. The flow field downstream is highly turbulent due to the vortex shedding caused by the bluff body.

Sjunnesson et al. conducted experimental results for this test case on a test rig at Volvo [1–3], and it has been added to the European Research Community On Flow, Turbulence, And Combustion (ERCOFTAC) knowledge base as a combustion application challenge. For this study, combustion has not been investigated, and only the inert flow experiment is considered for this analysis. Johansson et al. [4] conducted one of the earlier studies of the inert flame holder using the k-ε turbulence model and found very good agreement with respect to the streamwise velocity profile at the centerline, the profiles for velocity, and turbulent kinetic energy. However, being a RANS simulation, it was only able to recover time-averaged results.

Hasse et al. [5] performed a DES involving an Unsteady Reynolds-Averaged Naiver–Stokes (URANS) SST model and Large Eddy Simulation (LES) combination. Their results generally agree better with the experimental results [1–3] than the URANS simulations, particularly for the length of the recirculation zone.

Lysenko et al. [6,7] conducted a comprehensive numerical study for the inert and reactive flow conditions of this experimental test case using Unsteady Reynolds-Averaged

Naiver–Stokes (URANS), Scale-Adaptive Simulation (SAS), and Large Eddy Simulation (LES). The standard k-ε [8] model was used in the URANS simulations. For SAS, the model of Menter and Egorov [9] is based on the k-ω SST [10] with updated coefficients [11] and High-Wave-Number damping [12] based on the Wall-Adapting Local Eddy-viscosity (WALE) model [13]. Finally, for the LES calculations, the k-equation eddy viscosity subgrid-scale model [14] and the Smagorinsky model [15] were used. Results were compared with the earlier work of Hasse et al. [5], who compared Detached Eddy Simulation (DES) and LES. The results of these studies show that for the inert flow, the recirculation length (L_{rc}) behind the triangular bluff body was predicted with a reasonable level of accuracy between all models, as was streamwise velocity and turbulent kinetic energy.

Spalart et al. [16] first coined the term DES in 1997. The central idea is that it is expensive to conduct LES in the attached boundary layer due to the requirement that the grid cells for LES need to be isotropic. As the boundary layer becomes thinner at higher Reynold's numbers, the grid requirement scales accordingly, making LES prohibitively expensive for most practical applications. It is, however, inexpensive to solve the free flow turbulence away from the walls with LES. RANS does not have the same boundary layer requirement and, as such, is relatively cheap for solving thin boundary layers with a reliable prediction of separation. With this in mind, it can be seen that a model with LES for the areas of strongly separated flows, combined with a RANS model for the reduced grid requirements of the attached boundary layer, is an appealing combination.

This original proposal used the one-equation model of Spalart and Allmaras [17] with a modification to a DES-length scale. This means that in the boundary layer, the model will operate in a RANS mode, and outside the turbulent boundary layer, it will operate in LES.

Spalart foresaw a key weakness in the original DES concept where too fine grid resolution in the boundary layer would cause the length scale to switch from RANS to LES. This would result in LES being solved on a grid not sufficiently fine enough to resolve the turbulent scales and the modeled stress dropping below RANS levels, causing a phenomenon known as Modeled Stress Depletion (MSD). The result of MSD is a strong under-prediction of the skin friction, which can lead to very poor prediction of the flow separation, known as Grid-Induced Separation (GIS) [18]. Menter et al. [19] first proposed shielding the boundary layer regardless of grid size with a DES variant based on the k-ω SST turbulence model [11,20], and the first major modification to the methodology was published by Spalart et al. [21], called "Delayed" DES (DDES). This is now the default over the initial DES model and has been shown to greatly improve its robustness.

While the intention of DDES is to maintain RANS in the boundary layer and LES in regions of massive separation, there was a view that DES could work as a Wall-Modeled Large Eddy Simulation (WMLES) [22]. WMLES aims to resolve the largest structures of the boundary layer and, as such, is considered a higher fidelity method compared with DES. Building from this, Shur et al. [23] worked on a new DES variant, which blended DDES with WMLES, called Improved Delayed Detached-Eddy Simulation (IDDES). This model uses empirical blending function to switch between DDES and WMLES depending on the flow conditions and, like DDES, has been formulated to include the k-ω SST model [24].

In most engineering companies RANS using commercial code is the staple approach applied in CFD. More computer-intensive approaches of LES or even DNS can provide more accurate results; however, they are computationally more expensive and are typically reserved for academic or research rather than industrial applications. Indeed, it has been predicted that LES will not be ready for full-scale engineering applications until 2045 [25]. This means that companies that are limited in the time and money they can invest in CFD are limited to the less accurate RANS approach, and this is likely to remain the case even after 2045 due to the computational costs of LES. Recent advances such as IDDES offer a compromise between the efficiency of RANS and the accuracy of LES and offer an opportunity to replace industrial applications RANS in some areas. To this end, this paper will evaluate the accuracy that can be obtained using IDDES using a relatively modest mesh to ensure that the approach can be replicated within industry.

The objective of this study is to expand the DES methodology that has previously been conducted on this test case by Hasse et al. [5] to include the subsequent developments of the technique incorporated in IDDES, such as the inclusion of WMLES, so the use of IDDES in problems of this type involving vortex shedding at sharp corners was evaluated. This was performed using the bluff body, triangular cylinder geometry of the flame holder [1–3]. This study recreates the DDES study of Hasse et al. [5], the LES results of Lysenko et al. [6,7], and conducts a new study using IDDES methodology. Different grid resolution and numerical schemes are tested with the aim of developing a method for best practice for this type of test case. The results are validated against the experimental data from Sjunnesson et al. [1–3].

2. Numerical Methodology

For this study, the commercial code Star CCM+ version 12 is used for all simulations. The Navier–Stokes equations are discretized on a structured grid strategy using the finite volume method. All numerical simulations are incompressible implicit unsteady and are second-order accurate in time and space. A time step of 10^{-5} s ensures that the CFL number is below 1 in most of the domain. Both the Bounded Central (BC) [26] (a second-order central difference scheme blended with a second-order upwind scheme) and the Hybrid Bounded Central (HBC) numerical scheme are used within this study, where the HBC is defined as

$$F_{\text{HBC}} = (1-\sigma)F_{\text{CDS}} + \sigma F_{\text{UDS}} \tag{1}$$

where F_{CDS} and F_{UDS} are the central and upwind estimates, and σ is designed as a limiter based on a tanh function that changes rapidly from around 1 to around 0. When σ is approximately 1, an upwind scheme is applied for the RANS simulation, while when σ is around 0, a central difference scheme is applied for the LES simulation. A complete description of σ is neglected here for the sake of simplicity, but a full description is provided by Travin et al. [27].

All simulations were run on the SCIAMA HPC facility at the University of Portsmouth. Each simulation used 20 cores and at maximum had 135×10^3 cells per core.

2.1. Turbulence Modeling

Two turbulence models were tested within the present study, both of which are modifications of the original DES model by Spalart et al. [16] based on the one-equation model by Spalart and Allmaras [17]. The central idea behind DES, as explained in the section above, is to use a RANS model as the subgrid-scale model for LES by taking the minimum of the RANS turbulence length scale d and the cell length $\Delta = \max(\Delta x, \Delta y, \Delta z)$ so that

$$\tilde{d} = \min(d, C_{\text{DES}}\Delta) \tag{2}$$

the constant C_{DES} is usually equal to 0.65 and was calibrated from decaying isotropic turbulence [28]. This means that in the boundary layer where $d < C_{\text{DES}}\Delta$, the model will operate in a RANS mode, and outside the turbulent boundary layer where $d > C_{\text{DES}}\Delta$, it will operate in LES.

Travin et al. [27] then modified the original DES model for the two-equation k-ω SST model. This model behaves in the same way as the original DES model, but the formulation will solve RANS k-ω in the boundary layer and a blended mix of LES with RANS k-ε elsewhere, where the RANS turbulence length scale reads as

$$l_{k-\omega} = k^{1/2}/(\beta^{*}\omega) \tag{3}$$

when the RANS turbulence length scale is larger than the local grid spacing Δ, the model will switch to an LES model. Here, k is the turbulent kinetic energy, ω is the specific dissipation rate of turbulent kinetic energy, and $\beta^* = 0.09$ is a constant. The only term that requires modification for DES is the dissipative term in the k-transport equation for RANS:

$$D^k_{\text{RANS}} = \rho\beta^* k\omega = \frac{\rho k^{\frac{3}{2}}}{l_{k-\omega}}, \tag{4}$$

where ρ is the density of the fluid. This leads to a substitution of the length scale to modify the equation for DES:

$$D^k_{\text{DES}} = \rho k^{3/2} / \tilde{l} \tag{5}$$

where

$$\tilde{l} = \min(l_{k-\omega}, C_{DES}\Delta), \tag{6}$$

and the C_{DES} term is formulated using k-ω and k-ε models that are blended using Menter's blending function [10]:

$$C_{\text{DES}} = (1 - F_1)C^{k-\varepsilon}_{\text{DES}} + F_1 C^{k-\omega}_{\text{DES}} \tag{7}$$

and F_1 is a blending function between the k-ω and k-ε models. A complete description of F_1 is neglected here for the sake of simplicity, but a full description is provided by Menter [10].

The 'delayed' in DDES comes from an addition limiter that is introduced in order to reduce the risk of MSD and GIS. This modification to the equation is defined as

$$D^k_{\text{DES}} = \rho\beta^* k\omega \cdot F_{\text{DES}}, \tag{8}$$

where

$$F_{\text{DES}} = \max\left(\frac{l_{k-\omega}}{C_{\text{DES}}\Delta}(1 - F_{\text{SST}}), 1\right), \tag{9}$$

and F_{SST} can be taken as F_1 or F_2 [10] in the SST model.

The 'improved' in the IDDES of Shur et al. [23] builds another layer of complexity into the modeling strategy, as both WMLES and DDES branches are added to the code. The length scale l_{IDDES} is modified to accommodate these two branches, defined as

$$l_{\text{IDDES}} = \tilde{f}_d \cdot (1 + f_e) \cdot l_{\text{RANS}} + \left(1 - \tilde{f}_d\right) \cdot l_{\text{LES}} \tag{10}$$

The two blending functions complete the model. The first blends between DDES and WMLES and is defined as

$$\tilde{f}_d = \max[(1 - f_{dt}), f_b]. \tag{11}$$

The second blending function is aimed at preventing the excessive reduction in the RANS Reynolds stresses, which is a known problem between RANS and LES interfaces. This elevating function is considered instrumental in combating log-layer mismatch and is defined as

$$f_e = f_{e2} \cdot \max[(f_{e1} - 1.0), 0.0]. \tag{12}$$

Here, f_{dt}, f_b, f_{e1}, and f_{e2} are functions to switch between the WMLES and DDES. For a full derivation of these blending functions, the reader is encouraged to view the work of Shur et al. [23] and Gritskevich et al. [24].

In summary, there are three switches that occur within the IDDES simulation: between RANS and LES, between k-ω and k-ε, and between WMLES and DDES. The switch between RANS and LES is covered by the length scale in Equation (6), which depends on the k and ω but is primarily controlled by Δ, that is, the mesh spacing, with LES occurring where the mesh is sufficiently fine. Coupled with this, an upwind scheme is applied for RANS simulations that is too dissipative for LES. Thus, there is an associated switch from upwind to central difference, following Equation (1) [27]. The switch between k-ω and k-ε is determined by F_1 in Equation (7), with a full discussion and definition provided in [10].

The switch from this DDES model and WMLES is the final component of IDDES. This is performed through Equations (10)–(12), with full details provided in [23,24].

2.2. Domain and Boundary Conditions

The computational domain for this study is shown in Figure 1, following the experimental setup in [1–3]. The triangular flame holder, with a length L = 0.04 m at its rear face, spans the computational domain. The inlet is placed 25 L upstream of the $X = 0$ L position and is set as a velocity inlet with an average velocity of 16.6 m/s. The actual velocity distribution and turbulence profiles of the inlet are obtained from a precursor RANS simulation of a channel flow to match the experimental domain length. The outlet is placed 25 L downstream of the $X = 0$ L position and is set as an atmospheric pressure outlet, with a reference pressure of 101 kPa. The lateral boundaries in the Z direction have a span of 1 L and are set as symmetry planes with a periodic condition. The bottom-to-top boundary is at $Y = 0$ L and $Y = 3$ L, and these are set as no-slip walls, as is the flame holder itself. The boundary layer thickness on the top and bottom walls in the region of the flame holder and where the subsequent results are presented was found to be no greater than 0.4 L. The red lines are the positions where subsequent results are presented.

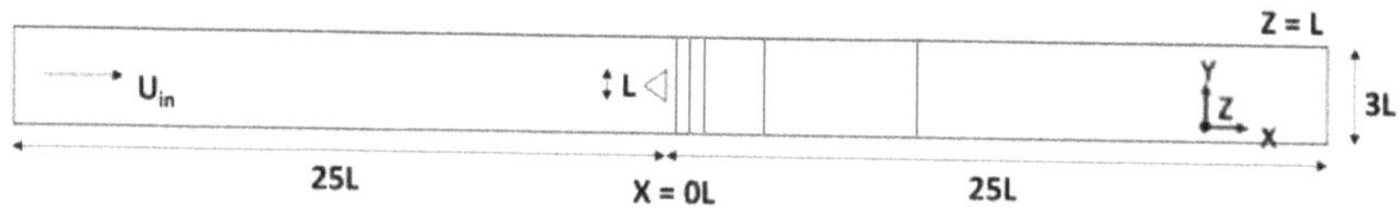

Figure 1. Computational domain. The red lines are the positions where subsequent results are presented.

2.3. Computational Mesh

In this study, all meshes are created within Star CCM+ version 12 using a 3D multi-block-structured grid strategy. Three different grids were designed using the best-practice guidance for DES from Spalart and Streett [29], with the coarsest grid having a cell count of 621,000 cells, a medium resolution grid consisting of 1,276,000 cells, and a fine grid with 2,700,000 cells. The turbulence length scale was obtained from the precursory RANS simulations following the methodology of turbulent flows by Pope [30], and the LES mesh size was selected to be 1/6 of this for the fine grid. The coarser meshes were also considered to evaluate the performance of the IDDES approach in cases with a lower computational cost. A zoomed-in cross-section of the coarsest grid is shown in Figure 2.

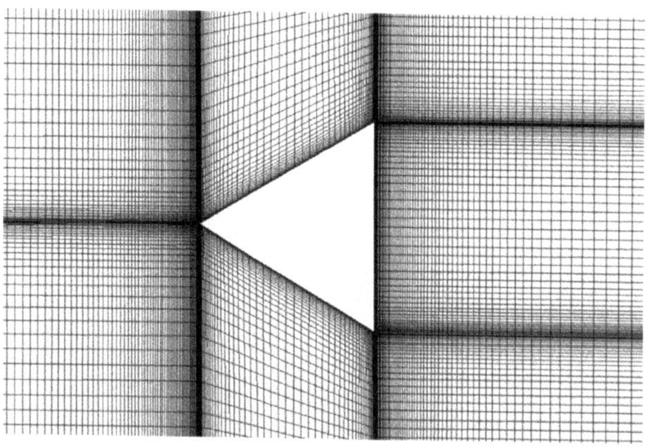

Figure 2. Cross-section of the structured mesh around the flame holder.

For all grids, a total of 45 grid cells are placed in the Z direction to ensure that the grid is as isotropic as possible in the shear layer region. These meshes were designed in order to resolve the boundary layer, with the cells close to the walls being refined to ensure that the boundary layer is suitably captured and a first cell wall normal distance of $y^+ < 1$.

3. Results and Discussion

The results are presented here in terms of the effect of the different models and the computational mesh refinement.

3.1. Effect of Turbulence Modeling

Table 1 shows the grid sizes for the whole mesh for various simulations, along with the predicted length of recirculation for the flame holder and the percentage difference between the experimental results. The experimental data from ERCOFTAC are provided without any experimental uncertainty. Subsequent IDDES results presented for the fine mesh LES simulations are generally considered accurate, are expected to be in excellent agreement with the experimental data, and are often used as a measure to compare approaches such as IDDES when experimental data are not available. However, here, the 'fine' mesh is not sufficiently fine to perform a standard LES simulation and is included for a direct comparison between IDDES and LES on the same mesh. DDES and IDDES simulations are presented on the same fine mesh to enable a comparison between the methods. One cannot mesh converge an LES study in the same way as a RANS simulation. The fine grid here performs LES in all domains, but it is not refined enough in the Y direction to give an accurate result. For the course grid, the resolution is not sufficient to perform a good LES simulation.

Table 1. Case descriptions.

Case Name	Numerical Scheme	Nx	Ny max	Nz	Ntotal ($\times 10^6$)	Lrc	% Difference
Experiment	-	-	-	-	-	1.33	-
LES	BC	450	150	45	2.70	1.14	19.00
DDES Fine	HBC	450	150	45	2.70	1.20	13.26
IDDES Coarse	HBC	270	60	45	0.62	1.20	12.94
IDDES Medium	HBC	360	90	45	1.28	1.27	6.00
IDDES Fine	HBC	450	150	45	2.70	1.16	17.23

When comparing the experimental results [1–3] against the LES, DDES, and IDDES results of this study, it can be seen in Figure 3 that all models manage to capture all of the main flow physics with a very good level of accuracy. Figure 3 also shows the recirculation length at the back of the flame holder, and it can be seen that both DDES and IDDES are nearly identical, which is to be expected as the IDDES model uses the DDES branch of the model. However, there are two main areas of difference between the two, the first being from X/L = 0 to X/L = 1, where the DDES model follows the experimental data slightly better than IDDES, and the opposite can be said for X/L = 4 to X/L = 6, where IDDES is in better agreement with the experimental data. In regards to recirculation length, as shown in Table 1, the DDES and IDDES models show a ~13% and ~17% difference, respectively. This gives the IDDES model a 4% overprediction in the recirculation length compared with the simpler DDES code. The LES model underpredicts the recirculation length by 19% compared with the experimental results; this is most likely related to the inlet conditions of the synthetic turbulence generator used and an under-resolved mesh that would not be considered suitable for a typical LES simulation.

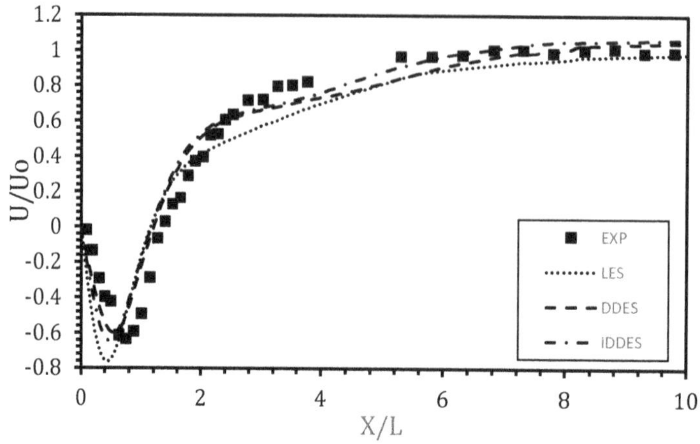

Figure 3. Time-averaged velocity at the centerline in the streamwise direction.

The velocity in the U direction and the resolved stresses for U and V are shown in Figures 4–6, respectively. These profiles show that IDDES and DDES have near identical results, which is to be expected, as discussed previously, and both match the experimental data with a very good level of accuracy. The LES results in these figures show that there is not a great difference between the hybrid modeling and the LES at this grid resolution.

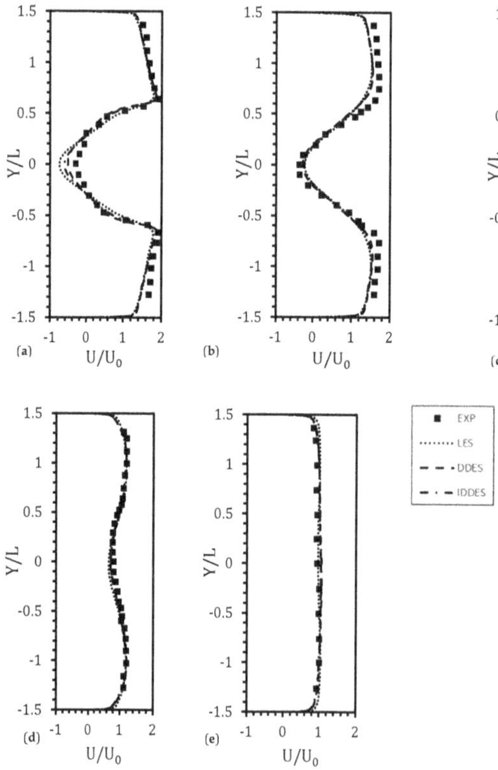

Figure 4. U velocity: (**a**) 0.375 L; (**b**) 0.95 L; (**c**) 1.53 L; (**d**) 3.75 L; (**e**) 9.4 L.

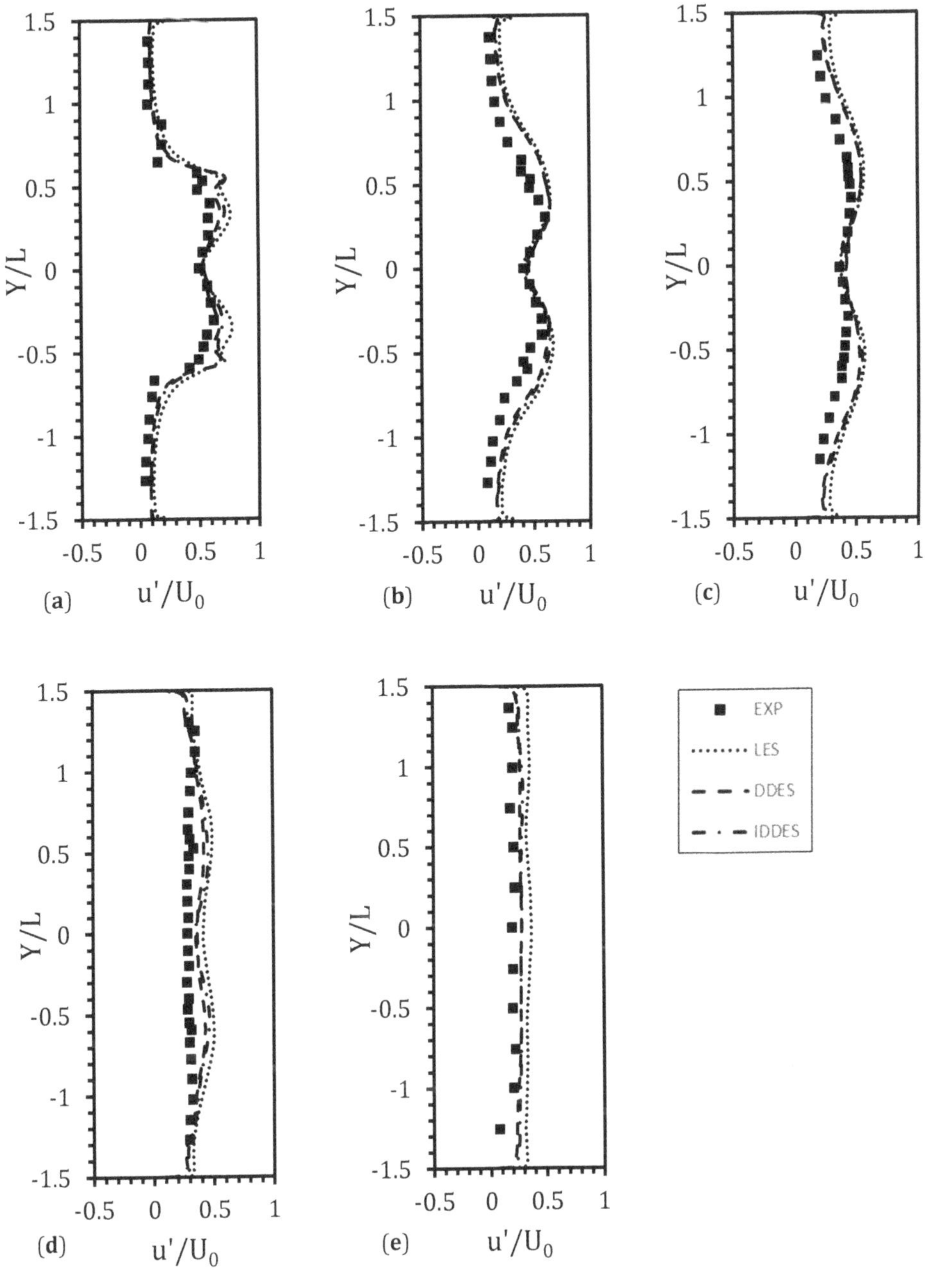

Figure 5. Velocity fluctuations (u′): (**a**) 0.375 L; (**b**) 0.95 L; (**c**) 1.53 L; (**d**) 3.75 L; (**e**) 9.4 L.

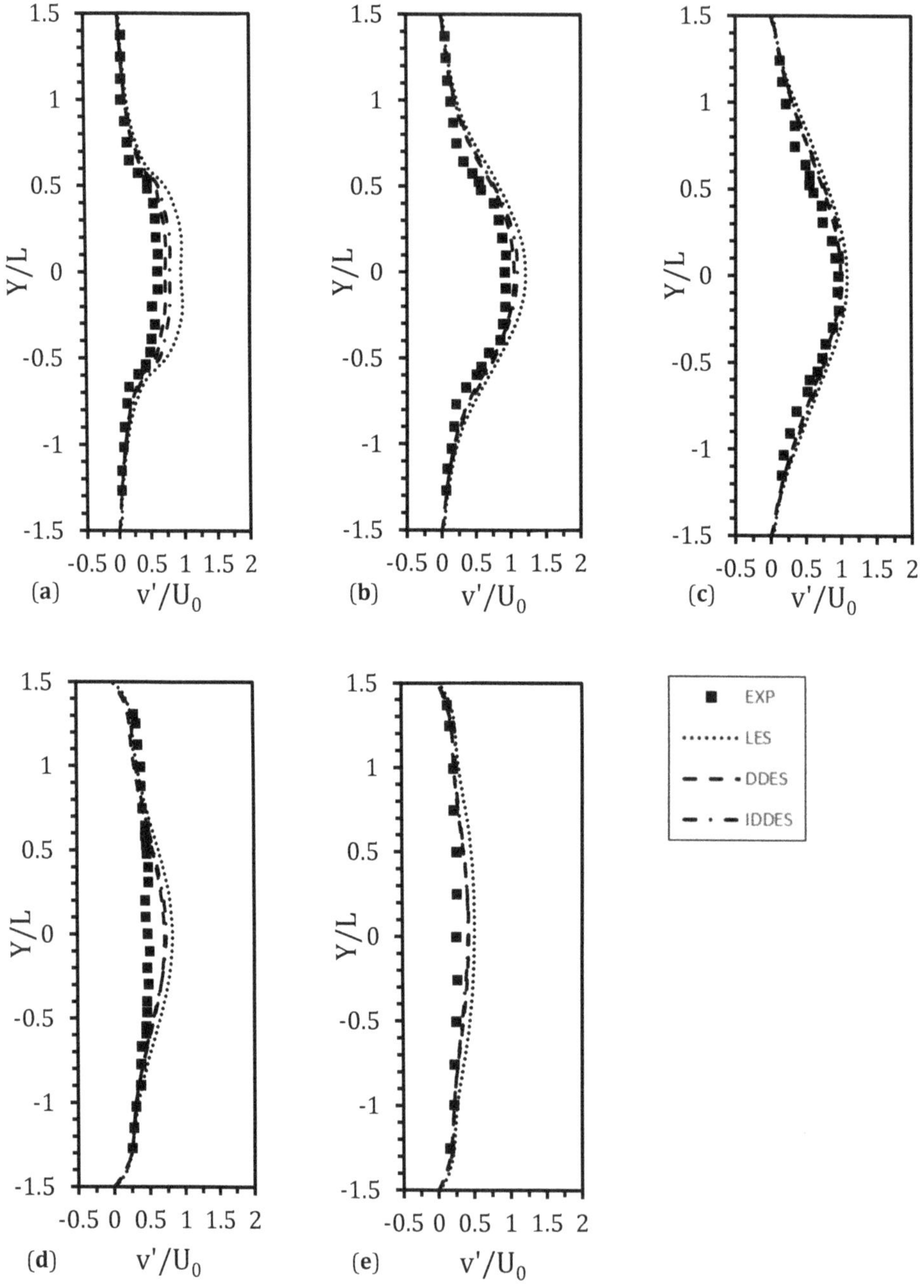

Figure 6. Velocity fluctuations (v'): (**a**) 0.375 L; (**b**) 0.95 L; (**c**) 1.53 L; (**d**) 3.75 L; (**e**) 9.4 L.

The Q criterion for both DDES and IDDES is shown in Figures 7 and 8. There is a noticeable difference in the breakdown of the turbulence, with the wake of the DDES model showing more turbulent structures than that of IDDES, and the vortex cores are being sustained further downstream of the triangular bluff body. Figures 9 and 10 show the

RANS-LES and upwind-bounded central blending functions for the DDES and IDDES simulations. When looking at the numerical scheme and how it is used within the domain, it can be seen by comparing the DDES in Figure 9 and IDDES in Figure 10 that with IDDES, there is more upwind within the wake region downstream, and this is causing the vortex cores to be damped out. It is also shown that the regions of LES and RANS have been calculated very differently: with IDDES, there is a large clear area of LES in the wake, which is being solved predominantly by an upwind scheme. The DDES model seems to be able to pair the numerical scheme with the blending function better than IDDES.

Figure 7. Iso surface of the vortex structures for DDES (**a**) around the flame holder (**b**) domain.

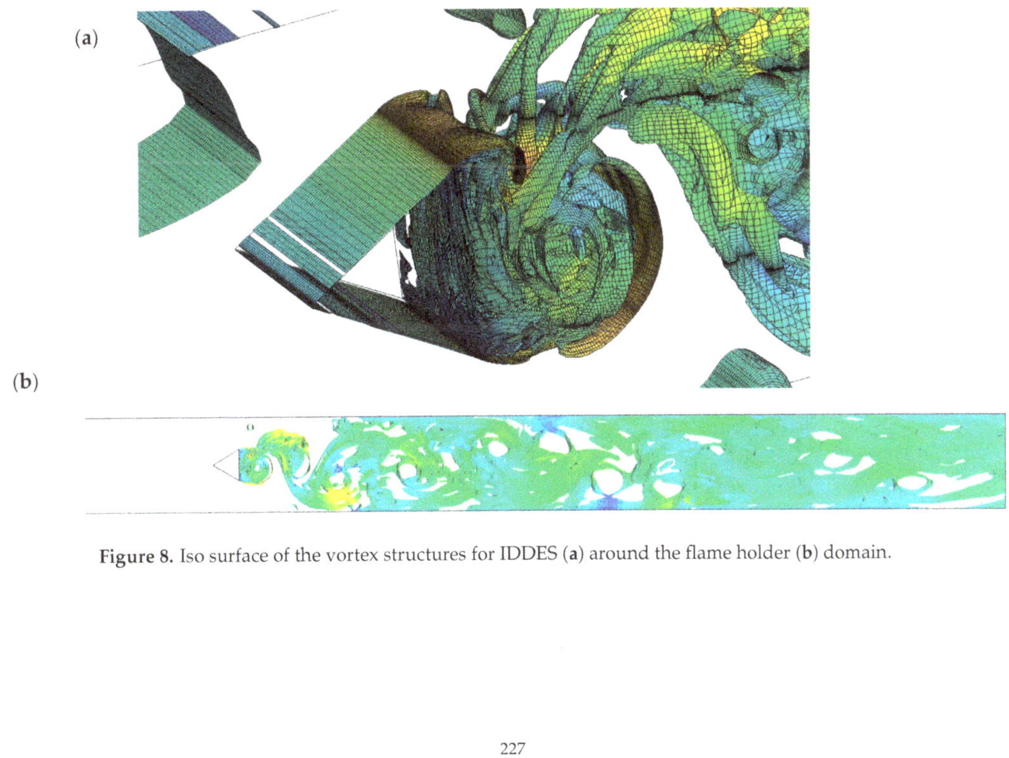

Figure 8. Iso surface of the vortex structures for IDDES (**a**) around the flame holder (**b**) domain.

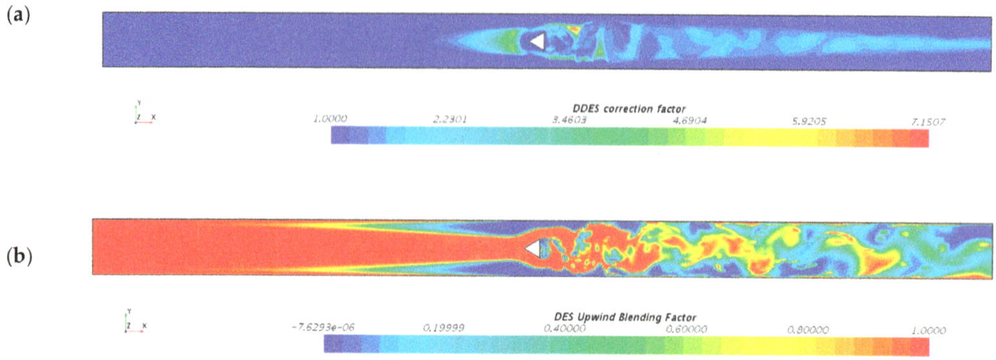

Figure 9. Blending factor for DES: (**a**) DDES correction factor, where 1 = RANS and >1 = LES; (**b**) upwind blending factor, where 1 = upwind and 0 = bounded central.

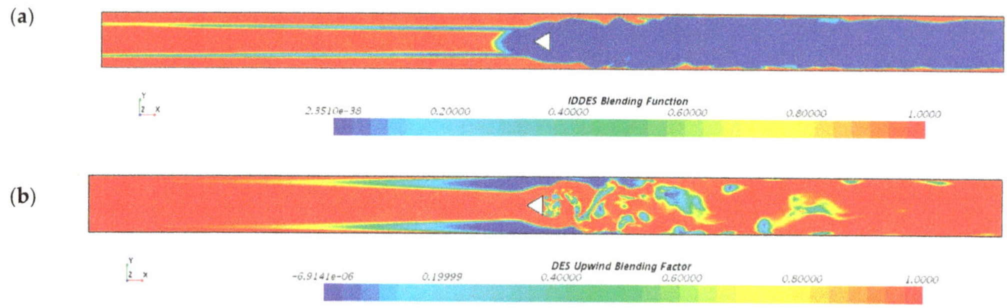

Figure 10. Blending factor for IDES: (**a**) IDDES blending function, where values close to 1 = RANS and values close to 0 = LES, with intermediate values blended; (**b**) upwind blending factor, where 1 = upwind and 0 = bounded central.

3.2. *Effect of Computational Mesh*

Three different computational meshes were tested to ascertain the performance of the IDDES model. These different mesh cases are shown in Table 1 and in the streamwise velocity results in Figure 11. It can be seen that all three meshes perform reasonably well and manage to capture the general flow physics. The fine mesh is the only one to reach the correct velocity for the minimum behind the bluff body, whereas the medium and coarse meshes overpredict this by ~30%, which would not be considered acceptable. We also note that refining the grid does not necessarily improve the results obtained, for example, the recirculation length in Table 1 is most closely predicted using the medium mesh. This could be due to the fine mesh not being sufficiently fine for a good-quality LES simulation, while the medium mesh restricts the regions where LES is performed or the inlet conditions of the synthetic turbulence generator.

Figures 12–14 show that for all meshes, there is very good agreement for the streamwise velocity and resolved stresses, with the only discrepancy being that the fine mesh overpredicts the resolved stresses on most of the profiles. Table 1 shows that the recirculation length for the course mesh is approximately the same as the DDES result, although the DDES result was able to better capture the magnitude of the velocity, which contributes to the overall accuracy. The medium mesh has the best recirculation length with a 6% difference from the experiments, but as with the coarse mesh, it still suffers from underprediction of the magnitude of the maximum velocity. The Power Spectral Density (PSD) of all three meshes is shown in Figure 15.

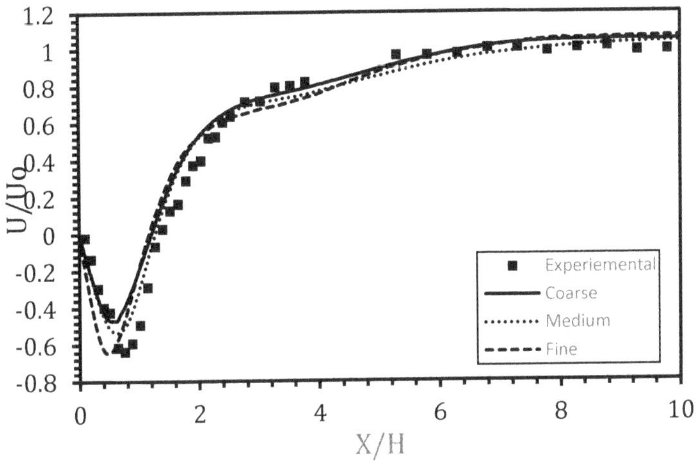

Figure 11. Time-averaged velocity at the centerline in the streamwise direction.

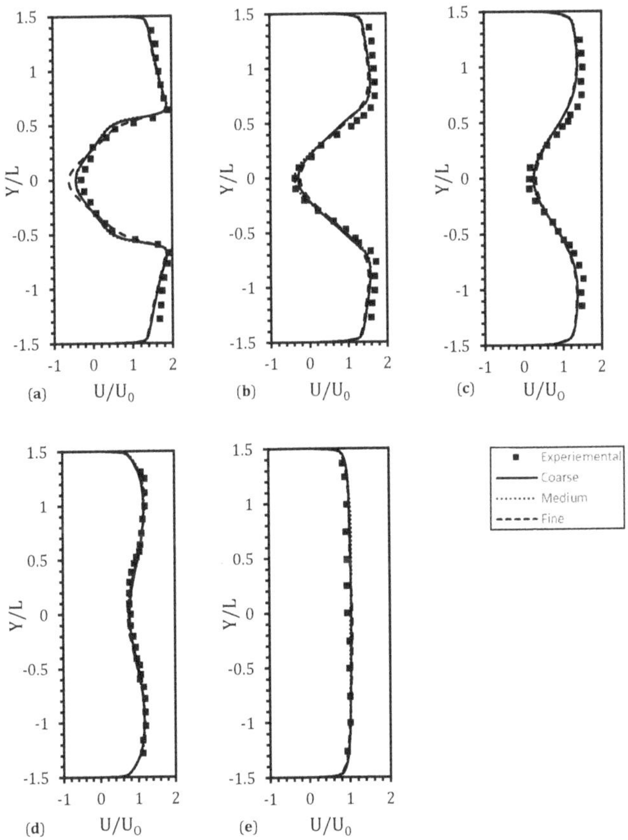

Figure 12. U velocity: (**a**) 0.375 L; (**b**) 0.95 L; (**c**) 1.53 L; (**d**) 3.75 L; (**e**) 9.4 L.

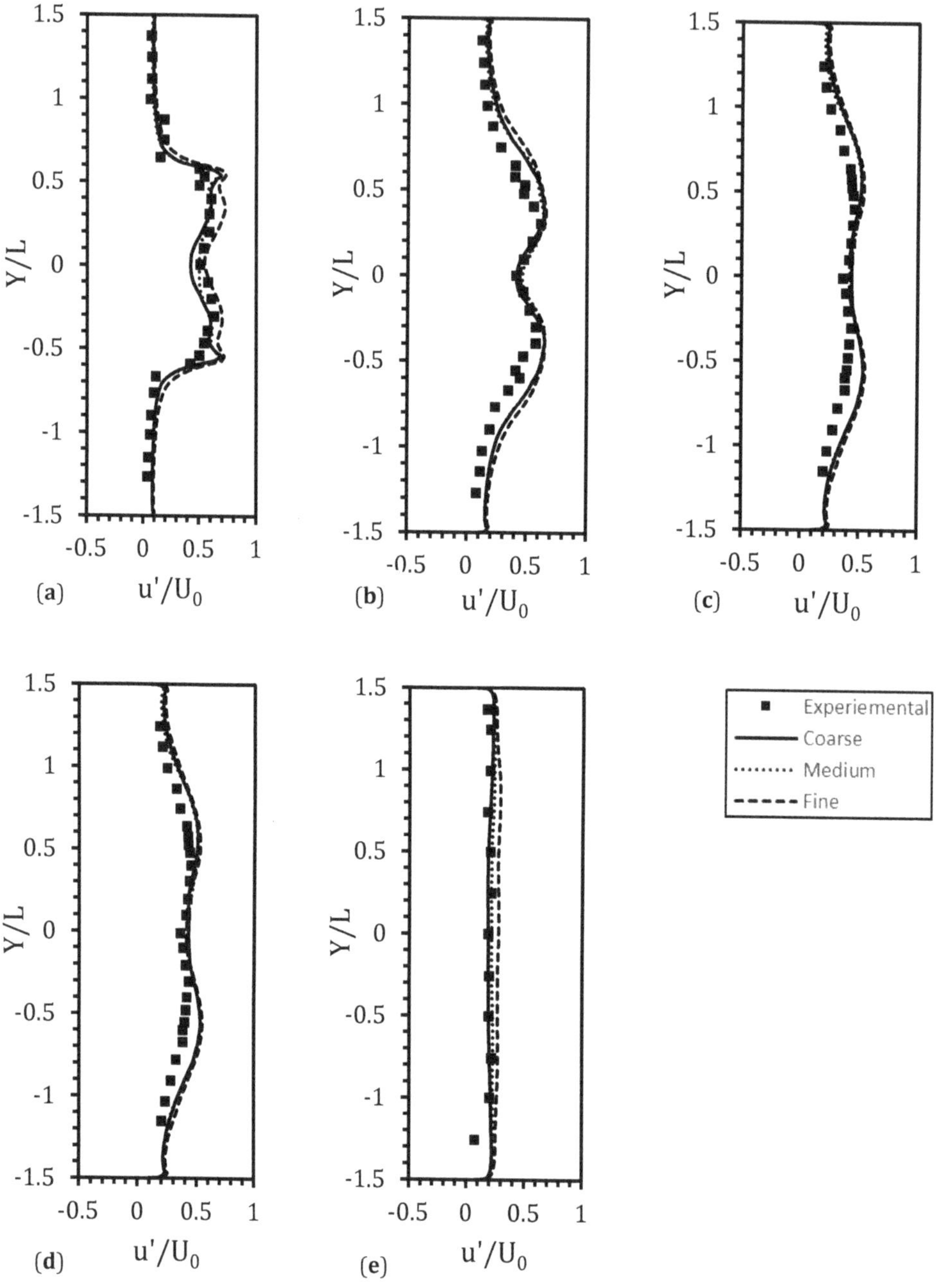

Figure 13. Velocity fluctuations (u′): (**a**) 0.375 L; (**b**) 0.95 L; (**c**) 1.53 L; (**d**) 3.75 L; (**e**) 9.4 L.

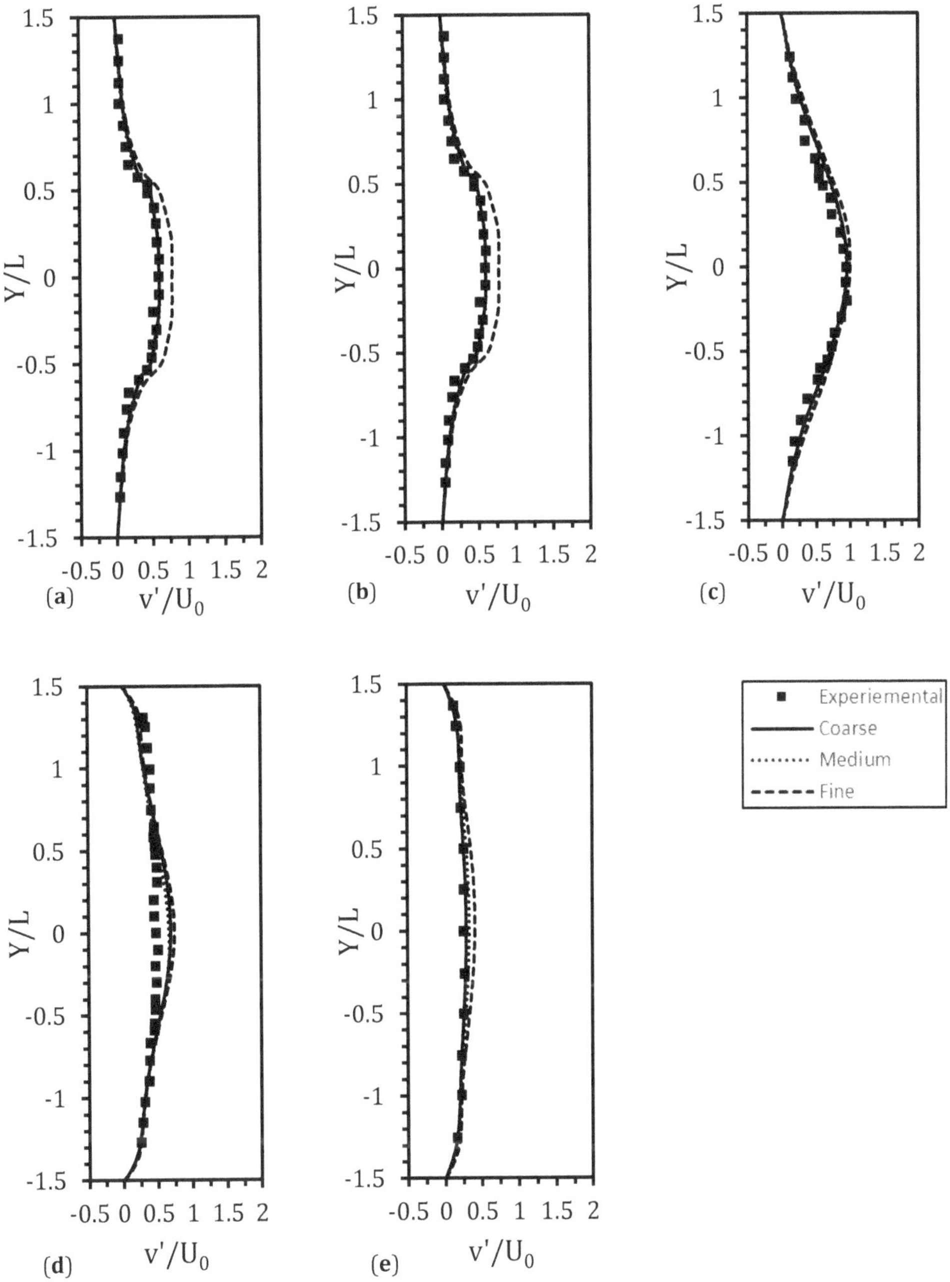

Figure 14. Velocity fluctuations (v′): (**a**) 0.375 L; (**b**) 0.95 L; (**c**) 1.53 L; (**d**) 3.75 L; (**e**) 9.4 L.

(a)

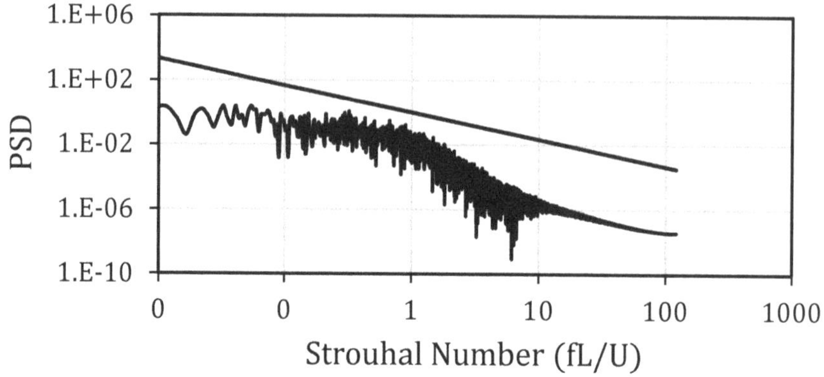

(b)

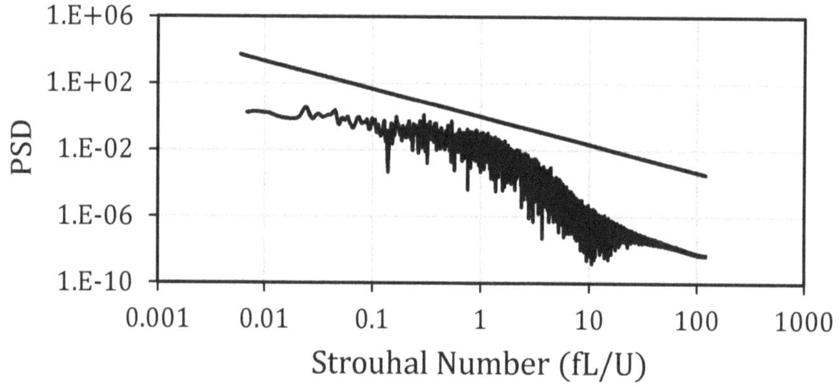

(c)

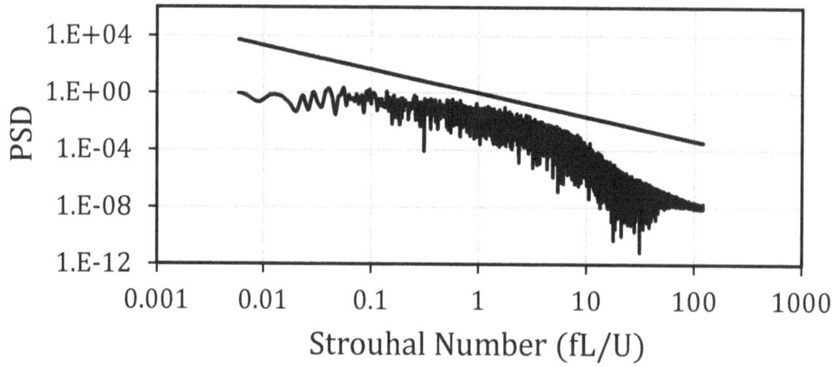

Figure 15. Power Spectral Density: (**a**) coarse; (**b**) medium; (**c**) fine.

In terms of computation effort required, there has always been a concern that as these types of models become more complex, they will inherently be more expensive to run. Table 2 shows typical run times for each code against a base RANS simulation. When comparing just the solver elapsed time per time step, it can be seen that there is a ~5% difference in DDES with a typical RANS solution and a ~8% difference with IDDES, which, considering the extra information included within the solution, could be considered a worthwhile expense. In regards to the difference between DDES and IDDES, the ~3% difference does not improve the solution, and in the test case, the DDES model outperforms IDDES in speed and accuracy. The LES results are considerably better than those of all models due to its simplicity, but the overall accuracy on this mesh is not considered to be sufficient, and the mesh would need to be further refined to improve the results to an acceptable level. Additionally, the time step would have to be reduced as a consequence, both of which will have the obvious effect of increasing the computational time.

Table 2. Computational time for different cases.

Case Name	Solver Elapsed Time per Time Step (s)	Total Time (hrs)	% Difference to RANS
Typical RANS	6.1	113	-
DDES	6.4	119.5	5.4
IDDES	6.6	123	8.1
LES	4.6	86	−31.4

4. Conclusions

The k-ω SST IDDES model used within this study has been shown to perform well against the experimental results, with very good agreement for all the main flow features. The IDDES model overpredicts the recirculation length by ~4% compared with DDES and is ~3% slower in computing the solution.

The testing of different grids shows that IDDES can still perform relatively well on a coarse mesh, with a generally good agreement found with the velocity and Reynolds stresses. The recirculation length shows an underprediction of the magnitude of the time-averaged velocity in the streamwise direction through the center of the domain as the grid is coarsened. We also note that the recirculation length is most accurately predicted by the medium mesh, rather than the fine mesh. This may be due to the different balance of LES and RANS on the two meshes or the synthetic turbulence in the inlet conditions; however, fully explaining this is a topic for future research.

The IDDES model is considered the default within Star CCM+ version 12, and although it can compute the solution with relatively good accuracy for this test case, it is not an improvement over the DDES model, and careful model selection should be considered before implementing a DES strategy.

Author Contributions: Conceptualization, M.R.M., J.K., and J.M.B.; methodology, M.R.M., J.K., and J.M.B. software, M.R.M.; validation, M.R.M.; formal analysis M.R.M., J.K., and J.M.B.; investigation, M.R.M.; resources, J.K. and J.M.B.; data curation, M.R.M.; writing—original draft preparation, M.R.M.; writing—review and editing, J.K. and J.M.B.; visualization, M.R.M.; supervision, J.K. and J.M.B.; project administration, J.K. and J.M.B. All authors have read and agreed to the published version of the manuscript.

Funding: This research received no external funding.

Data Availability Statement: Dataset available on request from the authors.

Acknowledgments: The authors would like to thank Siemens for the use of their code Star CCM+ version 12. Numerical computations were performed on the Sciama High-Performance Compute (HPC) cluster, which is supported by the ICG, SEPNet, and the University of Portsmouth.

Conflicts of Interest: The authors declare no conflicts of interest.

fluids MDPI

Article

The Use of Computational Fluid Dynamics (CFD) within the Agricultural Industry to Address General and Manufacturing Problems

Navraj Hanspal [1] and Steven A. Cryer [2,*]

[1] Crop Protection Discovery & Development, Corteva Agriscience, Indianapolis, IN 46268, USA; navraj.hanspal@corteva.com
[2] Advanced Modeling and Artificial Intelligence, Corteva Agriscience, Indianapolis, IN 46268, USA
* Correspondence: steven.cryer@corteva.com; Tel.: +1-765-481-7398

Abstract: Computational fluid dynamics (CFD) is a numerical tool often used to predict anticipated observations using only the physics involved by numerically solving the conservation equations for energy, momentum, and continuity. These governing equations have been around for more than one hundred years, but only limited analytical solutions exist for specific geometries and conditions. CFD provides a numerical solution to these governing equations, and several commercial software and shareware versions exist that provide numerical solutions for customized geometries requiring solutions. Often, experiments are cost prohibitive and/or time consuming, or cannot even be performed, such as the explosion of a chemical plant, downwind air concentrations and the impact on residents and animals, contamination in a river from a point source loading following a train derailment, etc. A modern solution to these problems is the use of CFD to digitally evaluate the output for a given scenario. This paper discusses the use of CFD at Corteva and offers a flavor of the types of problems that can be solved in agricultural manufacturing for pesticides and environmental scenarios in which pesticides are used. Only a handful of examples are provided, but there is a near semi-infinite number of future possibilities to consider.

Keywords: agriculture; CFD; manufacturing; general R&D

1. Introduction

Engineering principles have traditionally been divided between experimental and theoretical disciplines, with the former being dominated at agrochemical companies such as Corteva, Syngenta, BASF, and FMC. Experiments are often considered and required by regulatory bodies to understand objective details, but these experiments are often expensive, risky, and time consuming. Modeling and simulations based upon science, math, and computer science began in the 1950s and their use has been increasing recently due to the rapid evolvement of CPUs. Even today's computer desktops/laptops, running more than a single CPU (multi-processers), would have been classified as supercomputers several decades ago (a supercomputer can be thought of as a machine/server with multiple CPUs grouped into tens to thousands of computing nodes). This is often convenient for solving real-world problems that can be easily solved by theory vs. often expensive experimental analysis.

Computational Fluid Dynamics (CFD) is a modeling and simulation technique based upon fluid flow modeling and has been around for the past 50 years, providing a numerical solution to the governing equations of conservation of mass, momentum, and energy (Equations (1)–(3)). These governing conservation equations (e.g., theories) are solved numerically using various techniques such as Large Eddy simulation [1–6] and can offer insight when no experimental knowledge is available or is difficult to measure (i.e., discrete element modeling, finite difference, volume of fluid (VOF) method, etc.). Today's CPU horsepower has been used to solve complex problems that were difficult to solve only a decade ago. Thus, problems involving fluid flow, heat and mass transfer, reactive flow, multi-phase flow and combustion can now be simulated via various computational models. Some real-world examples of CFD use include multiphase flow (e.g., the breakup of formulations being sprayed in air, waves breaking on a beach, etc.), the design of airplanes/cars where drag can influence design, etc. CFD is often used when experiments are costly or cannot be carried out (i.e., experiments on humans, catastrophic events like sabotage to chemical facilities, manufacturing where changes in input conditions often are limited w/o interrupting production, and so on). In the EU and USEPA, even low mammal toxicity testing is coming under fire, so a future without any animal testing can be achieved (EPA by 2035, and the EU already banned testing for cosmetics in 2013). CFD is an ideal pathway used to simulate these results without the need for human and animal testing.

CFD replaces the continuous problem domain with a discrete domain using a grid. Thus, numerical methods are used to convert these partial differential governing equations to a coupled algebraic set of equations. Typically, the governing partial differential equations of fluid mechanics are first discretized into algebraic equations using specialist numerical techniques/methods such as Finite Differences, Finite Elements, Finite Volumes, Lattice Boltzmann, Smoothed Particle Hydrodynamics, etc. Specialist matrix solvers which can be both iterative (i.e., Conjugate Gradient) or direct (Multi-Frontal) are then used to solve this algebraic system of equations. Each variable is defined at every point in the continuous domain, while in the discrete domain, each variable is defined only at the grid points, as in Figure 1. CFD directly solves for flow variables only at the grid points, but interpolation techniques are used for values between the grid points. The discrete variables become a large system of coupled algebraic equations that involves iterative calculations amenable for computer computation. The reader can view many introductory writings on CFD found elsewhere [7–14]. Solving CFD problems involves three major steps that include (i) preprocessing (creation of geometry and meshing), (ii) solving the governing equations over this meshed model, and (iii) post-processing results to be analyzed. An example is defined at every point in the continuous domain converted into a discrete domain.

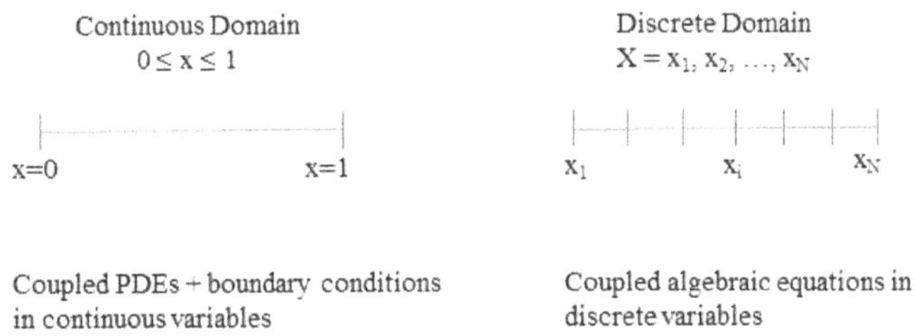

Figure 1. Example continuous to discrete domain in 1 dimension (i.e., discretizing).

2. Materials and Methods

Conservation of mass can be expressed as

$$\frac{\partial C}{\partial t} + \nabla \cdot (uC) = \nabla \cdot (D\nabla C) - \mu C \tag{1}$$

C = Concentration;
D = Diffusion coefficient;
u = Velocity vector.
The conservation of momentum equation can be written as

$$\underbrace{\rho(\frac{\partial u}{\partial t} + \mathrm{u} \cdot \nabla \mathrm{u})}_{1} = \underbrace{-\nabla p}_{2} + \underbrace{\nabla \cdot [\mu(\nabla u + (\nabla u)^T) - \frac{2}{3}\mu(\nabla \cdot u)]I}_{3} + \underbrace{F}_{4} \tag{2}$$

where (1) represents inertial forces, (2) pressure forces, (3) viscous forces, and (4) external forces applied to the fluid (i.e., gravity).

μ = viscosity;
V = kinematic viscosity = μ/ρ;
u = V = velocity vector;
ρ = fluid density;
I = identity matrix;
F = body force term;
p = pressure;
T = temperature;
τ = shear stress tensor;
D/Dt = material derivative (the rate of change following a moving fluid parcel);
k = material thermal conductivity;
$e_t = C_v\,T + u^2/2$ = specific internal energy (internal energy per unit mass);
f_G = external body forces acting on the fluid (such as gravity);
S_g = generation source term for energy (radiation effects, thermal heating from electrical current, etc.).

The conservation of energy can be written as

$$\partial\,(\rho\, e_t)/\partial t + \nabla\cdot \mathrm{u}\,(\rho\, e_t + p) = \nabla\cdot[k\,\nabla T + \tau\cdot \mathrm{u}] + S_g \tag{3}$$

and is used to generate temperatures for the scenario being investigated. Each term in the continuity, momentum, and energy equation (e.g., Equations (1)–(3)) can be found in any elementary treatise on the subject (Bird, Stewart, Lightfoot, 1960 [15]). All three governing equations are numerically solved to generate solutions using CFD. There are many opensource (free) CFD packages, such as OpenFOAM, SU2, Palabos, Fire Dynamics Simulator, and MFIX (OpenFOAM is most popular with users, with a tutorial, extended code, and programmers guide), along with many commercial packages users can utilize. A major advantage of commercial CFD packages is the ease of use via graphical user interfaces for pre- and post-processing, which reduces the hurdle associated with learning the package. CFD code became common during the 1950s–1970s. Computers necessary to run such code cost ~USD 10 million (2024 dollars) and key players were the Los Alamos National Lab, Douglas Aircraft, and NASA, with code written to model specific (not generic) examples. Off-the-shelf commercial code become available during the 1980s–2010s, and many examples of this work used the commercial CFD package COMSOL Multiphysics (https://www.comsol.com, accessed on 11 August 2024), but other packages exist, such as ANSYS FLUENT (https://www.ansys.com, accessed on 11 August 2024), STAR-CCM+ (https://plm.sw.siemens.com/en-US/, accessed on 11 August 2024), Altair (https://www.altair.com/altair-cfd, accessed on 11 August 2024), etc. These codes are designed to run on CPUs and often require extensive training and project time (weeks/months) and computers to run this software, costing ~USD 1–10 million (in 2024 dollars). More recently, GBU execution has evolved (2010s–present) and can be orders of magnitude faster than CPU code via software such as MstarCFD (https://mstarcfd.com accessed on 11 August 2024), Palabos (University of Geneva, https://palabos.unige.ch/publications/ accessed on 11 August 2024), and LB-2D (Harvard University, Randles and Kaxiras 2014 [16]). Project time is now measured in hours or days using computers costing ~USD $2000 (in 2024 dollars). Examples of various GBU algorithms are found in Figure 2.

Physics	Governing Equations	Algorithm
Fluid Mechanics	Transient Navier-Stokes	Lattice-Boltzmann[1]
Particle Transport	Newtons Second Law	BVH Trees[2]
Species Transport	Transient Advection-Diffusion	Explicit Coupling[3]
Conduction	Transient Heat Equation	Linear-Neighbor[4]

Figure 2. Examples of various GBU approaches in use. [1] https://www.sciencedirect.com/science/article/pii/S0898122111001064 (accessed on 11 August 2024), [2] https://www.sciencedirect.com/science/article/abs/pii/S092702561930206X (accessed on 11 August 2024), [3] https://papers.ssrn.com/sol3/papers.cfm?abstract_id=4368195 (accessed on 11 August 2024), [4] https://onlinelibrary.wiley.com/doi/abs/10.1002/num.22730 (accessed on 11 August 2024).

3. Results

What follows are specific R&D examples where CFD has been used at Corteva to offer the reader a flavor for the types of simulations and results that can be obtained. This section is broken into two parts, namely (1) General Corteva R&D Examples, and (2) Corteva Manufacturing Examples. It takes an agrochemical company approximately USD 250–300 million and over 10 years of studies to obtain registration of an active ingredient for commercial use. CFD is often used to "fill-in" our understanding of the manufacturing process and environmental fate of a molecule once placed in the environment. The examples provided illustrate the importance of CFD in both R&D and manufacturing at Corteva, and in many cases deal with predictions that could not be measured directly (e.g., testing on humans) or were cost-prohibitive to achieve.

3.1. General Corteva R&D Examples

3.1.1. Potential for Breathing in a Volatile Pesticide

We begin with an R&D example to understand the propensity for humans to breathe in and intake volatile/semi-volatile pesticides. A soil fumigant is a pesticide that when applied to soil forms a gas that subsequently emanates within the soil pores to control pests that ultimately impact crop production and yield. The soil fumigant 1,3-dichloropropene (1,3-D), sulfuryl fluoride (used to treat buildings and grain storage areas), and various metabolites of active pesticides (e.g., IN-VM862, a soil metabolite of Reklemel) are all examples of volatile pesticides. Semi-volatile pesticides include chlorpyrifos, various formulations of 2,4-D, dicamba, etc. The volatile soil fumigant 1,3-dichloropropene (1,3-D) used to be a USD 150 MM a year product for Dow AgroSciences, but went back to the Dow Chemical portfolio following the creation of Corteva in 2019. However, the following example highlights the power of CFD when experiments on humans could not be performed. 1,3-dicloropropene is injected into the ground as a liquid to control nematodes (primarily) and some weed species. The material vaporizes and begins to rise back towards the soil surface, while impacting soil-borne nematodes. There are several techniques to stop the material at the soil surface, such as the use of tarps or some types of chemical and/or an impermeable layer, but the potential of becoming airborne is a real possibility and thus can create potential inhalation risks for humans.

References for three dimensional CFD modeling for inhaled material transport are found elsewhere [17–20]. The following provides an example of the use of CFD for 1,3-D, which was undertaken about the time the product was returned to Dow when Corteva was formed. It is based upon an award-winning paper by Corley et al., 2012 [21] (Pacific Northwest National Lab). These researchers investigated the inhalation of a volatile material when smoking cigarettes. Files used by Corley et al., 2012 [21] were converted from OpenFoam (a freeware CFD package) to COMSOL Multiphysics. Now, we assume that a former Corteva product (1,3-dichloropropene) resulting from agricultural use was inhaled instead of cigarette smoke. COMSOL was used for human and rat inhalation scenarios; see Figures 3 and 4. This analysis bypassed the need for lower tier (and higher safety factor) approaches and allowed Corteva to assess the risk directly associated with inhaled vapor of 1,3-dichloropropene. Computed tomography (CAT) scans for a human and rat (corpses) were used, with a cylinder placed around the mouth/nose for CFD input; see Figure 3. The assigned breathing rates for a typical human and rat were used, partition coefficients and physicochemical properties for 1,3-D, etc., were assumed, and the determination of where 1,3-D was absorbed in the body was determined. The regional uptakes of 1,3-D in the human respiratory track were sensitive to air concentration levels, airflow rates, air-tissue partition coefficients, tissue thickness, and metabolism rates and occurred early in the process, well before reaching the lung. CFD predictions of 1,3-D exposure and uptake to humans via the inhalation exposure route eliminated the simplistic back-of-the-envelope exposure approaches (with safety factors) often used to estimate dose and exposures for inhalation gas dosimetry, while providing a key scientific approach for the inhalation dosimetry of gasses. It was found that 1,3-D is absorbed rapidly early in the breathing process (e.g., in the trachea) and does not reach deep into the lungs when predicted losses of 1,3-D from the soil (vapor phase) are anticipated; see Figure 4. Computational Fluid Dynamics eliminates the need for lower-tier (and higher safety factor) approaches to allow the ability to predict risk associated with inhaled vapor and aerosol exposures of organic chemicals.

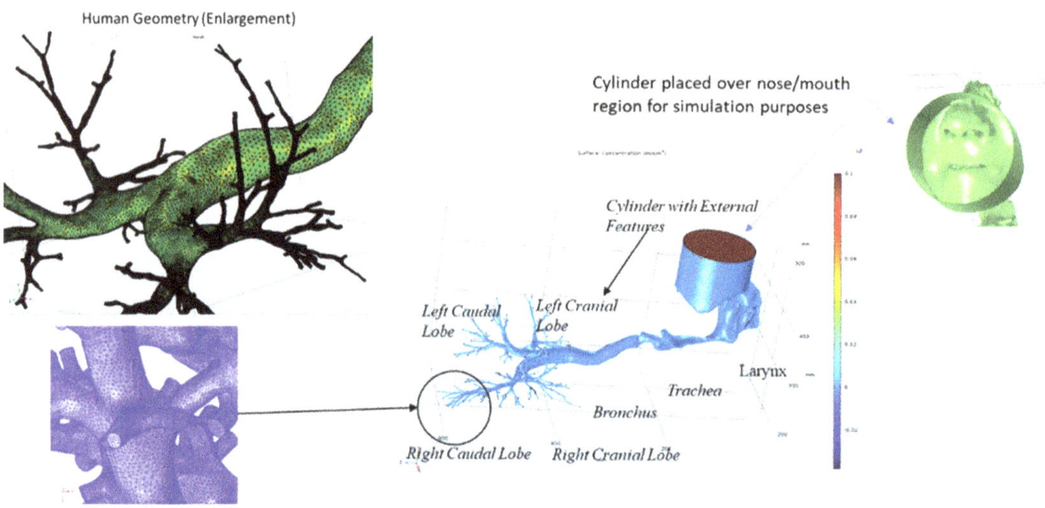

Figure 3. Example of using CFD modeling when experiments cannot be performed (human/rat breathing 1,3-D-laden air required for reregistration of 1,3-D).

3.1.2. Reactor Design

Several decades ago, CFD was used to investigate whether process conditions for the dichlorophenol reactor in Midland, MI (a precursor to 2,4-D synthesis) could be operated to enhance dichlorophenol yield. Several findings from this work summarized the filling of the reactor with fluid, the simulation of the transient liquid/gas interface as a steady state was considered, and operating conditions were looked at for the recycle stream as a product to be recycled back into the reactor to optimize performance; see Figure 5. As the velocity stream is slowed (slow recycle), the recycle region within the reactor is minimized, thus giving insight into the ideal conditions for operating the reactor.

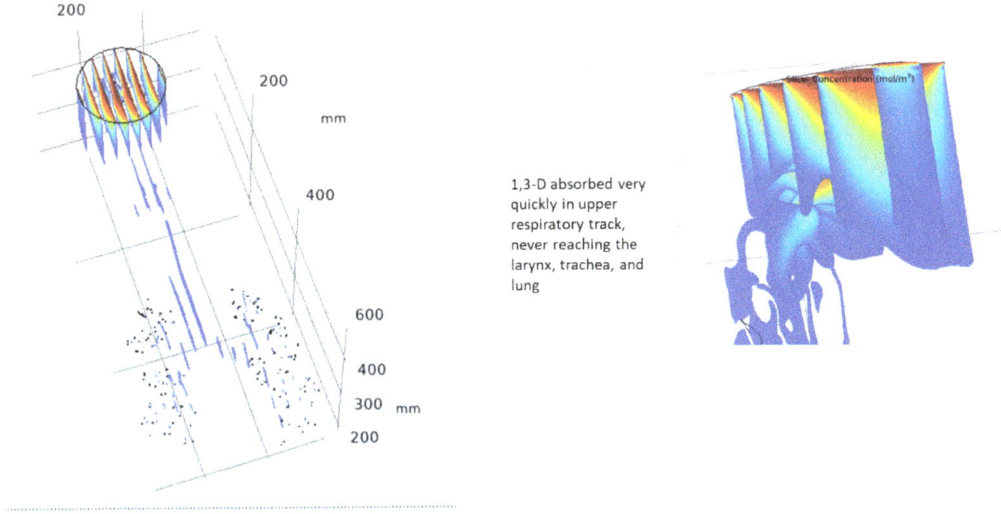

Figure 4. COMSOL CFD results [human, 1,3-D concentration results (mol/m^3)].

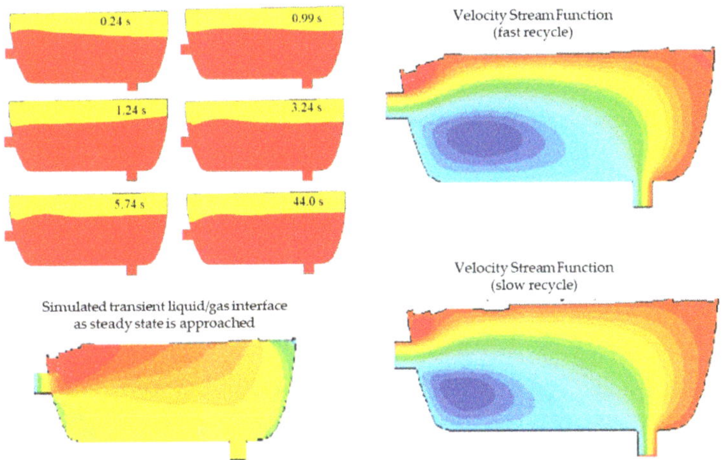

Figure 5. CFD results using the FLUENT software package to simulate the optimal practices for filling and operating the dichlorophenol reactor.

3.1.3. Home Fumigation

Sulfuryl fluoride is a gas that is pumped into a house/structure for around a 24 h period to kill insect pests. The structure is often covered with tarps during fumigation to keep the fumigant from leaking out. Following fumigation, the structure is vented to the outside using fans, open doors/windows, etc., so residents and workers can reenter safely. Several constraints need to be heeded. First is that the venting to the outside is slow enough that the outside air concentrations are below threshold levels for the safety of residents/passersby in proximity, and the second deals with when residents can safely return to the structure post fumigation and ventilation. CFD is an ideal tool to address this scenario; see Figure 6.

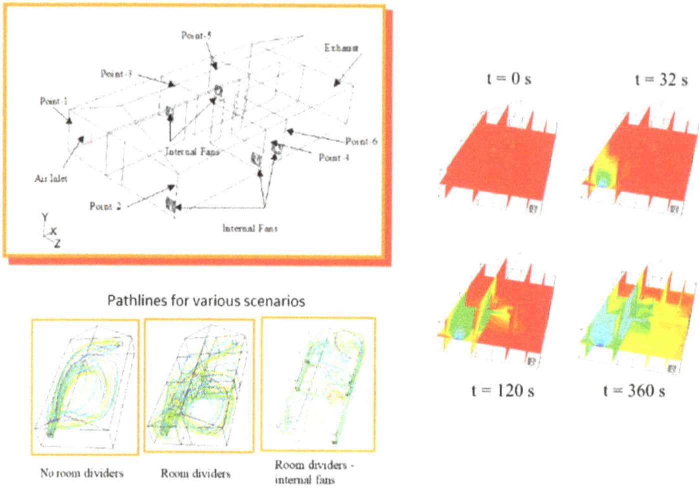

Figure 6. Home aeration following pest fumigation (e.g., when can a resident safely reenter a treated structure?) Graphic on the right gives 2D slices in the 3D simulation domain as the home is aerated with fresh air.

Outside air concentrations were simulated following mill fumigation using Sulfuryl Fluoride, where the gas was released into the environment; see Figure 7. Best management practices could be deciphered to minimize bystander exposure during the clearing process for the mills, such that operators could safely return [22,23]. The work documented in this Biosystems Engineering paper, "Estimating Outside Air Concentrations Surrounding Fumigated Grain Mills," won the EurAgEng outstanding paper award from Biosystems Engineering in 2008, the only nonacademic paper to do so [24].

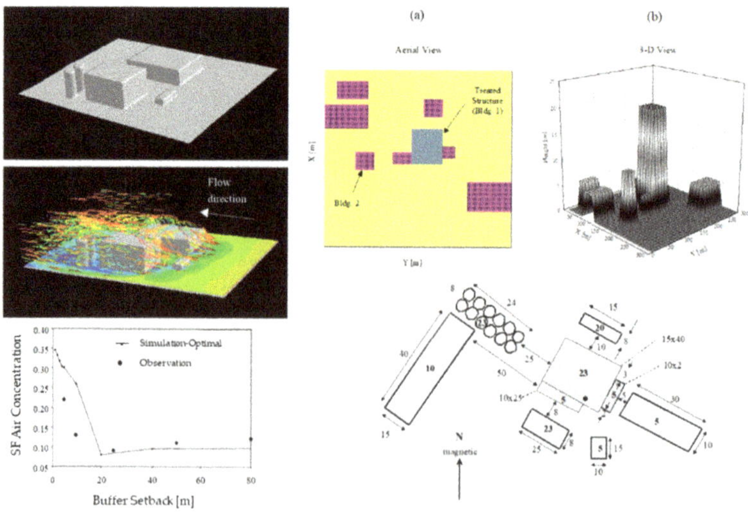

Figure 7. Mill fumigation and aeration with potential exposure to sulfuryl fluoride, with results compared to experimental results for different large-scale milling facilities. (**a**) Aerial View, (**b**) 3-D View.

3.1.4. Quantifying Efficacy and Bystander Exposure from Expanded Uses of Sulfuryl Fluoride

Grain is often stored on the ground and covered with a tarp for extended periods of time before being taking to market in regions such as Australia and New Zealand. A fumigant such as sulfuryl fluoride is often used to control pests over this time interval. An applicator walks along the tarp and injects fumigant into the grain below and repeats this process until the entire pile has been treated. CFD was used to simulate the resulting concentration profile in the grain as a function of depth and location, such that advice about application practices (and amount of chemical used) can be made to maximize efficacy; see Figure 8.

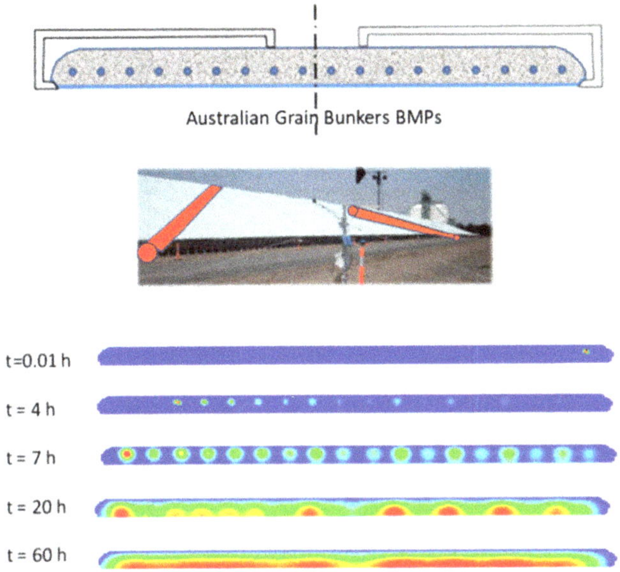

Figure 8. An Australian grain bunker treated with sulfuryl fluoride to control insect pests, where CFD was used to simulate the amount, location, and duration (e.g., best management practices (BMP)) for the pesticide such that efficacy estimates could be made.

3.1.5. Drift Loading to Stream and Backwater Regions: Refinement of Acute Exposure

Salmon are an endangered species in California, USA. In many areas of California, agricultural fields often border streams where small salmon live during part of their lifecycle, before heading back to the ocean to grow, ultimately returning to these streams to spawn and continue this lifecycle. There is a concern that pesticide sprays from neighboring farms can enter these surface waters and injure the small salmon fry that live there. This work looked at three different regions within the Mokelumne River that have backwater regions where these young salmon fry live. Thus, work was undertaken to see if a stream received pesticide loading from a spray drift, and then to see how long the concentrations remained in the stream, especially in these backwater regions where the salmon live. CFD was an ideal tool to simulate the residence time in this California stream for a pesticide to be cleared following the original insult; see Figure 9. These transient concentrations could be compared with acute toxicity data for the organism in question.

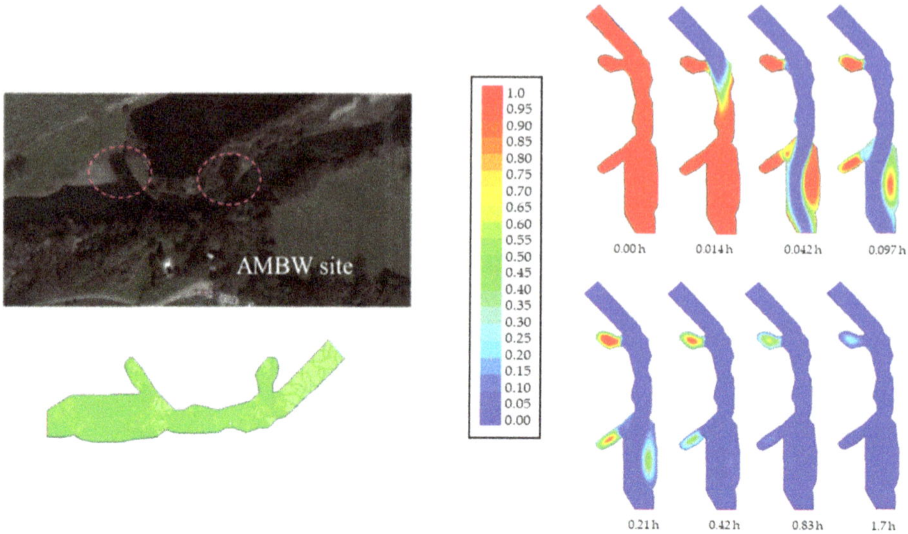

Figure 9. Drift loading to stream and backwater regions: 2D geometry of the stream section was discretized by hand, where backwater regions are clearly seen. Graphic on the right is the transient stream mass fraction following a chemical insult predicted by CFD.

3.1.6. Predicting Pesticide Volatility through Coupled above/below Ground Multiphysics Modeling

A large CFD simulation that involves both mass transfer for chlorpyrifos (semi-volatile) from plant surfaces and chlorpyrifos air concentrations once in the air was conducted, such that volatile airborne concentrations of chlorpyrifos could be addressed. The flow field and streamlines for geometries associated with row/furrow agriculture for a peach tree are found in Figure 10 (tree only). The air velocity is much lower near the plant/crop due to the crop canopy, due to obstructions created by the plant (Figure 10). The highly curved streamlines (Figure 10) lead to effective mixing of pesticide vapor within the canopy. Interested readers can find more detail elsewhere (Mao et al., 2018 [25], Wolters et al., 2004 [26]; Yates et al., 2002 [27]).

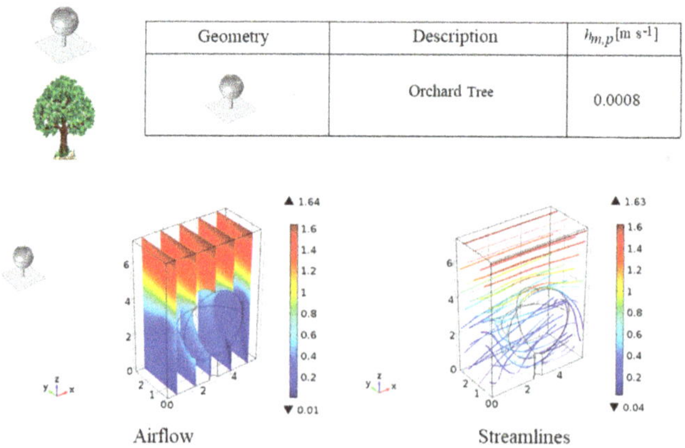

Figure 10. Turbulent flow past a modeled tree (**top**); air flow (**left**) and airflow streamlines (**right**). Specific details can be found in Mao et al., 2018 [25].

3.1.7. Why Holes Form in Oil-in-Water-Sprayed Liquid Sheets, Leading to Sheet Breakup

The breakup for a spray sheet consisting of an oil-in-water formation (e.g., atomization of a continuous sheet emanating from an agricultural spray nozzle) was also investigated using CFD. The breakup pattern for oil-in-water formulation starts with holes first forming within the spray sheet. These holes grow, collide, and eventually lead to sheet breakup. The oil drop within the water makes it way to the air/water interface and creates a disturbance; see Figure 11. These disturbances can grow and lead to holes within the sheet that are responsible for the sheet's demise and the resulting atomization drop sizes that ensue. Details of the modeling approaches for oil-in-water formulation breakup patterns emanating from agricultural nozzles mounted along a spray boom can be found elsewhere (Altieri et al., 2014 [28], Cryer and Altieri, 2017 [29], Altieri and Cryer, 2018 [30], Krishnan et al., 2022 [31].

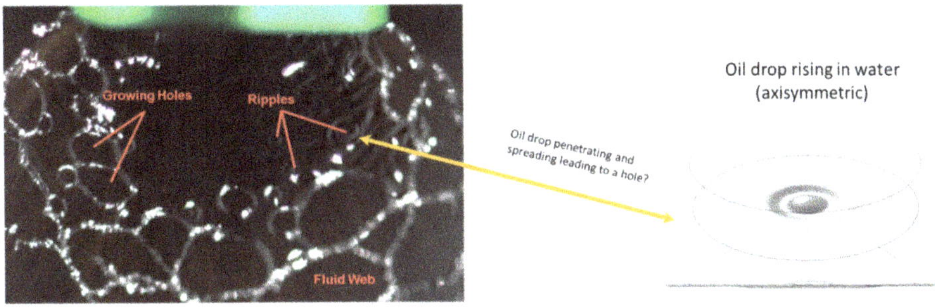

Figure 11. CFD results for the "dimple" that that forms in an oil-in-water formulation being sprayed from an agricultural nozzle as an oil drop penetrates the water film surface, compared to an observation of the film breakup process.

3.2. Corteva Manufacturing Examples

As discussed in the introduction section, CFD involves the use of computer modeling to study a system or process equipment to visualize the fundamental fluid flow-based physics associated with the chemical process. The biggest advantage of using CFD modeling in manufacturing is that it does not rely on any scale-up or scale-down approaches, due to its reliability on fundamental governing equations of fluid flow (i.e., mass, momentum, and energy balances).

Additional benefits of using CFD in manufacturing are listed below:

- Achieving a better understanding of the physical phenomena in mixing-based processes.
- Prototyping mixing designs at manufacturing scale cost-effectively compared to experiments.
- Evaluating impacts of design/parameter changes through computer simulations.
- Understanding transient effects and mixing heterogeneities during scale-up.

CFD can be generically used for

- Troubleshooting and hypothesis testing;
- Feasibility studies;
- Engineering design, redesign, and optimization;
- Flow-physics-based visualization.

The traditional approach to characterizing mixing and associated physics, mass transfer, and chemical reactions in mixing systems has been to apply engineering correlations based on a combination of first principles and empirical relationships. This is a logical first step to understanding a mixing process. The investment is low, and the information is quite useful. Furthermore, the process of gathering the specific information necessary for this approach is essential for understanding the process. However, the challenge for the mixing experts is to navigate the vast amount of literature that is typically written for specific systems and then apply it to their unique mixing process.

Our present computational methodology to model, simulate, and understand mixing-based multi-physics processes within Corteva includes using a Lattice Boltzmann LES framework on GPU architecture [32–34]. This strategy inherently alleviates the limitations posed by Reynolds Averaged Navier–Stokes (RANS)-equation-based approaches by resolving the transients in fluid flow, turbulence, and additional physics associated with scalar species transport, gas–solid, gas–liquid mass transfer, liquid–liquid immiscible interactions, non-Newtonian rheology, bubble coalescence/break-up, bubble/particle motion tracking, and rotating impeller/shaft geometries within the stirred tank reactor vessels or continuous flow reactor systems.

Corteva in-house applications of CFD in manufacturing (specifically related to mixing) include:

- Flows in batch and continuous processes.
- Mixing and mass transfer in stirred tank reactors;
    - Mixing/blend time calculations;
    - Power requirement calculations;
    - Vortexing and splashing computations;
    - Non-Newtonian fluid mixing;
    - Multiple miscible fluid blending;
    - Gas–liquid mixing and mass transfer.
- Flow in static/in-line mixers.
- Flows in jet mixers/tanks/non-standard geometries.
- Free surface flows in centrifugation.
- Tank recirculation flows.
- Gas entrainment in recirculation flows;
    - Mixing and hydrodynamics in crystallization;
    - Mixing and mass transfer in fermentation and chlorination;
    - Chemical reactions coupled with mixing and mass transfer.

Although there are a diverse range of manufacturing (mixing-related) examples in which CFD has been applied, for sake of brevity only a few will be discussed here.

3.2.1. Mixing and Hydrodynamics in Crystallization

Both empirical and CFD modeling were used to compare mixing in different vessels at different sites in the manufacturing process. CFD modeling helped identify agitation conditions at site-1, in Cernay, which would be closer to site-2 performance. CFD was also used to quantify the effect of impeller geometry and agitation speed on the following:

- Mixing hydrodynamics and statistics;
- Power number;
- Mixing time.

The modeling work was utilized to guide configuration changes to the site-1 crystallizer to best represent the site-2 process; see Figure 12. Corteva was able to increase the agitation performance from 63 to 88% relative to site-1, which was a minor change without any significant capital investment.

3.2.2. Reactor Yield Mixing-Based Investigation

CFD was used to investigate the impact of mixing on yield during the production of a fine chemical using the first commercial campaign. Here, both empirical and CFD studies were undertaken to compare the agitation profiles within glass-lined and stainless steel reactors at a third-party manufacturing site.

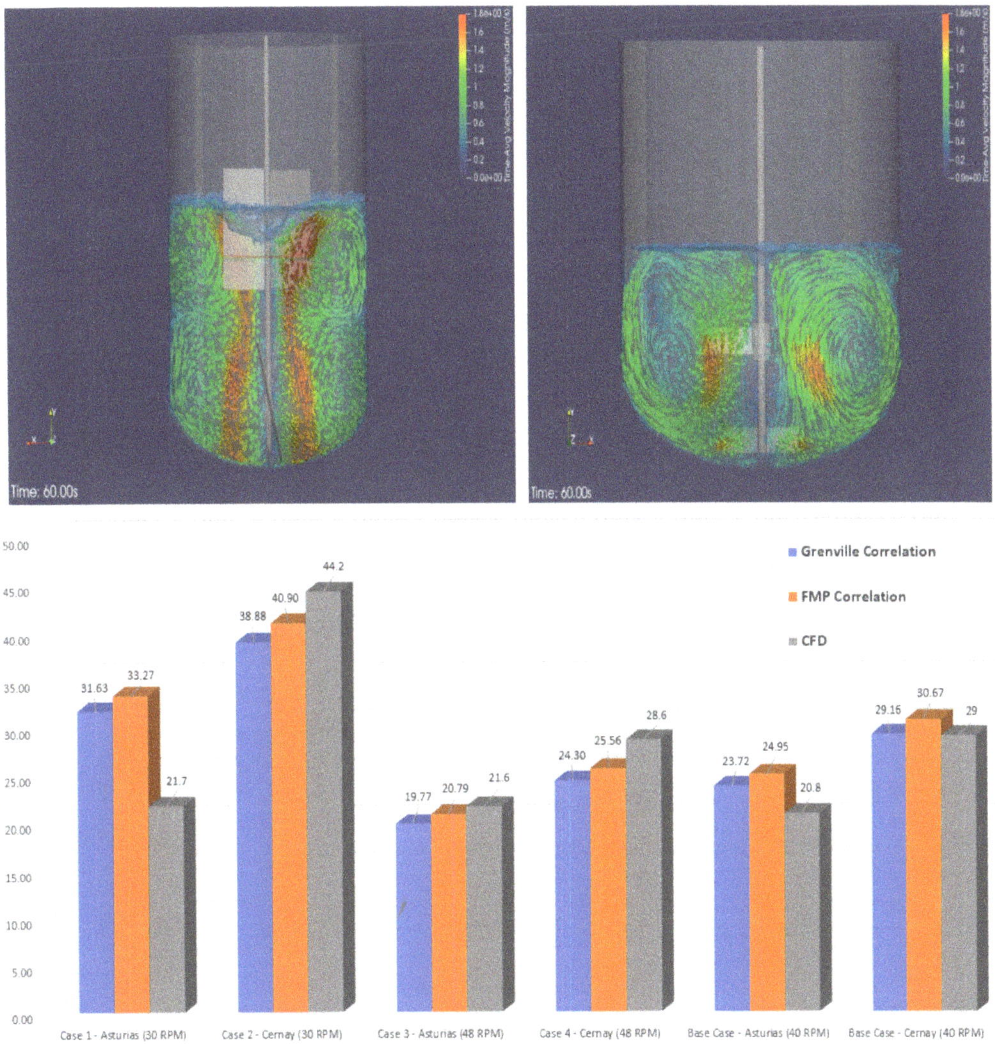

Figure 12. Mixing hydrodynamics and blend times.

Key differences in the two different mixing vessels were identified, which include:

- Fluid flow and hydrodynamics (i.e., vortex formation);
- Mixing performance (dead velocity zones, flow recirculation regions).

This case study, as illustrated in Figure 13, demonstrated how CFD modeling could be used to

- Evaluate the impact of mode of addition of reactants;
    - Top surface vs subsurface loading provided the opportunity to increase yield, batch size, improve cycle time and increase production rates.
- Investigate new reactor designs for upcoming/new commercial campaigns when there is a question about yield performance for mixing sensitive reactions, impeller location, retrofit tank shape, or optimal batch size.
- Understand key mixing scenarios affecting the yield performance of the cyclization step of the chemical manufacturing process.

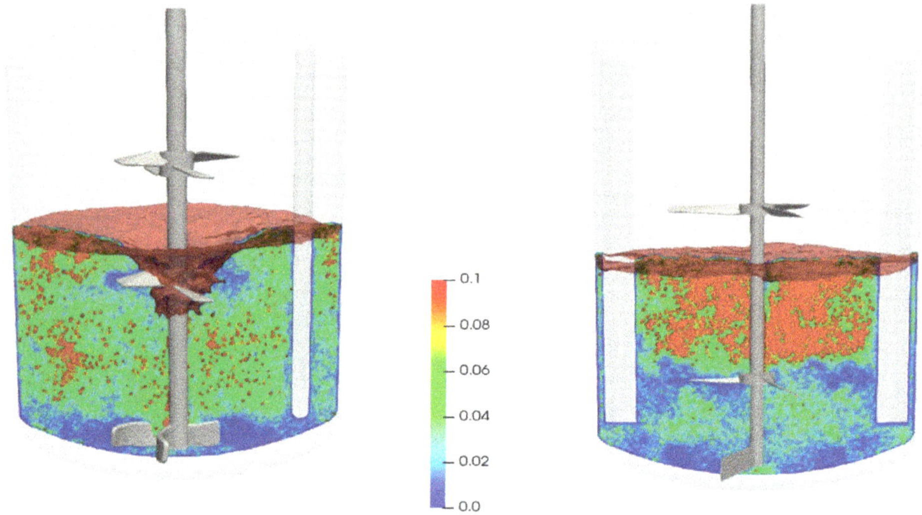

Figure 13. Reactor yield investigation (glass-lined vs. stainless steel), free surface velocity, and micro-mixing times.

Field observations illustrated that yield performance directly correlated to the mixing performance identified through CFD simulations.

3.2.3. Non-Newtonian Miscible Fluid Mixing in Formulations

This next case-study is for a formulations process at Valdosta, GA, where poor flow at the top liquid surface resulted in extended blend/mixing times. As a result, the batch processing volumes reduced from 9k->7k gal, resulting in extended processing/cycle times. Here, CFD modeling was used for the following:

- Proposing optimal impeller designs (narrow blade hydrofoils) and optimal impeller placement with original batch size.
- Investigating single phase and miscible non-Newtonian fluid blending.
- Quantifying the effect of miscible fluid blending on velocity profiles, blend times and power consumption.

Figure 14 illustrates how CFD helped in the following:

- Modeling single phase and miscible non-Newtonian fluid blending.
- Quantifying the effect of non-Newtonian flow on velocity profiles.
- Quantifying the effect of miscible fluid blending on the following:
    - ○ Velocity profiles;
    - ○ Blend times;
    - ○ Power consumption.
- Determining the optimal impeller placement for blend time reduction.

3.2.4. Agitation Design Modifications in Formulations

This example is for a formulation solid–liquid mixing case in a gel tank that did not incorporate xanthan gum powder into water as expected. As a result, powder was floating on the liquid surface and was not being drawn down by the sawtooth blade shear impeller. This led to production delays and forced the gel to be prepared in another area of the plant and transferred via totes. This resulted in added costs for the fleet of totes, adjusting the schedules of other production units, and concerns about the operation of the gel tank. CFD modeling was used here to propose agitation design modifications, enhancing the tank and subsequent formulation processes up to design rates (by as much as 15%).

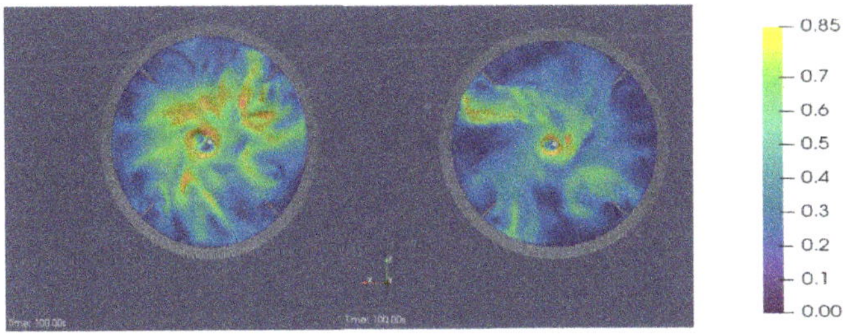

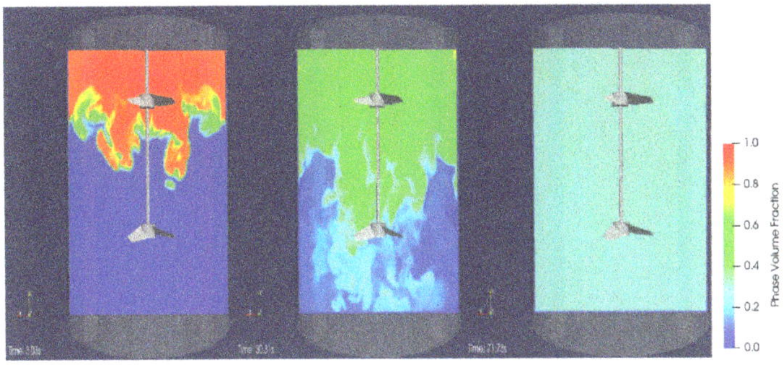

Phase Volume Fractions @ 3.03 s, 30.31 s and 71.73 s respectively

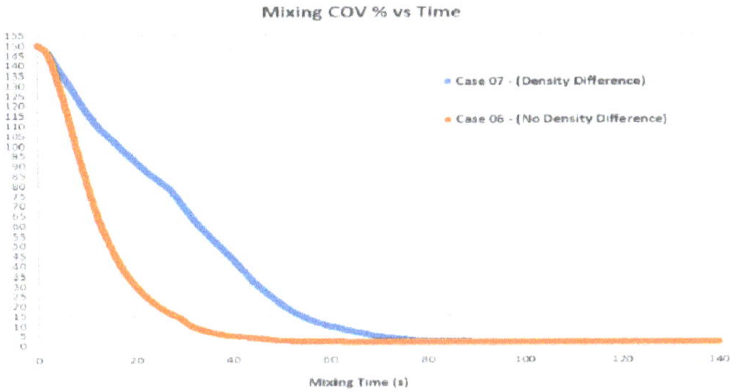

Figure 14. Non-Newtonian fluid mixing, top free surface velocity, miscible phase volume fraction, and blend time comparison.

Figure 15 illustrates how CFD was used in the following:

- Validation of modifications to agitator design with the addition of a second hydrofoil impeller.
- Comparison of current and modified designs.

The base design showed no vortexing during simulations, in comparison to the modified design, which illustrated a strong vortex development 18 s into the simulation. A higher velocity was observed in the entire tank in the modified design, as compared to the base scenario.

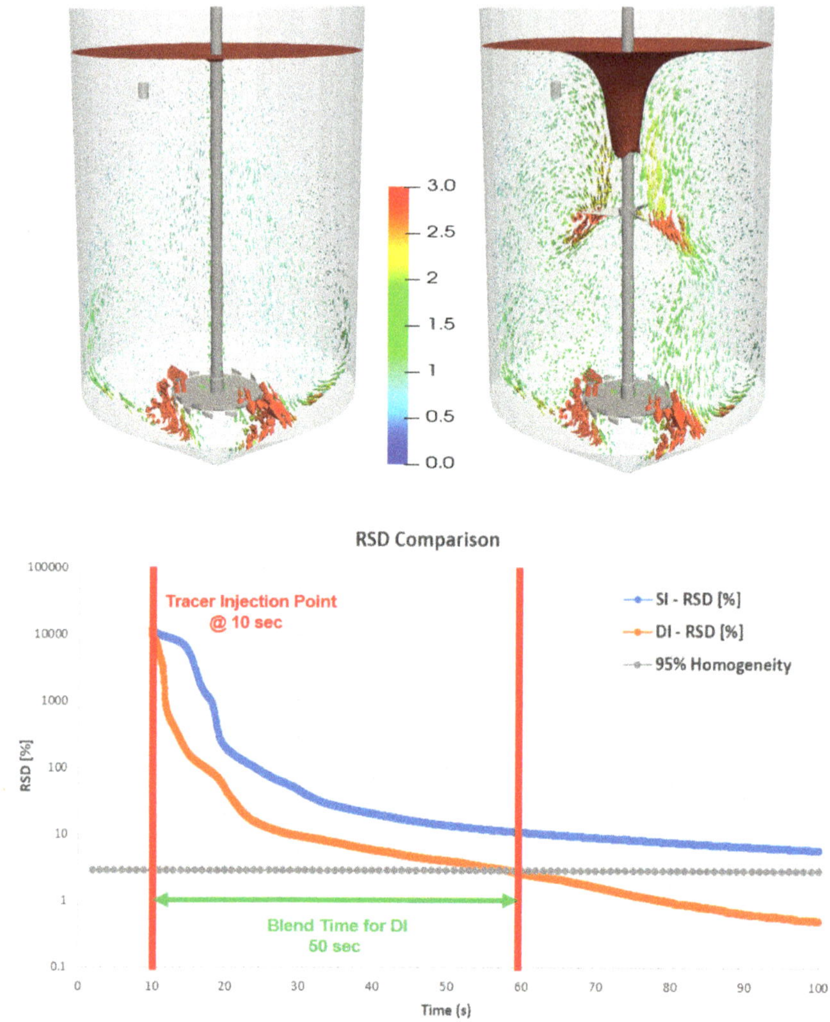

Figure 15. Gel tank agitator design modifications, and the effect on flow and blend times.

3.2.5. Bi-Phasic Immiscible Mixing CFD Model

This example corresponds to the acidic hydrolysis of an organic toluene bi-phasic system in which two immiscible liquids sit on top of each other. CFD was utilized to predict the range of agitation speeds at which the two liquid layers would break apart and mix with each other, resulting in an increased interfacial surface area. CFD simulations helped to predict the dispersed speed and the contact interfacial surface area between the immiscible phases.

Figure 16 illustrates how CFD was used for this immiscible liquid–liquid mixing scenario. The following were noticed:

- Increasing the thermowell ID from 0.125″ to 0.375″ at 100 RPM does not help break up the toluene–HCl layer.
- Introducing baffles with thermowell and pH probes does not help the liquid layer to break up when operating RPM < Njd (just dispersed speed).

- Increasing agitation to 500 RPM breaks up the two liquid layers, inducing maximum interfacial surface area.

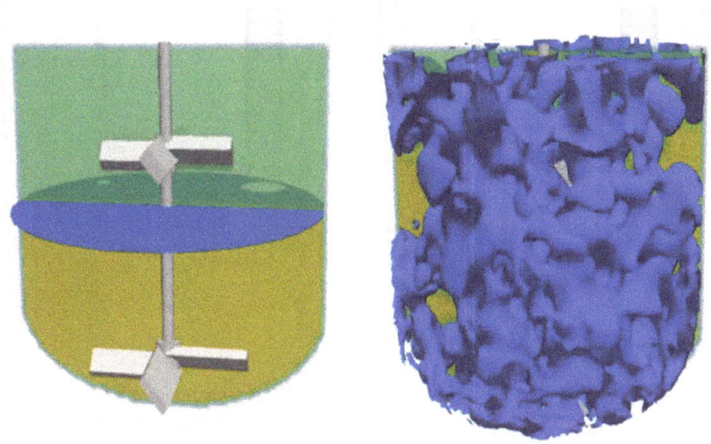

Figure 16. Bi-phasic immiscible mixing model and interfacial liquid–liquid layer at two different operating RPMs.

CFD can therefore help with predicting dispersed speed and interfacial surface area, and in designing lab scale experiments.

3.2.6. Solid–Liquid Mixing in Formulations

The case study focuses on the solid–liquid mixing formulation process to incorporate floating powder particles within the bulk liquid.

Figure 17 illustrates how CFD was used for the following:

- To determine if the operating RPM was sufficient to entrain the powder particles in the bulk liquid.
- To quantify the effect of powder incorporation on velocity, blend times, power consumption, and cloud heights.

3.2.7. Static Mixer Flow Modeling for Nozzle Design Comparison

Next, we have an example that relates to a drying unit operation which experienced plugging/fouling. CFD was utilized to investigate 12 different static mixer nozzle designs and identify one which promotes the highest droplet break-up. A specific laser-scanned nozzle provides a higher turbulent dissipation rate and a higher pressure drop, indicating a better droplet breakup capability and finer particle size while reducing the possibility of fouling.

CFD, as illustrated in Figure 18, was used for the following:

- To understand the impact of key dimensions in the static mixer nozzle design.
- To quantify the impact of nozzle dimensions on the energy dissipation rate and droplet break-up through the simulation of turbulent energy dissipation rates.
- To compare twelve different nozzle variants for reducing fouling.

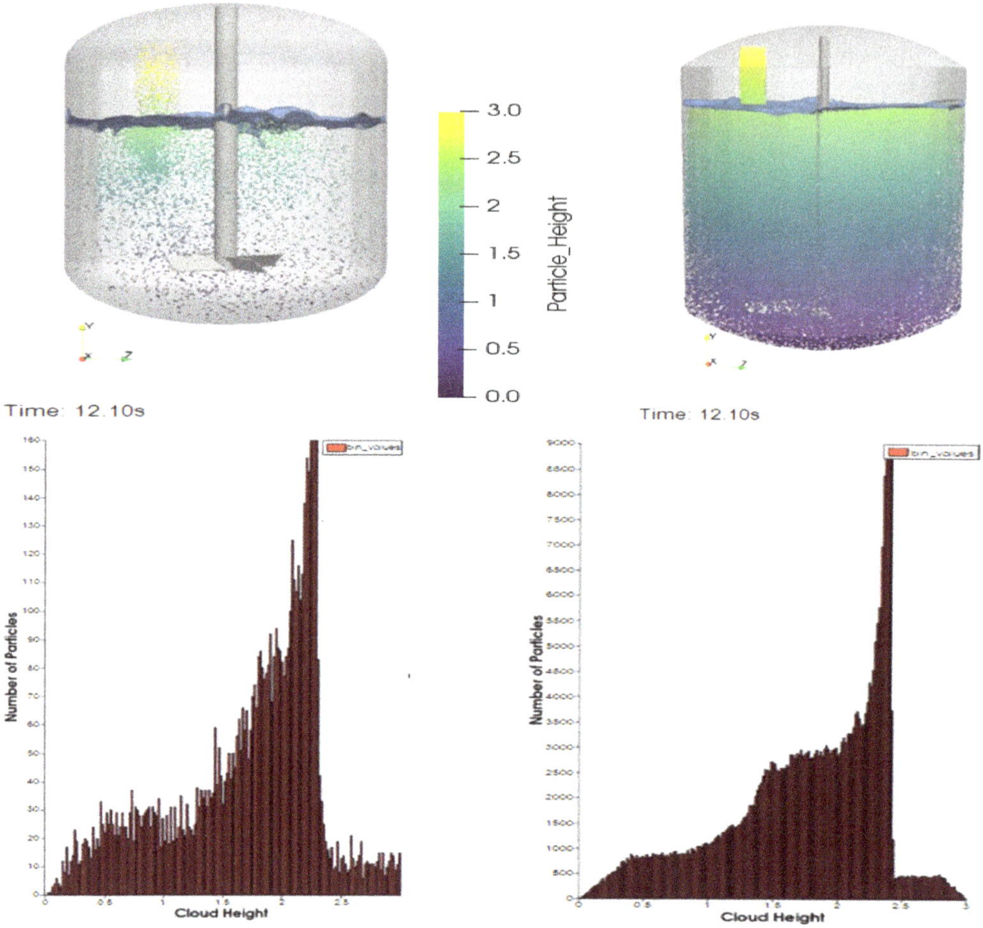

Figure 17. Solid–liquid mixing for formulations and particles colored by velocity-Y component and cloud height.

3.2.8. Jet Mixing Modeling

This example refers to computing blend times for a jet mixing process in non-standard vessel geometries for which empirical correlations are not available. We first built a CFD model to simulate the transient jet mixing hydrodynamics in a vessel with a flat bottom and compared the blend times predicted using CFD to those obtained using empirical correlations. Once we were confident in the CFD predictions, we further extended our CFD modeling approach to compute jet mixing blend times in dished vessels, as illustrated in Figure 19.

CFD, as illustrated in Figure 19, was used for the following:

- To model flow hydrodynamics in continuous flow systems like jet mixers with flat and dished bottom heads.
- To quantify the effect of inlet nozzle flow and tank bottom head geometry on the following:
    - Mixing hydrodynamics;
    - Mixing/blend times;
    - Spatial/transient tracer concentration.

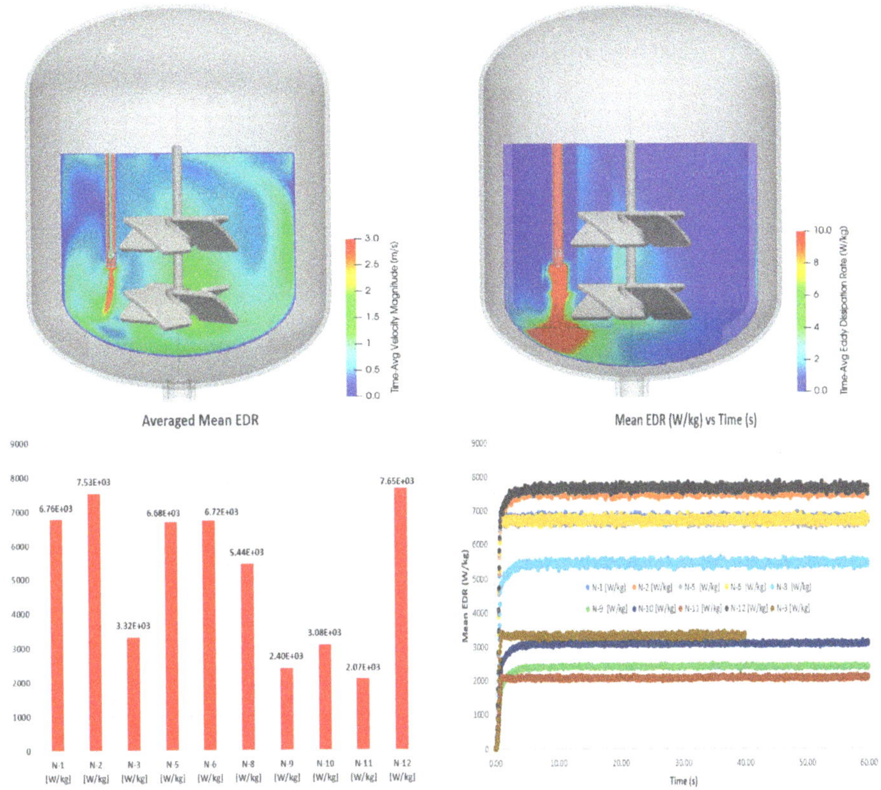

Figure 18. Static mixer nozzle flow model; base case mixing and nozzle design comparison.

3.2.9. Continuous Flow Process CFD Modeling

A reaction was being carried out in a coiled reactor. Optimizing the heat exchanger fluid velocity profile through the reactor shell was critical to maintaining the appropriate temperature profile and heat removal rate. As such, agitating nozzles needed to be designed with an optimized nozzle size, no. of nozzles, and vertical position(s) in the shell considering fluid and flow properties, to provide a spiral flow and hence turbulence.

CFD was leveraged to provide insights into the velocity profile as a function of the variables mentioned above. As illustrated in Figure 20, CFD was used for the following:

- To quantify the (%) volume of annular flow region within given thresholds.
- To guide the optimal nozzle placement in a continuous flow reactor for minimizing dead zones associated with low velocity thresholds.
- To quantify the effect of viscosity on velocity profiles, and in turn its effect on the wall heat transfer coefficient.

3.2.10. Fermentation, Integrating Biochemical Kinetics with CFD

Finally, we highlight an approach to integrate the unstructured biochemical reaction kinetics with our existing Lattice Boltzmann-based CFD modeling framework (Hanspal et al., 2023 [32], Hanspal et al. [33], Hanspal and Thomas [34], Yeoh et al., 2004, 2005 [35,36], Sirasitthichoke et al., 2022 [37]). This case study involves the application of a Euler–Lagrange CFD approach for the simulation of gluconic acid production based on Contois chemical kinetics published in the literature. Sample results illustrate the time evolution of species concentrations, including glucose, dissolved oxygen, gluconic acid, and biomass, which have been simulated using CFD and compared against the ordinary differential equation (ODE)-based solution implemented in MATLAB (R2020a). Illustrated in Figure 21 are the spatially varying heterogeneous distributions of gluconic acid in the system that can result in an overall yield loss for the fermentation process.

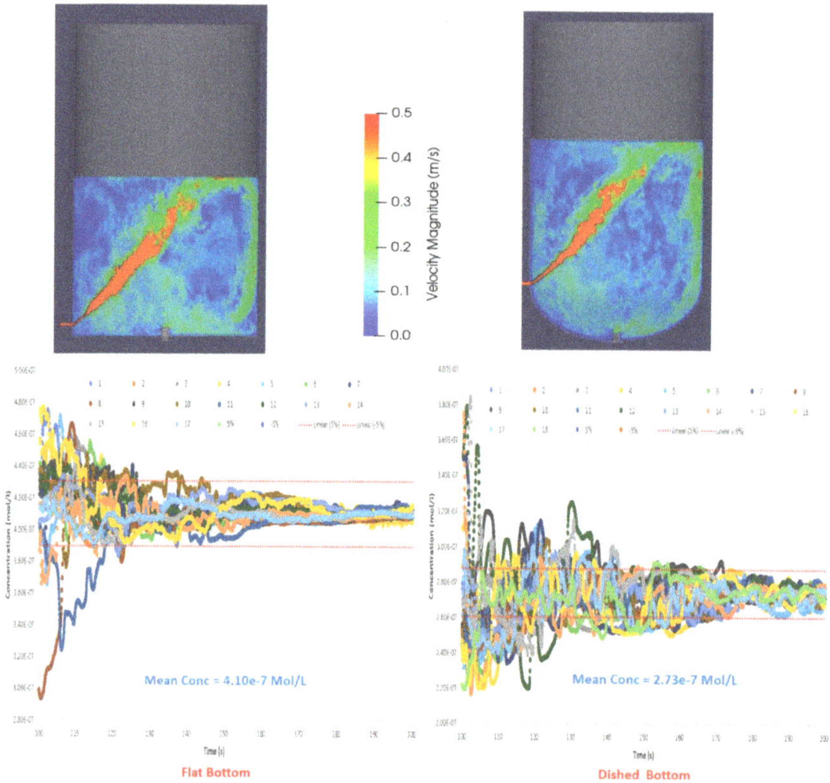

Figure 19. CFD modeling of jet mixing, velocity profiles, and tracer concentrations for blend time determination.

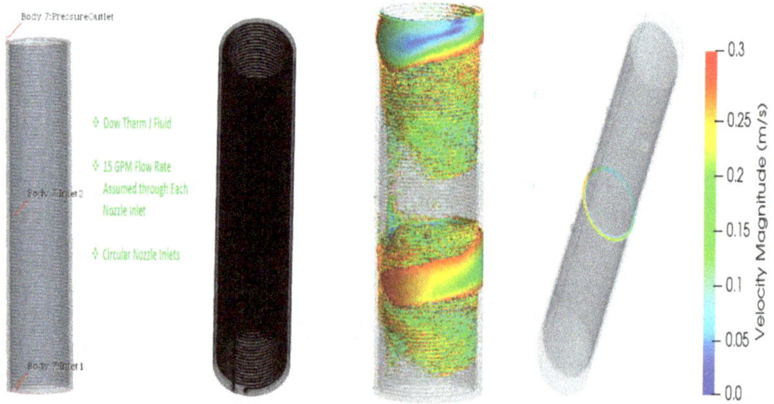

Figure 20. Continuous flow processes, velocity profile in volume, and cross-section.

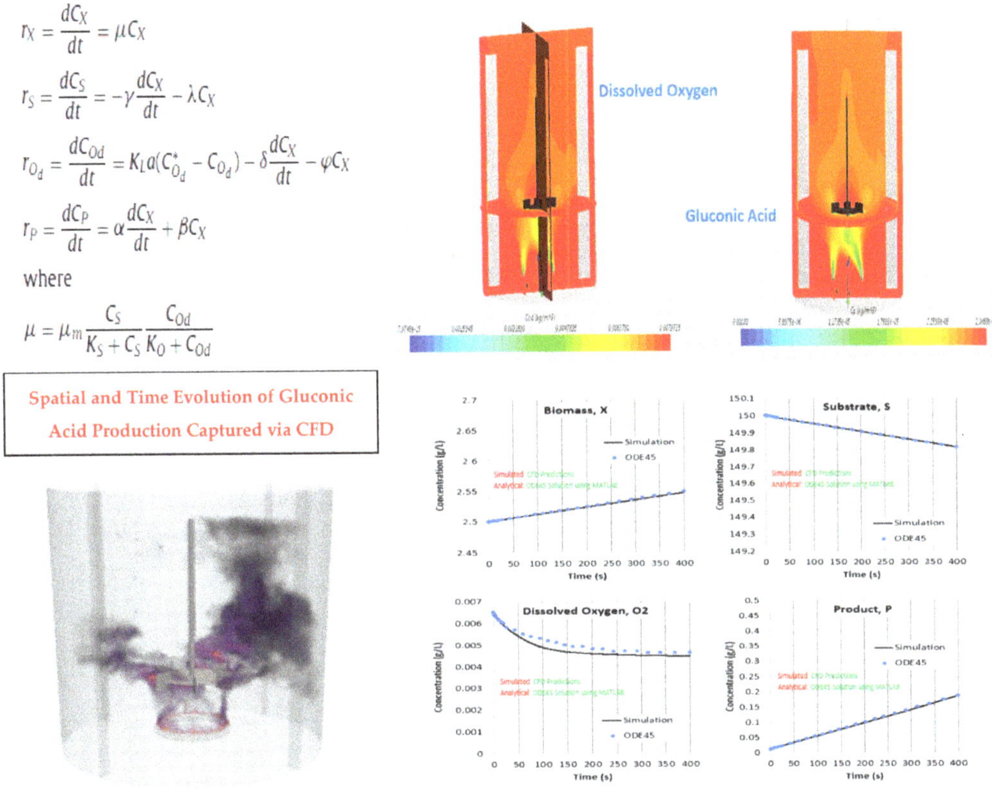

Figure 21. Fermentation, integrating biochemical kinetics with CFD.

A CFD modeling approach, including the unstructured kinetics (illustrated in Figure 21), was used to simulate the following (Hanspal et al. [33]):

- Heterogeneous distribution of gluconic acid that can result in an overall yield loss.
- Slight differences in DO concentration between CFD and perfectly mixed ODE solution.
- Differences in the results caused by the underlying spatial variations in the gas–liquid mass-transfer coefficient across fluids.

4. Discussion

CFD is the numerical solution to the Navier–Stokes equation (conservation of momentum), continuity equation (conservation of mass), and energy equation (all are partial differential equations—several analytical solutions with various limits, but most applications need to be solved numerically). CFD is used to solve these governing equations and finite volume/element/difference techniques. Many examples in manufacturing and throughout R&D have been provided to highlight the types of problems that can be investigated. Since CFD is based on theory, there are a vast array of other problems that can be explored.

As shown in this work, CFD has been used in a plethora of R&D projects. For example, COMSOL was used to estimate pesticide uptake in humans via inhalation when little or no prior knowledge is available and/or experiments cannot be performed (since humans cannot be tested on experimentally). It was found the pesticide 1,3-D readily adsorbed immediately during inhalation and never reached deep into the lung tissue, which could never have been found out without CFD. These CAT scan files for humans and rats exist and can be used for any volatile material, including volatile pesticide metabolites. These computational approaches using CFD can help avoid the 10x safety factor assigned by regulatory bodies in going from rat to human.

Subsequently, Corteva's Global Active Process Technology Expertise Center (GAPTEC) group acquired the CFD tool MStarCFD, which is based on the Lattice Boltzmann approach and has been successfully used to study and investigate a variety of continuous flow and batch processes involving mixing as the key unit operation. The key outcomes when applying MStarCFD have been in troubleshooting existing processes, retrofitting existing agitation designs, and proposing new mixing systems/designs that can increase the efficiency, enhance the productivity, and reduce batch cycle or processing times, leading to increased cost savings and revenue.

As illustrated through diverse examples presented in this manuscript, CFD is quite valuable in the following areas:

- For extrapolating field study observations to different and diverse scenarios or when testing on mammals such as humans cannot be performed.
- For achieving a better understanding of the physical phenomena in mixing-based processes.
- For prototyping mixing designs at manufacturing scale cost-effectively, as compared to experiments.
- Understanding transient effects and mixing heterogeneities during scale-up.

Henceforth, CFD has led to many engineering solutions for Corteva problems, as documented by the many publications and examples listed, while increasing our understanding of experimental observations and mixing processes. This paper provides examples in industrial manufacturing and R&D to offer a flavor of the types of simulations that have been historically performed and to offer insights for other projects to pursue.

5. Future Work

Scientists at Corteva have been on the cutting edge of CFD advancement as relevant to the agricultural industry, utilizing various techniques to overcome the numerical problems and to increase speed as problems continue to grow in size. Architecture is now focused on multi-core cloud or grid-computing CPUs and GPUs. In addition, the coupling of machine learning approaches with CFD is coming to the forefront (Vinuesa and Brunton 2022 [38]; Panchigar et al., 2022 [39]), and many commercial software companies offer this coupling in their most recent versions (e.g., SmartUQ, COMSOL, Ansys, Altari, etc.).

Author Contributions: N.H. contributed the manufacturing examples using MstarCFD, STAR-CCM+ and MATLAB, and S.A.C. contributed to the general R&D section using COMSOL Multiphysics and Fluent CFD software. N.H. and S.A.C. were responsible for conceptualization, methodology, use of commercial software packages, validation, formal analysis, and writing this manuscript. All authors have read and agreed to the published version of the manuscript.

Funding: This research received no external funding and was entirely supported by Corteva Agrisciences.

Data Availability Statement: Data are unavailable due to commercial confidential privacy restrictions.

Acknowledgments: All work was supported by Corteva Agriscience.

Conflicts of Interest: Navraj Hanspal and Steven A. Cryer were employed by Corteva at the time of this publication, and this information went through Corteva's Information Release process before being published. The research was conducted in the absence of any commercial or financial relationships that could be construed as a potential conflict of interest.

Nomenclature

C	concentration [g·m^{-3}]
D	diffusion coefficient [m^2·s^{-1}]
u = V	velocity vector [m·s^{-1}]
p	pressure [kg·m^{-1}·s^{-2}]
F = f_G	external body forces acting on the fluid (such as gravity) [kg·m·s^{-2}]
μ	dynamic viscosity [kg·m^{-1}·s^{-1}]
ζ	kinematic viscosity [m^2·s^{-1}]
ρ	fluid density [kg·m^{-3}]
I	identity matrix
T	temperature [°K]
τ	shear stress tensor [kg·m^{-1}·s^{-2}]
D/Dt	material derivative (the rate of change following a moving fluid parcel) [s^{-1}]
k	material thermal conductivity [W·m^{-1}·°K^{-1}]
e_t	specific internal energy (internal energy per unit mass) [J·kg^{-1}]
S_g	generation source term for energy (radiation effects, thermal heating from electrical current, etc.) [W·m^{-1}]
C_v	heat energy absorbed/released per unit mass (constant volume) [J·kg^{-1}·°K^{-1}]

References

1. Yuan, X.; Chai, Z.; Wang, H.; Shi, B. A generalized lattice Boltzmann model for fluid flow system and its application in two-phase flows. *Comput. Math. Appl.* **2020**, *79*, 1759–1780. [CrossRef]
2. Lallemand, P.; Luo, L.S.; Krafczyk, M.; Yong, W.A. The lattice Boltzmann method for nearly incompressible flows. *J. Comput. Phys.* **2021**, *431*, 109713. [CrossRef]
3. Succi, S. *The Lattice Boltzmann Equation for Fluid Dynamics and Beyond*; Oxford University Press: Oxford, UK, 2001.
4. Shu, C.; Guo, Z. *Lattice Boltzmann Method and Its Application in Engineering*; World Scientific: Hackensack, NJ, USA, 2013.
5. Kruger, T.; Kusumaatmaja, H.; Kuzmin, A.; Shardt, O.; Silva, G.; Viggen, E. *The Lattice Boltzmann Method: Principles and Practice*; Springer: Cham, Switzerland, 2017.
6. Sukop, M.; Thorne, D. *Lattice Boltzmann Modeling: An Introduction for Geoscientists and Engineers*; Springer: Berlin/Heidelberg, Germany, 2006.
7. Pachpute, S.N. An Introduction to Computational Fluid Dynamics (CFD). Basics of CFD Modeling for Beginners CFD Flow Engineering. Available online: https://cfdflowengineering.com/basics-of-cfd-modeling-for-beginners/ (accessed on 11 August 2024).
8. Ray, B.; Bhaskaran, R.; Collin, L. *Introduction to CFD Basics*; Cornell University: New York, NY, USA, 2012. Available online: https://dragonfly.tam.cornell.edu/teaching/mae5230-cfd-intro-notes.pdf (accessed on 11 August 2024).
9. Tu, J.; Yeoh, G.H.; Liu, C.; Tao, Y. *Computational Fluid Dynamics: A Practical Approach*, 3rd ed.; Butterworth-Heinemann, Elsevier Science: Amsterdam, The Netherlands, 2018; 498p, ISBN 9780081012444.
10. Tey, W.Y.; Asako, Y.; Sidik, N.A.C.; Goh, R.Z. Governing Equations in Computational Fluid Dynamics: Derivations and A Recent Review. *Prog. Energy Environ.* **2017**, *1*, 1–19.
11. Bournet, P.-E.; Rojano, F. Advances of Computational Fluid Dynamics (CFD) applications in agricultural building modelling: Research, applications and challenges. *Comput. Electron. Agric.* **2022**, *201*, 107277. [CrossRef]
12. Kanti, P.K.; Chandran, S.A. A Review Paper on Basics of CFD and Its Applications. 2016. Available online: https://api.semanticscholar.org/CorpusID:212454797 (accessed on 11 August 2024).
13. Hosain, M.L.; Fdhila, R.B. Literature Review of Accelerated CFD Simulation Methods towards Online Application. *Energy Procedia* **2015**, *75*, 3307–3314. [CrossRef]
14. Soodmand, A.M.; Azimi, B.; Nejatbakhsh, S.; Pourpasha, H.; Farshchi, M.E.; Aghadasinia, H.; Mohammadpourfard, M.; Heris, S.Z. A comprehensive review of computational fluid dynamics simulation studies in phase change materials: Applications, materials, and geometries. *J. Therm. Anal. Calorim.* **2023**, *148*, 10595–10644. [CrossRef]
15. Bird, R.B.; Stewart, W.E.; Lightfoot, E.N. *Transport Phenomena*; John Wiley & Sons: New York, NY, USA, 1960; ISBN 978-0-470-11539-8.
16. Randles, A.; Kaxiras, E. Parallel in time approximation of the lattice Boltzmann method for laminar flows. *J. Comput. Phys.* **2014**, *270*, 566–586. [CrossRef]
17. Kabilan, S.; Lin, C.L.; Hoffman, E.A. Characteristics of airflow in a CT-based ovine lung: A numerical study. *J. Appl. Physiol.* **2007**, *102*, 1369–1482. [CrossRef] [PubMed]
18. Longest, P.W.; Holbrook, L.T. In silico models of aerosol delivery for the respiratory tract- Development and applications. *Adv. Drug Deliv Rev.* **2011**, *64*, 296–311. [CrossRef]
19. Garcia, G.J.; Schroeter, J.D.; Segal, R.A.; Stanek, J.; Foureman, G.L.; Kimbell, J.S. Dosimetry of nasal uptake of water-soluble and reactive gases: A first study of interhuman variability. *Inhal. Toxicol.* **2009**, *21*, 607–618. [CrossRef]

20. Schroeter, J.D.; Kimbell, J.S.; Gross, E.A.; Willson, G.A.; Dorman, D.C.; Tan, Y.M.; Clewell, H.J., III. Application of physiological computational fluid dynamics models to predict interspecies nasal dosimetry of inhaled acrolein. *Inhal. Toxicol.* **2008**, *20*, 227–243. [CrossRef]
21. Corley, R.; Kabilan, S.; Kuprat, A.P.; Carson, J.P.; Minard, K.R.; Jacob, R.E.; Timchalk, C.; Glenny, R.; Pipavath, S.; Cox, T.C.; et al. Comparative Computational Modeling of Airflows and Vapor Dosimetry in the Respiratory Tracts of Rat, Monkey, and Human. *Toxicol. Sci.* **2012**, *128*, 500–516. [CrossRef]
22. Quinn, A.D.; Wilson, M.; Reynolds, A.M.; Couling, S.B.; Hoxey, R.P. Modeling the dispersion of aerial pollutants from agricultural buildings-an evaluation of computational fluid dynamics (CFD). *Comput. Electron. Agric.* **2001**, *20*, 219–235. [CrossRef]
23. Cryer, S. Predicted Gas Loss of Sulfuryl Fluoride and Methyl Bromide During Structural Fumigation. *J. Stored Prod. Res.* **2008**, *44*, 1–10. [CrossRef]
24. Cryer, S.A.; Barnekow, D.E. Estimating Outside Air Concentrations surrounding Fumigated Grain Mills. *Biosyst. Engr.* **2006**, *94*, 557–572. [CrossRef]
25. Mao, M.; Cryer, S.A.; Altieri, A.; Havens, P. Predicting Pesticide Volatility Through Coupled Above- and Belowground Multi-physics Modeling. *Environ. Model. Assess.* **2018**, *23*, 560–582. [CrossRef]
26. Wolters, A.; Leistra, M.; Linnemann, V.; Klein, M.; Schäffer, A.; Vereecken, H. Pesticide volatilization from plants: Improvement of the PEC model PELMO based on a boundary-layer concept. *Environ Sci Technol.* **2004**, *38*, 2885–2893. [CrossRef] [PubMed]
27. Yates, S.; Wang, D.; Papiernik, S.K.; Gan, J. Predicting pesticide volatilization from soils. *Environmetrics* **2002**, *13*, 569–578. [CrossRef]
28. Altieri, A.L.; Cryer, S.A.; Acharya, L. Mechanisms, Experiment and Theory of Liquid Sheet Breakup and Drop Size from Agricultural Nozzles. *At. Sprays* **2014**, *24*, 695–721. [CrossRef]
29. Cryer, S.A.; Altieri, A.L. Use of large inhomogeneity's to initiate sprayed sheet demise during atomization. *Biosyst. Eng.* **2017**, *163*, 103–115. [CrossRef]
30. Altieri, A.L.; Cryer, S.A. Break-up of sprayed emulsions from flat-fan nozzles using a hole kinematics model. *Biosyst. Eng.* **2018**, *169*, 104–114. [CrossRef]
31. Krishnan, G.; Cryer, S.A.; Turner, J.; Rajan, N.S. Spray Atomisation in Multiphase Flows with References to Tank Mixes of Agricultural Products. *Biosyst. Eng.* **2022**, *223*, 232–248. [CrossRef]
32. Hanspal, N.; Chai, N.; Allen, B.; Brown, D. Applying multiple approaches to deepen understanding of mixing and mass transfer in large-scale aerobic fermentations. *J. Ind. Microbiol. Biotechnol.* **2020**, *47*, 929–946. [CrossRef]
33. Hanspal, N.; Thomas, J.; DeVincentis, B. Modeling Multiphase Fluid Flow, Mass Transfer, and Chemical Reactions in Bioreactors using Large-Eddy Simulation. *Eng. Life Sci.* **2023**, *23*, e2200020. [CrossRef] [PubMed]
34. Hanspal, N.; Thomas, J. CFD Modeling of Two-Phase Stirred Bioreaction Systems: Applications of Large-Eddy-Simulation (LES) Simulation. In Proceedings of the 2020 American Institute of Chemical Engineers (AIChE) Annual Meeting, San Francisco, CA, USA, 16–20 November 2020; American Institute of Chemical Engineers: New York, NY, USA, 2020.
35. Yeoh, S.L.; Papadakis, G.; Yianneskis, M. Determination of mixing time and degree of homogeneity in stirred vessels with large eddy simulation. *Chem. Eng. Sci.* **2005**, *60*, 2293–2302. [CrossRef]
36. Yeoh, S.L.; Papadakis, G.; Yianneskis, M. Numerical Simulation of Turbulent Flow Characteristics in a Stirred Vessel Using the LES and RANS Approaches with the Sliding/Deforming Mesh Methodology. *Chem. Eng. Res. Des.* **2004**, *82*, 834–848. [CrossRef]
37. Sirasitthichoke, C.; Teoman, B.; Thomas, J.; Armenante, P.M. Computational prediction of the just-suspended speed, Njs, in stirred vessels using the Lattice Boltzmann method (LBM) coupled with a novel mathematical approach. *Chem. Eng. Sci.* **2022**, *251*, 117411. [CrossRef]
38. Vinuesa, R.; Brunton, S.L. Enhancing computational fluid dynamics with machine learning. *Nat. Comput. Sci.* **2022**, *2*, 358–366. [CrossRef] [PubMed]
39. Panchigar, D.; Kar, K.; Shukla, S.; Mathew, R.M.; Chadha, U.; Selvaraj, S.K. Machine learning-based CFD simulations: A review, models, open threats, and future tactics. *Neural Comput. Applic.* **2022**, *34*, 21677–21700. [CrossRef]

fluids MDPI

Article

Deeper Flow Behavior Explanation of Temperature Effects on the Fluid Dynamic inside a Tundish

Enif Gutiérrez [1], Saul Garcia-Hernandez [1,*], Rodolfo Morales Davila [2] and Jose de Jesus Barreto [1]

1 Metallurgy Graduate Center, TecNM-Instituto Tecnologico de Morelia, Av. Tecnologico No. 1500, Morelia 58120, Mexico; enif.gg@morelia.tecnm.mx (E.G.); jose.bs@morelia.tecnm.mx (J.d.J.B.)
2 Departament of Metallurgy and Materials Engineering-ESIQUIE, Instituto Politécnico Nacional, Ed. 7, UPALM, Col. Zacatenco, Ciudad de México 07738, Mexico; rmorales@ipn.mx
* Correspondence: saul.gh@morelia.tecnm.mx

Abstract: The continuous casting tundish is non-isothermal due to heat losses and temperature variation from the inlet stream, which generate relevant convection forces. This condition is commonly avoided through qualitative fluid dynamic analysis only. This work searches to establish the conditions for which non-isothermal simulations are mandatory or for which isothermal simulations are enough to accurately describe the fluid dynamics inside the tundish by quantifying the buoyant and inertial forces. The mathematical model, simulated by CFD software, considers the Navier-Stokes equations, the realizable k-ε model for solving the turbulence, and the Lagrangian discrete phase to track the inclusion trajectories. The results show that temperature does not significantly impact the volume fraction percentages or the mean residence time results; nevertheless, bigger velocity magnitudes under non-isothermal conditions than in isothermal conditions and noticeable changes in the fluid dynamics between isothermal and non-isothermal cases in all the zones where buoyancy forces dominate over inertial forces were observed. Because of the results, it is concluded that isothermal simulations can accurately describe the flow behavior in tundishes when the flow control devices control the fluid dynamics, but simulations without control devices or with a weak fluid dynamic dependence on the control devices require non-isothermal simulations.

Keywords: mathematical simulation; non-isothermal model; isothermal model; ideal volume fractions; inclusion removal rates

1. Introduction

The tundish is an essential reactor of the continuous casting machine. Its primordial goal is to feed the mold with steel that is homogeneous in temperature and chemical composition. Additionally, the tundish tries to minimize the new inclusion generation and improve the inclusion removal, implying that its good operation guarantees high-quality steel production. Focusing on tundish temperature control, the slag layer on the tundish top provides thermal insulation to prevent the molten steel from losing heat to the atmosphere. However, heat losses exist throughout steel contact with the refractory walls, flow control devices, and even more from heat to heat. Because of that, in recent years, more researchers have considered the temperature variable in their simulations [1–26] versus most previous works, which ignore their effects [27–45]. Both approaches have generated debate about the importance of temperature on flow patterns among researchers [19–26]. Some assure that temperature effects strongly impact the flow patterns inside the tundish. For example, Miki et al. [19], studying a tundish with one strand and a dam, found a steel temperature drop of five grades inside the tundish, generating notorious changes in the fluid dynamics compared with the isothermal case due to the thermal convection, which also induces a stronger inclusion removal rate for the non-isothermal case. Alizadeth et al. [20], studying a tundish with two strands, a turbulence inhibitor, and dams, analyze the RTDs for isothermal and non-isothermal conditions, determining that the mixed fraction volume is lower for the isothermal case than the non-isothermal case due to the buoyancy force forcing the flow towards the top surface. Sun et al. [21], studying a single-strand tundish with a turbulence inhibitor, dam, weir, and stopper rod, confirm the results reported by Alizadeth [20] and additionally note that the RTD-s for both cases have the same tendency, concluding that although the isothermal simulation cannot reflect in detail the flow patterns, they can determine the optimal design for the flow controls. Chaterjee et al. [22], studying a tundish with four strands and an impact pad, found that the flow patterns far from the inlet zone can change drastically in a non-isothermal case. The temperature induces a strong inclusion coupling with the flow, which induces an increment in the removal inclusion rate. Zhu et al. [23] use a four-strand tundish with baffles, finding that control devices can cushion but not eliminate the temperature impact.

In contrast, other researchers affirm that this variable is unimportant, disdaining the temperature effects for considering them insignificant. For example, Morales et al. [24], using a two-strand tundish with a weir, dam, and turbulence inhibitor, determined that the thermal disturbances due to the ladle change modify the fluid dynamics immediately after the input. Still, the flow behaves similarly to an isothermal state after a short period. Using a four-strand tundish, Chattopadhyay et al. [25] found a strong influence of the natural convection if the tundish is empty; however, using an impact pad, the temperature effects on the fluid dynamics and removal rates vanish. Sousa Rocha et al. [26], using a two-strand tundish with different configurations for the flow control devices, did not find substantial differences between the isothermal and non-isothermal cases.

The previously mentioned works studied and clarified the differences and similitudes between isothermal and non-isothermal simulations on the steel flow patterns in the tundish. Nevertheless, all these efforts are qualitative, and it is still undetermined if the temperature does or does not substantially affect the flow patterns and if the flow control devices can or cannot cancel out the temperature effects on the fluid dynamics. Therefore, this work aims to establish the conditions for which non-isothermal simulations are mandatory or for which isothermal simulations are enough to accurately describe the fluid dynamics inside the tundish by quantifying the buoyant and inertial forces. To achieve this, a mathematical model will be validated using a scaled water model to later employ the validated model to study the effect of temperature on fluid dynamics by flow pattern analysis and by quantifying the changes in the variation of the non-ideal flows inside the tundish.

2. Numerical Methods

The commercial ANSYS Fluent® CFD code (Fluent 16.0, Ansys Inc., Centerra Resource Park, Lebanon, 2009) solved the equations simultaneously using the following assumptions and considerations:

2.1. Main Assumptions and Considerations

Three Cartesian coordinates consider isothermal and non-isothermal conditions. The gravity force acts vertically in the incoming flow direction. The typical non-slip conditions apply to solid surfaces. The molten steel behaves as an incompressible Newtonian fluid. The simulation considers an air zone of 100 mm and a slag zone of 70 mm. The non-isothermal simulation considers a tundish wall of magnesia. The physical properties of steel, slag, air, and magnesia are in Table 1.

The inlet and outlet temperatures were 1800 K and 1793 K, respectively. Heat losses from the bottom, walls, and flow control devices are in Table 2. The inlet and outlet velocities were calculated based on the volumetric flow of 297.8 L/min.

The simulated inclusions were assumed to be of spherical, rigid shape and to have the physical properties of alumina (density = 3960 kg/m^3). No interaction among the inclusions was considered; therefore, agglomeration and collision were not simulated. The inclusion-removal mechanism considered was Stokes flotation. The boundary conditions for inclusion removal were as follows: any inclusion that reached the interphase slag steel was considered removed, and the rest was considered escaped. Inclusion trajectories were calculated using a Lagrangian particle-tracking approach, which solves a transport equation for each inclusion as they travel through the previously calculated velocity field of liquid steel. This approach assumes that the interaction between steel and the inclusion is one-way coupled; that is, only the steel affects the trajectories of inclusions, but these do not affect the steel flow. Once isothermal and non-isothermal cases achieve the quasi-steady state, 2000 inclusions of a single size 1, 20, 40, 60, 80, 100, 130, 150, 180, and 200 μm were fed as an impulse at the ladle shroud surface entry.

Table 1. Physical properties of steel, slag, air, and magnesia.

Property	Steel	Slag	Air	Magnesia
ρ (kg/m^3)	$8580 - 0.883T$	2600	1.225	3580
C_p(J/(kg·K))	1200	628	1006.93	937.86
TC (w/(m·K))	35	1.1	0.0242	55
μ (kg/(m·s))	0.0064	0.09	1.7894×10^{-5}	---

Table 2. Heat losses in the simulation.

Wall	Heat Loss
Bottom	1400 w/m^2
Back and front	3200 w/m^2
Right and left	3800 J/(kg·K)
Control flow devices	1750 J/(kg·K)

2.2. *Mathematical Model Equations*

The mathematical model is based on the Navier-Stokes equations for incompressible flows. The model solves simultaneously the following continuity equations and momentum conservation equations:

$$\frac{\partial \rho}{\partial t} + \frac{\partial}{\partial x_i}(\rho V_i) = 0 \tag{1}$$

$$\frac{\partial}{\partial t}(\rho V_i) + \frac{\partial}{\partial x_i}(\rho V_i V_j) = -\frac{\partial P}{\partial x_i} + \frac{\partial}{\partial x}\left[\mu_{eff}\left(\frac{\partial V_i}{\partial x_j} + \frac{\partial V_j}{\partial x_i}\right)\right] + \rho\beta\Delta T g \tag{2}$$

where ρ is the fluid density. V is the fluid velocity, P is the pressure, μ_{eff} is the effective viscosity given by $\mu_{eff} = \mu + \mu_t$, with μ the fluid viscosity and μ_t the turbulent viscosity, β is the coefficient of volumetric thermal expansion, T is the temperature, and g is the gravity acceleration.

2.2.1. The Standard k-ε Realizable Turbulence Model

The realizable k-ε proposed by Shih et al. [46] model contains a new formulation for the turbulent viscosity and the dissipation rate ε.

$$\mu_t = \rho C_\mu \frac{k^2}{\varepsilon} \tag{3}$$

where C_μ is no longer a constant.

This model is more accurate for flows involving rotation. Boundary layers under strong adverse pressure gradients, separation, and recirculation.

The equations that describe the transport of the turbulent energy k and dissipation rate ε are given by:

$$\frac{\partial}{\partial t}(\rho k)+\frac{\partial}{\partial x_j}(\rho kju_j)=\frac{\partial}{\partial x_i}\left[\left(\mu+\frac{\mu_t}{\sigma_k}\right)\frac{\partial k}{\partial x_j}\right]+G_k+G_b-\rho\varepsilon-Y_M+S_k \quad (4)$$

$$\frac{\partial}{\partial t}(\rho\varepsilon)+\frac{\partial}{\partial x_i}(\rho\varepsilon u_i)=\frac{\partial}{\partial x_j}\left[\left(\mu+\frac{\mu_t}{\sigma_\varepsilon}\right)\frac{\partial \varepsilon}{\partial x_j}\right]+\rho C_1 S_\varepsilon-\rho C_2\frac{\varepsilon^2}{k+\sqrt{\nu\varepsilon}}+C_{1\varepsilon}\frac{\varepsilon}{k}C_{3\varepsilon}G_b+S_\varepsilon \quad (5)$$

In this equation G_k represents the generation of turbulence kinetic energy due to the mean velocity gradients. G_b is the generation of turbulence kinetic energy due to buoyancy Y_M represents the contribution of the fluctuating dilatation in compressible turbulence to the overall dissipation rate, and C_2, $C_{1\varepsilon}$ are constants, σ_k and σ_ε are the turbulent Prandtl numbers. S_k and S_ε are user-defined source terms.

2.2.2. The Lagrangian Discrete Phase Model

The Lagrangian model [47] solves a transport equation for each inclusion as it travels inside the flow field by integrating the force balance acting on the particle. The force balance is as follows:

$$\frac{du_p}{dt}=F_d(V-u_p)+F_B+F_G+F_{VM}+F_{PG}+F_S \quad (6)$$

The drag force per unit particle mass $F_d(V-u_p)$ tends to induce the inclusion to follow the steel flow trajectory. The buoyant (F_B) and gravity forces (F_G), act only in the vertical direction, inducing ascending or descending moves. The term (F_S) corresponds to the Saffman force, which promotes radial and lifting effects. The virtual mass force (F_{VM}) quantifies the required force to accelerate the fluid around the inclusions. The last force in Equation (6) is the gradient force (F_{PG}) generated by the pressure differences from one point to another.

2.3. Numerical Procedure

The simulations are transient following the next sequence: the isothermal case requires simulating 300 s using the implicit formulation and then 60 s using the explicit formulation using 0.01 s as the time step size to achieve the quasi-steady state. For the non-isothermal case, the base was the previously described isothermal case using a time step of 0.005, turning on the temperature. Reaching the quasi-steady state in the explicit formulation for the non-isothermal conditions after the 1920s. The simulation convergence criterion requires residuals smaller than 10^{-4}. All the images and calculations for the isothermal and non-isothermal cases were taken once the simulations achieved the quasi-steady state.

The geometries and dimensions of the scaled one-third model of the tundish and the flow control devices are in Figures 1 and 2. The computational grid shown in Figure 3 consists of 1,500,000 structured cells. The discretization of the governing equations uses the finite volume technique through the commercial CFD code. The non-linear momentum equations were linearized using the implicit approach, and the discretization requires the Second Order Upwind scheme. The pressure interpolations use the PRESTO scheme and the PISO algorithm for the pressure-velocity coupling. The volume of fluid (VOF) [47] method models the multiphase steel-slag-air system.

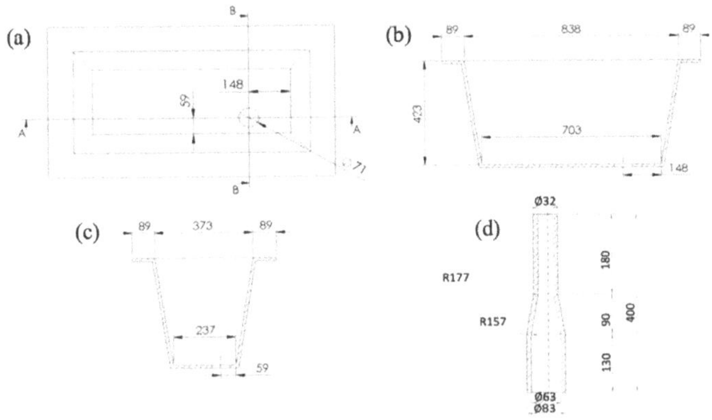

Figure 1. Geometry of the one-third scaled tundish and its flow control devices (mm). (**a**) Upper view; (**b**) frontal view; (**c**) lateral view; and (**d**) ladle shroud.

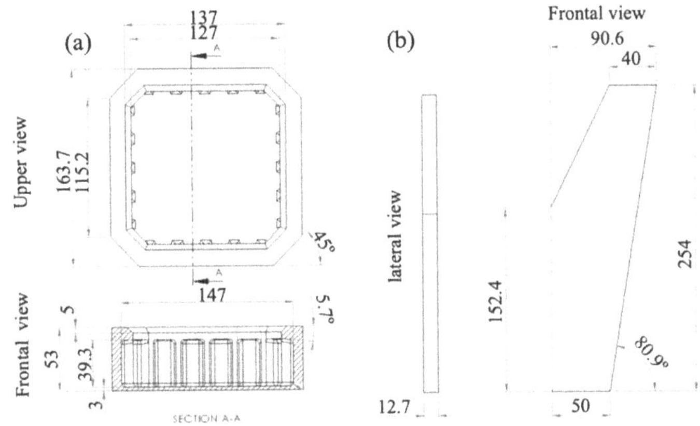

Figure 2. Geometry and dimensions of the one-third scaled (**a**) turbulence inhibitor and (**b**) vortex killer.

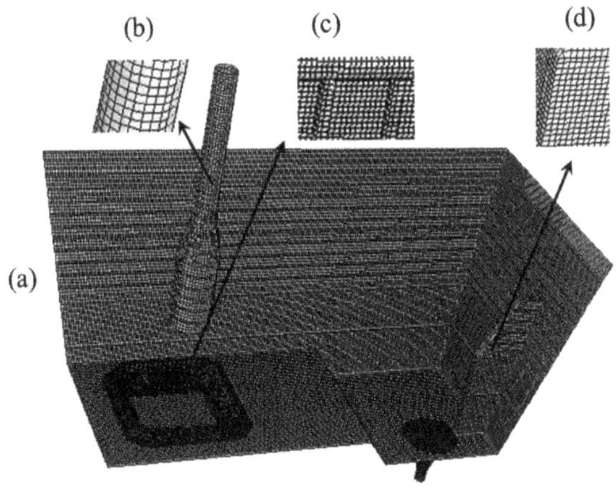

Figure 3. Tundish mesh: (**a**) general view, close up for the (**b**) ladle shroud, (**c**) turbulence inhibitor, and (**d**) vortex killer.

2.4. Physical Model

Physical modeling is a powerful tool for understanding and studying the fluid dynamics occurring inside industrial tundishes, with the advantage of being low-cost. In water physical modeling, the most common scale condition is the Froude similarity criterion, where the ratio between inertial and gravitational forces in the model (m) and prototype (p) must be equal to:

$$\left(\frac{VL\rho}{\mu}\right)_m = \left(\frac{VL\rho}{\mu}\right)_p \quad (7)$$

V and L are the characteristic velocity and length, and ρ is the fluid density. The Froude similarity criterion provides relationships between the variables in the scaled water model and the prototype isothermal tundish. With these relationships, if the variable values are known in the water model, the prototype variable values can be predicted. Some of these relationships, for example, for the length, volumetric flow rate, volume, velocity, and time, are:

$$\begin{aligned} (L)_m &= \lambda (L)_p \\ (Q)_m &= \lambda^{5/2} (Q)_p \\ (Vol)_m &= \lambda^3 (Vol)_p \\ (V)_m &= \sqrt{\lambda}\,(V)_p \\ (t)_m &= \sqrt{\lambda}\,(t)_p \end{aligned} \quad (8)$$

where λ is the scale factor.

Experimental water modeling was carried out on a $1/3$ scaled model from the prototype, obeying the Froude criterion. The scaled tundish walls, the turbulence inhibitor, and the vortex killer were built of transparent 12-mm-thick plastic. The scaled ladle shroud and the tundish outlet were made with acrylic block. The working fluid was water at room temperature under isothermal conditions. The tundish was collocated on a metallic structure over a water pit, which has an immersed water pump to transport water through a vertical pipe until the entry of the ladle shroud. The entry flow rate was regulated using a flowmeter, and the tundish model bath level was controlled using a slide gate. Once the scaled model was set, the model was left working for at least 5 min to achieve the steady state of the fluid dynamics, and at this point, the experiments started. A red dye tracer was used to analyze the fluid dynamics of the model; the dye was injected at the top of the ladle shroud, and the dye dispersion movement was recorded with a video camara to make its posterior image analysis.

3. Results and Discussion

3.1. Mathematical Model Validation

Frames from a recorded video of an impulse injection tracer are used to study and analyze the fluid dynamics inside this water model under isothermal conditions. At the same time, predictions of the impulse injection tracer behavior by mathematical modeling are obtained considering again isothermal conditions in a full-scale model. Figure 4 shows the results of tracer behavior for the physical modeling at three representative times (3, 22, and 35 s) and their corresponding times for the mathematical model (5.2, 38, and 61 s) following the equation $(t)_m = \sqrt{\lambda}\,(t)_p$. Figure 4a,b show that the tracer does not enter the tundish homogenously, having a faster entry for the right side. Once the tracer impacts the turbulence inhibitor, it moves upward until it reaches the bath level. Later, it shows two preferential streams: the first moves to the right side, crashing with the lateral tundish wall to continue a descending movement (see Figure 4c,d); the second stream tracer remains moving parallel to the top surface and starts to fall at the half distance between the ladle shroud entry and the right lateral tundish wall; at the same time, another small tracer stream moves close to the tundish floor (see Figure 4e,f). According to the previous results, it can be considered that the physical and the mathematical models have a good agreement, and consequently, the mathematical model is validated.

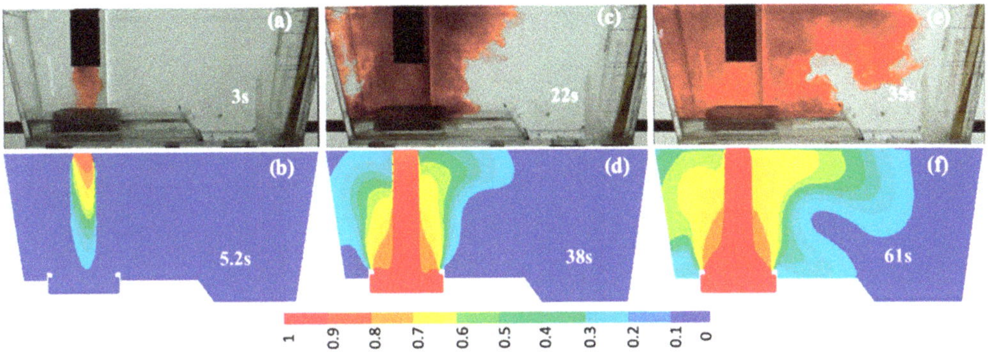

Figure 4. Comparison of the tracer behavior between the 1:3 scale water model (**a**,**c**,**e**) and the full-scale mathematical model (**b**,**d**,**f**).

To demonstrate if a scaled mathematical model can accurately predict the flow patterns and volume fractions as a full-scale model, a 1:3 mathematical model was also simulated. Since the full-scale mathematical model is validated, the following comparison will be made between only the mathematical results: Figure 5a,b exhibits the velocity vector fields in a longitudinal plane at the ladle shroud entry axis. The velocity range is from 0 to 0.023 for the reduced scale case and from 0 to 0.04 m/s, satisfying the expected relationship $(V)_m = \sqrt{\lambda}\,(V)_p$. The similarities between both cases are evident, showing a flow recirculation on the left side of the ladle shroud, a short circuit on the right side, and a second recirculation between the short-circuit position and the tundish floor. Consequently, both mathematical results predict qualitatively the same flow patterns. Nevertheless, it is crucial to verify if the quantifiable parameters also agree. Then, Figure 5c contains the RTD curves for both cases, and Table 3 has the piston, mixed, and dead volume fraction percentages and the non-dimensional residence time for both cases. The same tendency of both curves is visible in Figure 5c. The residence time for the scaled case is 222 s and 372 s for the full-scale case, which satisfy the relation $(t)_m = \sqrt{\lambda}\,(t)_p$ with a difference of 3.7%. The reported values in Table 3 are close for both cases, with a maximum difference of 4% for the mixed volume fraction. Therefore, the above-discussed results demonstrate that the scaled and full-scale models predict the same flow patterns, volume fractions, and residence time, with less than 4% differences following the Froude criteria. Consequently, both mathematical models are equally validated, and either can reasonably well predict the flow patterns of the physical model. Since the current mesh quality and mathematical model predict with good agreement the fluid dynamics of the physical model, mesh sensitivity analysis can be avoided. The present research considers a full-scale model for further analysis since the simulation will include the temperature variable.

Table 3. Volume fraction percentages for the scaled and non-scaled models.

Scale	V_P/V	V_M/V	V_d/V	θ
1:3	46	40	14	0.9
1:1	50	38	12	0.92

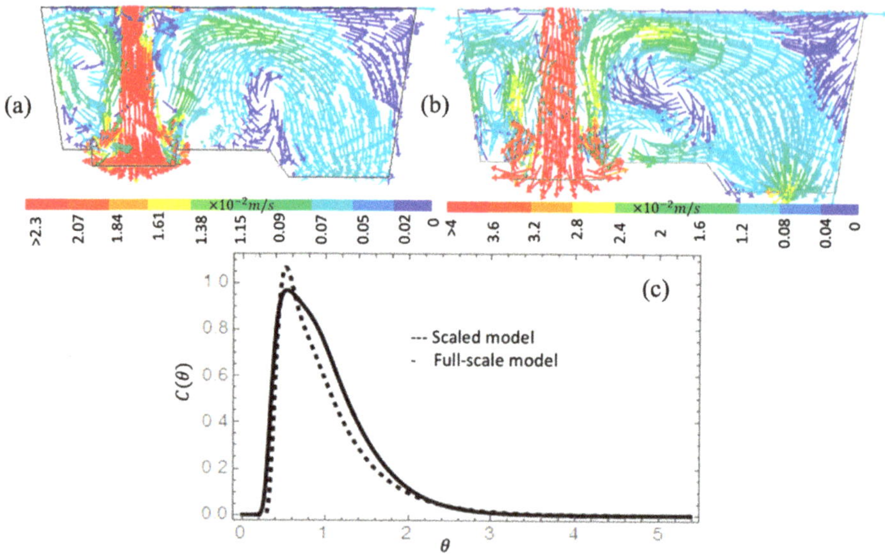

Figure 5. Fixed velocity vectors for (**a**) scaled tundish, (**b**) full-scale tundish, and (**c**) RTD curves for both cases.

3.2. Comparison between Isothermal and Non-Isothermal Cases

To establish the effect of the temperature on the tundish flow patterns, Figure 6 shows tracer concentration contours for three representative times to contrast the tracer behavior for the isothermal (Figure 6a–c) and non-isothermal (Figure 6d–f) cases. Figure 6a,d shows that the tracer behaves similarly in the entry zone for both cases, noting that the flow on the upper right side of the tundish presents a slightly longer trajectory for the non-isothermal case. However, comparing Figure 6b,c, and Figure 6e,f, the tracer shows significant differences in its flow behavior at the top tundish since, for the non-isothermal case, the flow tends to move near the bath level for a longer distance, while for the isothermal case, the tracer presents the same stream with a descending trajectory. Looking for a deeper analysis of the flow patterns, Figure 7 shows the fixed velocity vector fields in three planes: a longitudinal plane at the center of the ladle shroud, a longitudinal plane at the exit, and a horizontal plane at the bath level position.

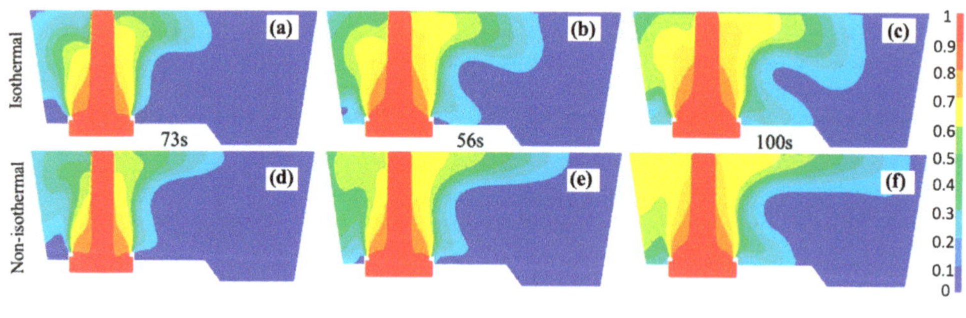

Figure 6. Tracer concentration contours at the central symmetrical plane of the ladle shroud for the isothermal (**a**–**c**) and non-isothermal (**d**–**f**) cases.

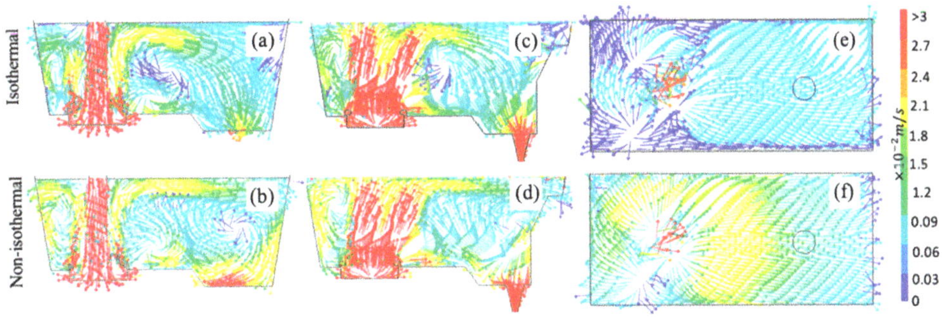

Figure 7. Fixed velocity vector fields in three planes: (**a**,**b**) center of the ladle shroud, (**c**,**d**) center of the exit, and (**e**,**f**) bath level position.

These figures show substantial differences between both cases. For example, the short-circuit previously mentioned in the isothermal case disappears in the non-isothermal case, and the recirculation observed on the right side of the ladle shroud remains in both cases but with different positions and intensities. Also, the non-isothermal case increases the velocity magnitudes for all the planes, being more evident at the bath-level plane. To explain the detected velocity increment, we need the Maxwell-Boltzmann velocity distribution function, which is:

$$f(V)d^3V = \left(\frac{m}{2\,\pi\,k_B\,T}\right)^{3/2} Exp\left[-\frac{m\,V^2}{2\,k_B\,T}\right]d^3V \tag{9}$$

where k_B is the Boltzmann constant, m is the mass particle, T is the temperature, and $f(V)$ is a probability distribution function. The function $f(v)$ indicates the probability of finding a particle with a velocity between V and $V + dV$. Spherical coordinates will be used to find the expression that lets us calculate its value:

$$f(V_x,\,V_y,\,V_z)dV_x dV_y dV_z = f(V_x,\,V_y,\,V_z)V^2 dV\,sin\theta d\theta d\varphi \tag{10}$$

Then, after its integration, the Equation (9) becomes:

$$f(V) = \left(\frac{m}{2\,\pi\,k_B\,T}\right)^{1/2} V^2 Exp\left[-\frac{m\,V}{2\,k_B\,T}\right] \tag{11}$$

By choosing two arbitrary temperatures, $T_2 > T_1$, and plotting their corresponding $f(v)$ function in Figure 8, it is possible to understand the effect of the temperature on the velocity distribution function. This figure demonstrates that for the higher temperature T_2, the curve flattens out, the maximum value of the curve diminishes, the maximum value shifts to the left, and the curve becomes narrower in contrast with the curve of T_1. The behavior of this probability distribution function indicates that flows with higher temperatures will have a range of higher velocities. This asseveration applies for the non-isothermal simulation, which will induce that the fluid dynamics present velocity variations because of temperature gradients, being, on average, faster for the hot streams and slower for the cold ones; such velocity variations will not be present in the isothermal case since there are no temperature gradients. Following this analysis and associating velocity increments observed in the non-isothermal case with the previous discussion, it is necessary to study the temperature gradients in the tundish. Figure 9 shows temperature and buoyancy magnitude forces contours. This figure shows that higher temperatures exist in the entry zone and along the bath level. The high velocities at the entry are because of the entry jet. Compared with the isothermal case, the velocity increment observed at the bath level follows the Maxwell-Boltzmann velocity distribution function explanation.

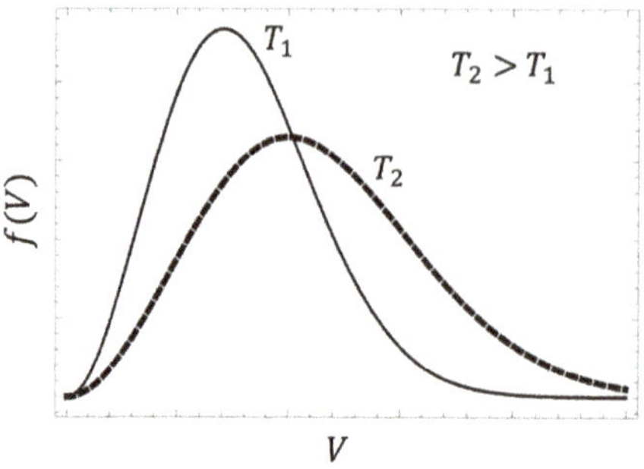

Figure 8. Maxwell–Boltzmann distribution function for two arbitrary temperatures.

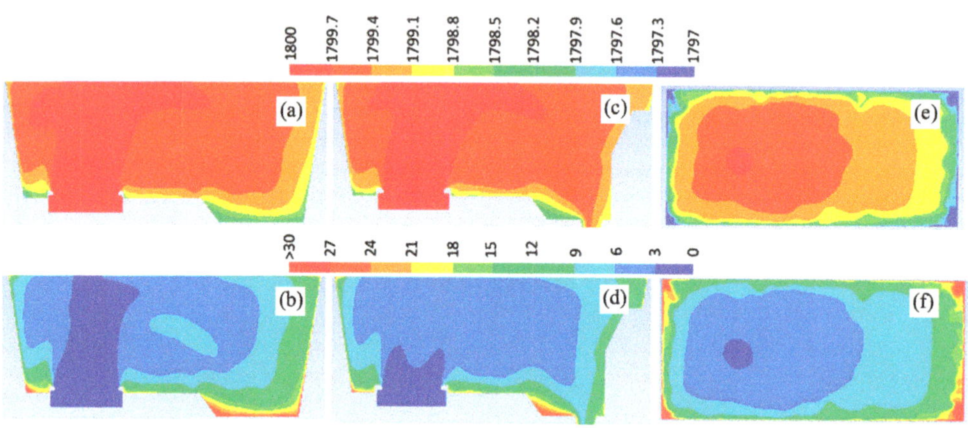

Figure 9. Temperature contours (K) and buoyancy forces (N) at three planes (**a**,**b**) center of the ladle shroud, (**c**,**d**) center of the exit, and (**e**,**f**) bath level position.

Although the Maxwell-Boltzmann velocity distribution function explains velocity magnitude variations between isothermal and non-isothermal models, it cannot explain why the short-circuit disappears under non-isothermal conditions. The buoyancy and inertial forces are studied since these forces are associated with the parameters of temperature and velocity, respectively, to analyze this point.

The buoyancy force depends on temperature variations inside the tundish, and consequently, it only applies to the non-isothermal case and is given by:

$$F_B = \rho\, g\, L^3 \beta\, \nabla T \tag{12}$$

where β is the thermal fluid expansion that arises due to the differences in the fluid density inside the tundish as follows:

$$\beta = -\frac{1}{\rho_{ref}}\left(\frac{\partial \rho}{\partial T}\right)_p = 1.27 \times 10^{-4} K^{-1} \tag{13}$$

Then, given the strong dependency of the buoyancy force on the temperature gradients, the zones that exhibit higher temperature gradients will produce higher buoyancy force values. The results indicate that the main effect of the buoyancy forces occurs in zones near the walls, being less significant in the rest of the tundish, especially at the entry zone. Furthermore, inertial force is another essential force to consider acting in the flow. The inertial force magnitude $F_i = \rho\, V^2 L^2$ is in Figure 10, Figure 10a–c for the isothermal case, and Figure 10d–f for the non-isothermal case. As expected, these figures reflect that inertial force magnitudes are meaningful throughout the entire tundish. However, the non-isothermal case has more zones where the inertial forces act with higher values than the isothermal case for all the shown planes, especially at the bath level.

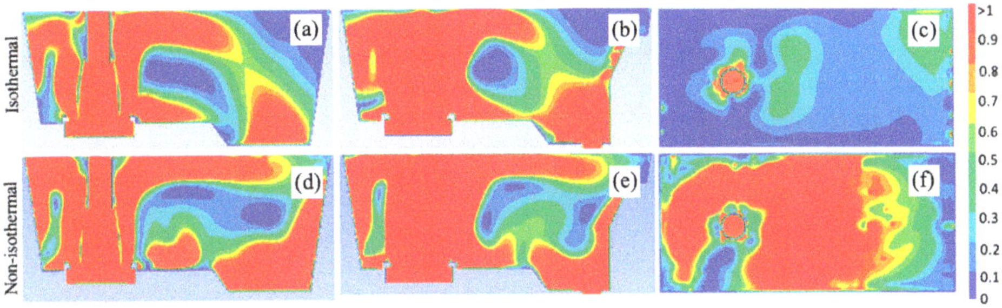

Figure 10. Inertial force magnitude (N) at three planes, for the isothermal and non-isothermal cases, (**a**,**d**) center of the ladle shroud, (**b**,**e**) center of the exit, and (**c**,**f**) bath level position.

To observe which of the two analyzed forces dominates each region of the tundish, the Richardson number (Ri) will be employed. This number defines the ratio of buoyancy force to inertial force as follows:

$$\mathrm{Ri} = \frac{gL\beta\Delta T}{V^2} \tag{14}$$

Then, if Ri tends to zero, the inertial force is predominant over the buoyancy force, and it is opposite when the ratio is higher than one. Figure 11 shows the contours of the Ri values for the non-isothermal case in the three analyzed planes. The results show that the inertial force dominates over the buoyancy force in the entry zone, around the ladle shroud at the bath level, and very close to the outlet; in contrast, buoyancy force dominates in the detected recirculation zones (points ① and ②) and dead flow zones (points ③, ④, and ⑤). It is crucial to notice that zones where Ri tends to zero correspond to zones with less appreciable flow pattern changes between isothermal and non-isothermal cases.

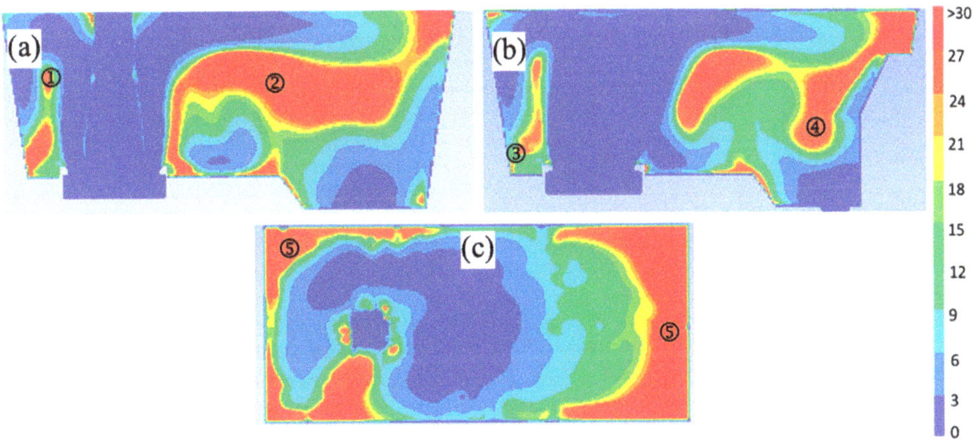

Figure 11. Richardson number magnitude at three planes. (**a**) center of the ladle shroud; (**b**) center of the exit nozzle; and (**c**) upper view of the tundish at the bath level.

Considering the above discussion from Figures 6–10, the flow variations observed between the isothermal and non-isothermal cases have now been supported. The upward stream flow reaching the bath surface from the turbulence inhibitor has a higher temperature than the rest of the bulk flow. This hotter flow must follow the Maxwell-Boltzmann velocity distribution function, which, together with the buoyancy forces acting upwards below this stream, induces the following changes: (1) at the left ladle shroud side, the recirculation moves upward with an increment in the flow velocity around it; (2) at the right ladle shroud side, the mainstream, moving parallel to the bath level towards the right tundish wall and traveling near the metal-slag interphase, has higher velocity, which reaches a longer distance than in the isothermal case. The second implication forces the short circuit to disappear, moving the recirculation below it toward the right tundish side. These changes are even when the temperature variations inside the tundish are as modest as 3 K. Until this point, the fluid dynamic changes between both cases have been analyzed and explained; however, it is necessary to determine their effect on variables such as volume fractions and inclusion removal rate quantitatively.

Figure 12 shows the RTD curves for both cases. The figure reveals that the curves present similarities in the first appearance of the tracer, θ_{min}, and when the tracer's maximum concentration occurs, θ_{max}. However, it is noticeable that the non-isothermal curve is the highest and slightly thinner than the isothermal one. These differences are quantifiable by the volume fraction calculation presented in Table 4.

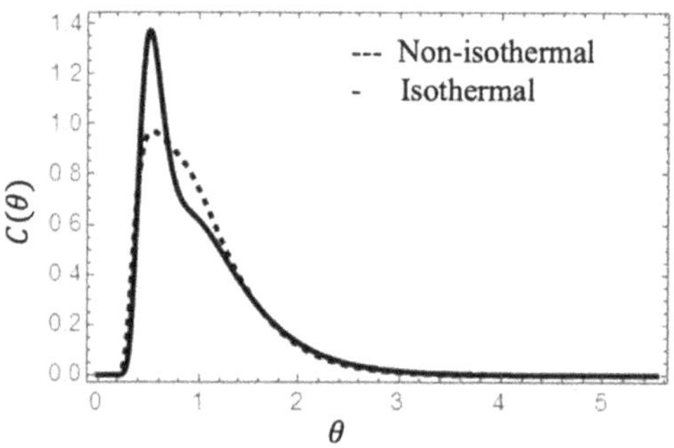

Figure 12. RTDs for isothermal and non-isothermal cases.

Table 4. Volume fraction percentages for the isothermal and non-isothermal models.

Case	V_P/V	V_M/V	V_d/V	θ
Isothermal	50	38	12	0.93
Non-isothermal	46	39	15	0.91

The results indicate that the volume fraction percentages are similar with a flow moving mainly as a mixed flow. In addition, there is an increment in the piston flow of 1%, a decrement in the mixed flow of 4%, an increment in the dead volume of 3%, and a decrement in the non-dimensional residence time of 2.1% for the non-isothermal case concerning the isothermal case.

To study if the temperature variable affects the behavior of the inclusions inside the tundish, first the changes in the number of removed inclusions were analyzed for both cases. To do this, for a single inclusion size, 2000 inclusions were fed at the ladle shroud entry as an impulse once the fluid dynamics acquired the quasi-steady state condition, and through the Lagrangian model, which uses the current flow patterns, the trajectories of the 2000 inclusions were calculated using the removal boundary condition previously indicated. As a result, the Lagrangian model indicates the number of inclusions that were removed in the steel-slag interphase and the number of inclusions that have reached the tundish exit. With this information, the inclusion removal percentage was calculated using Equation (15):

$$inclusion\ removal\ percentage = \frac{amount\ of\ removed\ inclusions}{2000} \times 100 \quad (15)$$

This procedure was the same for each of the inclusion sizes. Figure 13a presents the inclusion removal percentage calculated by the DPM for the inclusion sizes considered in this work, for isothermal and non-isothermal cases. The results show that the removed inclusions in a range of 1 to 40 μm are at least 35% of the fed inclusions, and the inclusion removal percentage, for both cases, is almost 100% for inclusions bigger than 130 μm. The non-isothermal case presents an increment of 5% in the inclusion removal percentages in comparison with the isothermal case for inclusions ranging from 1 to 60 μm; this was the most significant difference between the results for both cases.

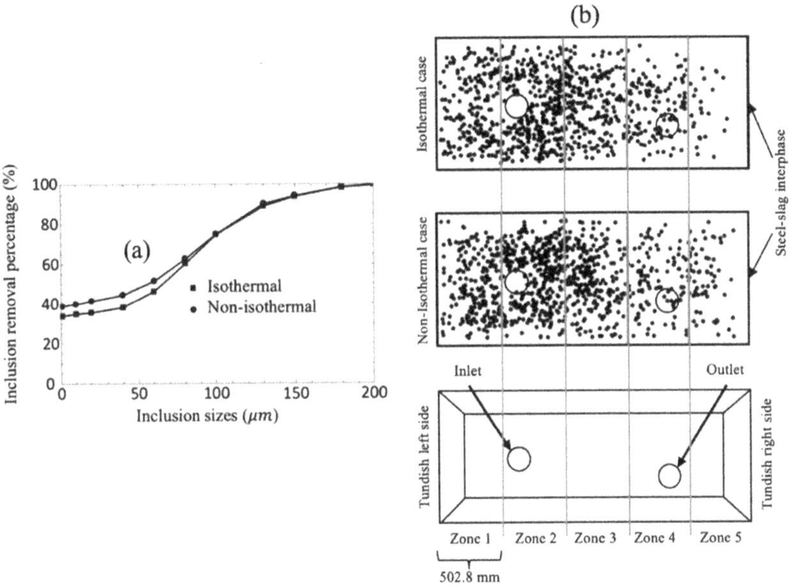

Figure 13. (**a**) Inclusion removal percentages for many inclusion sizes. (**b**) Sample of the removed inclusion distribution at the steel-slag interphase using 40 μm.

Previously, it has been demonstrated that the temperature barely affects the global tendency of the removal of inclusion percentages; however, it could be possible that the temperature affects the way inclusions are distributed when they are removed. To obtain the exact position where each inclusion was removed at the steel-slag interphase, a user-defined function (UDF) was employed within the CFD software. Figure 13b shows the removed inclusion distribution at the steel-slag interphase for an inclusion size of 40 μm in both cases. The results indicate that both removed inclusion distributions are quite similar, highlighting some characteristics: Most of the inclusions are removed around the ladle shroud zone, and towards the outlet position, the number of removed inclusions decreases. This behavior is attributed to the fluid dynamics induced by the turbulence inhibitor and the absence of more flow control devices between the turbulence inhibitor and the tundish outlet. These results apparently indicate that the temperature does not influence the removed inclusion distribution at the steel-slag interphase. In order to do a comparison between the number of inclusions and their removed position for each inclusion size, the steel-slag interface was divided into five zones of equal length. Subsequently, the removed inclusion positions were determined for each inclusion size, and the number of removed inclusions at each zone was calculated as a percentage for isothermal and non-isothermal cases. Figure 14 shows the removed inclusion percentage for each inclusion size calculated for the zones indicated in Figure 13b for isothermal and non-isothermal cases. The results of Figure 14 indicate that the removed inclusion distribution is similar for all the inclusion sizes regarding the non-isothermal or isothermal conditions. Therefore, the flow velocity increment in the non-isothermal case and the short circuit in the isothermal case do not significantly impact the inclusion removal rate and its removed distribution at the bath level.

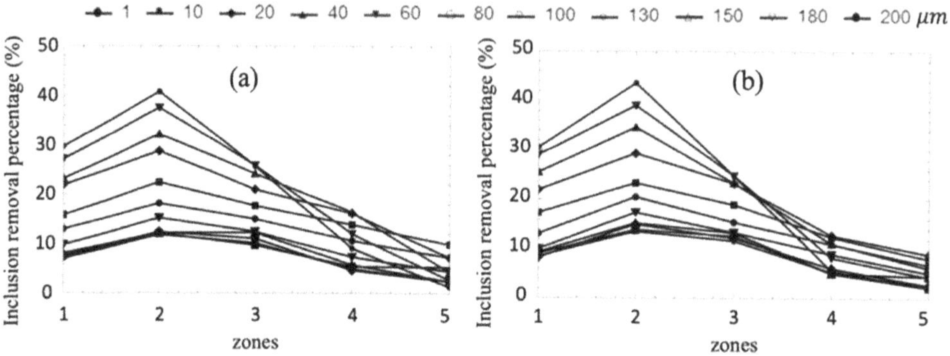

Figure 14. Inclusion removal percentages by zones in the tundish bath level (**a**) isothermal case, (**b**) non-isothermal case.

4. Conclusions

This work investigates the temperature effect on the fluid patterns and inclusion removal rates in a continuous casting tundish using mathematical simulation. From the previously discussed results, the following conclusions are drawn:

(1) The quantification of the differences between the results of 1:3 scaled and full-scale models, which have a maximum difference of 4% on the volume fractions percent and residence time, demonstrate that scaled and full-scale models can be used reliably to predict the flow patterns of an isothermal physical model following the Froude criteria.

(2) Temperature gradients inside the tundish induce variations in the flow velocity magnitude; if the stream has a higher temperature than its surrounding flow, its velocity will increase because of the Maxwell-Boltzmann velocity distribution function; this supports why hot streams, e.g., at the bath level, under non-isothermal conditions, have bigger velocity magnitudes than the same streams but under isothermal conditions.

(3) The quantification of the ratio between inertial and buoyancy forces demonstrates that inertial forces dominate over buoyancy forces at the entry zone because turbulence inhibitors strongly control the fluid dynamics in such an area. In contrast, buoyancy forces take more relevance than inertial forces in the recirculation and dead flow zones, inducing noticeable changes in the fluid dynamics between isothermal and non-isothermal cases far from the entry zone.

(4) Although the temperature induces substantial fluid dynamic changes between the analyzed cases, this variable does not significantly impact the volume fraction percentages or the mean residence time results, and it only increases the inclusion removal percentage by 5% for the non-isothermal case.

(5) The effect of the temperature on the flow patterns is not significant when the flow control devices strongly rule the fluid dynamics; nevertheless, when the flow control devices effects are not dominant, the temperature takes substantial importance because of the buoyancy forces and the increment of flow velocity because the Maxwell Boltzmann velocity distribution function changes the fluid dynamics in comparison with the obtained from the isothermal conditions. Consequently, isothermal simulations can accurately describe the flow behavior in tundishes, where flow control devices control the fluid dynamics. However, the simulation of tundishes without control devices or with a weak fluid dynamic dependence on the control devices requires non-isothermal simulations.

5. Future Work

Since temperature variance could depend on the turbulence model, and as a consequence, these differences can produce density variations and affect the flow, a further analysis of the differences between isothermal and non-isothermal models must be conducted, considering different turbulence models and many tundish designs.

Author Contributions: Conceptualization, R.M.D. and J.d.J.B.; methodology, E.G.; software, E.G. and S.G.-H.; validation, R.M.D. and E.G.; formal analysis, E.G. and S.G.-H.; investigation, E.G.; resources, E.G. and R.M.D.; data curation, E.G.; writing—original draft preparation, E.G. and S.G.-H.; writing—review and editing, E.G., S.G.-H. and J.d.J.B.; visualization, E.G.; supervision, R.M.D.; project administration, E.G. All authors have read and agreed to the published version of the manuscript.

Funding: This research received no external funding.

Data Availability Statement: Data are contained within the article.

Acknowledgments: The authors thank TecNM-ITM and SNI institutions for their permanent support of the Academic Research Group.

Conflicts of Interest: The authors declare no conflicts of interest.

Nomenclature

ρ	Fluid density (kg/m^3)
C_p	Specific heat (J/(kg K))
TC	Thermal conductivity (w/(m K))
μ	Dynamic viscosity (kg/(m s))
V	Fluid velocity (m/s)
u_p	Particle velocity (m/s)
P	Pressure (Pa)
t	Time (s)
T	Temperature (K)
μ_{eff}	Effective Viscosity (kg/(m s))
g	Gravity acceleration (m/s^2)
β	Coefficient of volumetric thermal expansion (K^{-1})
ε	Turbulent dissipation rate (m^2/s^3)
k	Kinetic energy (m^2/s^2)
G_k	Generation of turbulent kinetic energy due to the mean velocity gradients
G_b	Generation of turbulent kinetic energy due to buoyancy
Y_M	Contribution of the fluctuating dilatation in compressible turbulence to the dissipation rate
$F_d(V\text{-}u_p)$	Drag force per unit mass (N/kg)
F_B, F_G, F_S, F_{VM}, F_{PG}	The buoyant, gravitational, Saffman, virtual mass, and gradient force per unit mass (N/kg)
$C_{1\varepsilon}$, $C_{2\varepsilon}$, $C_{3\varepsilon}$	Model constants
σ_k, σ_ϵ	Turbulent Prandtl numbers
S_k, S_ϵ	User-defined source terms
L	Characteristic length (m)
λ	Scale factor
V_P/V	Piston volume fraction
V_M/V	Mixed volume fraction
V_d/V	Dead volume fraction
θ	Non-dimensional residence time
k_b	Boltzmann constant
K	Kelvin
Ri	Richardson number

References

1. Sheng, D.; Jönsson, P. Effect of Thermal Buoyancy on Fluid Flow and Residence-Time Distribution in a Single-Strand Tundish. *Materials* **2021**, *14*, 1906. [CrossRef] [PubMed]
2. Wang, Q.; Liu, Y.; Huang, A.; Yan, W.; Gu, H.; Li, G. CFD Investigation of Effect of Multi-hole Ceramic Filter on Inclusion Removal in a Two-Strand Tundish. *Metall. Mater. Trans. B* **2020**, *51*, 276–292. [CrossRef]
3. Cwudzinski, A. Physical and mathematical simulation of liquid steel mixing zone in one strand continuous casting tundish. *Int. J. Cast. Met. Res.* **2017**, *30*, 50–60. [CrossRef]
4. Agarwal, R.; Singh, M.K.; Kumar, R.; Ghosh, B.; Pathak, S. Extensive Analysis of Multi Strand Billet Caster Tundish Using Numerical Technique. *J. Mech.* **2019**, *9*, 29–51. [CrossRef]
5. Xing, F.; Zheng, S.; Liu, Z.; Zhu, M. Flow Field, Temperature Field, and Inclusion Removal in a New Induction Heating Tundish with Bent Channels. *Metals* **2019**, *9*, 561. [CrossRef]
6. Ramirez, O.D.; Torres-Alonso, E.; Ramos-Banderas, A.; Villa, S.A.; Bocanegra, C.H.; Martínez, J.T. Thermal and Fluid-Dynamic Optimization of a Five Strand Asymmetric Delta Shaped Billet Caster Tundish. *Steel Res. Int.* **2018**, *89*, 1–10. [CrossRef]
7. Tang, H.; Guo, L.; Wu, G.; Xiao, H.; Yao, H.; Zhang, J. Hydrodynamic Modeling and Mathematical Simulation on Flow Field and Inclusion Removal in a Seven-Strand Continuous Casting Tundish with Channel Type Induction Heating. *Metals* **2018**, *8*, 374. [CrossRef]
8. Neves, L.; Tavares, R. Analysis of the mathematical model of the gas bubbling curtain injection on the bottom and the walls of a continuous casting tundish. *Ironmak. Steelmak.* **2017**, *44*, 559–567. [CrossRef]
9. Sahai, Y. Tundish Technology for Casting Clean Steel: A Review. *Metall. Mater. Trans. B* **2016**, *47*, 2095–2106. [CrossRef]
10. Wang, Q.; Li, B.; Tsukihash, F. Hydrodynamic Problem in Continuous Casting Tundish with Channel Type Induction Heating. *ISIJ Int.* **2014**, *54*, 311–320. [CrossRef]
11. Ni, P.; Jonsson, L.T.I.; Ersson, M.; Jönsson, P.G. Turbulent Flow Phenomena and Ce_2O_3 Behavior during a Steel Teeming Process. *ISIJ Int.* **2013**, *53*, 792–801. [CrossRef]
12. Qu, T.; Liu, C.; Jiang, M.; Zu, L. Numerical Simulation for Effect of Inlet Cooling Rate on Fluid Flow and Temperature Distribution in Tundish. *J. Iron Steel Res. Int.* **2012**, *19*, 12–19. [CrossRef]
13. Ling, H.; Zhang, L. Numerical Simulation of the Growth and Removal of Inclusions in the Molten Steel of a Two-Strand Tundish. *JOM* **2013**, *65*, 1155–1163. [CrossRef]
14. Singh, V.; Ajmani, S.; Pal, A.; Singh, S.; Denys, M. Single strand continuous caster tundish furniture comparison for optimal performance. *Ironmak. Steelmak.* **2012**, *39*, 171–179. [CrossRef]
15. Braun, A.; Warzecha, M.; Pfeifer, H. Numerical and Physical Modeling of Steel Flow in a Two-Strand Tundish for Different Casting Conditions. *Metall. Mater Trans. B* **2010**, *41*, 549–559. [CrossRef]
16. Vargas-Zamora, A.; Morales, R.D.; Díaz-Cruz, M.; Palafox-Ramos, J.; Demedices, L.G. Heat and mass transfer of a convective-stratified flow in a trough type tundish. *Int. J. Heat Mass Transf.* **2003**, *46*, 3029–3039. [CrossRef]
17. Joo, S.; Han, J.; Guthrie, R. Inclusion behavior and heat-transfer phenomena in steelmaking tundish operations: Part II. Mathematical model for liquid steel in tundishes. *Metall. Mater Trans. B* **1993**, *24*, 767–777. [CrossRef]
18. Chakraborty, S.; Sahai, Y. Effect of Varying Ladle Stream Temperature on the Melt Flow and Heat Transfer in Continuous Casting Tundishes. *ISIJ Int.* **1991**, *31*, 960–967. Available online: https://www.jstage.jst.go.jp/article/isijinternational1989/31/9/31_9_960/_pdf (accessed on 4 January 2024). [CrossRef]
19. Miki, Y.; Thomas, B. Modeling of inclusion removal in a tundish. *Metall. Mater. Trans. B* **1999**, *30*, 639–654. [CrossRef]
20. Alizadeh, M.; Edris, H. Prediction of RTD Curves for Non-isotherm steel Melt Flows in Continuous Casting Tundish. In Proceedings of the International Symposium on Liquid Metal Processing and Casting, LMPC, Nancy, France, 2–5 September 2007; pp. 167–172.
21. Sun, H.; Yan, B.; Zhang, J. Effect of Thermal Buoyancy Force on the Flow, Temperature Distribution and Residence Time Distribution of Molten Steel in the Slab Casting Tundish. In *CFD Modeling and Simulation in Materials Processing*; The Minerals, Metals, & Materials Society: Pittsburgh, PA, USA, 2012; pp. 327–334. [CrossRef]
22. Chatterjee, S.; Chattopadhyay, K. Transient steel quality under non-isothermal conditions in a multi-strand billet caster tundish: Part II. Effect of a flow-control device. *Ironmak. Steelmak.* **2016**, *44*, 413–420. [CrossRef]
23. Zhu, M.; Peng, S.; Jiang, K.; Luo, J.; Zhong, Y.; Tang, P. Fluid Flow and Heat Transfer Behaviors under Non-Isothermal Conditions in a Four-Strand Tundish. *Metals* **2022**, *12*, 840. [CrossRef]
24. Morales, R.; Lopez-Ramirez, S.; Palafox-Ramos, J.; Zacharias, D. Mathematical simulation of effects of flow control devices and buoyancy forces on molten steel flow and evolution of output temperatures in tundish. *Ironmak. Steelmak.* **2001**, *28*, 33–43. [CrossRef]
25. Chattopadhyay, K.; Isac, M.; Guthrie, R. Modelling of Non-isothermal Melt Flows in a Four Strand Delta Shaped Billet Caster Tundish Validated by Water Model Experiments. *ISIJ Int.* **2012**, *52*, 2026–2035. [CrossRef]
26. de Sousa, J.; Barros, E.; Marcondes, F.; de Castro, J. Modeling and computational simulation of fluid flow, heat transfer and inclusions trajectories in a tundish of a steel continuous casting machine. *J. Mater. Res. Technol.* **2019**, *8*, 4209–4220. [CrossRef]
27. Neumann, S.; Asad, A.; Schwarze, R. Numerical Simulation of an Industrial-Scale Prototypical Steel Melt Tundish Considering Flow Control and Cleaning Strategies. *Adv. Eng. Mater.* **2020**, *22*, 1900658. [CrossRef]

28. Mishra, R.; Mazumdar, D. Numerical Analysis of Turbulence Inhibitor Toward Inclusion Separation Efficiency in Tundish. *Indian Inst. Met.* **2019**, *72*, 889–989. [CrossRef]
29. Gutiérrez, E.; Garcia-Hernández, S.; de Barreto, J. Mathematical Analysis of the Touching Inclusions Parameters at the Tundish Free Surface to Predict More Realistic Inclusion Removal Rates. *Steel Res. Int.* **2019**, *90*, 1900328. [CrossRef]
30. Gutiérrez, E.; Garcia-Hernández, S.; de Barreto, J. Mathematical Modeling of Inclusions Deposition at the Upper Tundish Nozzle and the Submerged Entry Nozzle. *Steel Res. Int.* **2016**, *87*, 1406–1416. [CrossRef]
31. Gutiérrez, E.; Garcia-Hernández, S.; de Barreto, J. Mathematical Analysis of the Dynamic Effects on the Deposition of Alumina Inclusions inside the Upper Tundish Nozzle. *ISIJ Int.* **2016**, *56*, 1394–1403. [CrossRef]
32. Lei, H.; Zhao, Y.; Geng, D. Mathematical Model for Cluster-Inclusion's Collision-Growth in Inclusion Cloud at Continuous Casting Mold. *ISIJ Int.* **2014**, *54*, 1629–1637. [CrossRef]
33. Liu, Z.; Li, B.; Jiang, M. Transient Asymmetric Flow and Bubble Transport Inside a Slab Continuous-Casting Mold. *Metall. Mater. Trans. B* **2014**, *45*, 675–697. [CrossRef]
34. Liu, Z.; Li, B.; Jiang, M.; Tsukihashi, F. Euler-Euler-Lagrangian Modeling for Two-Phase Flow and Particle Transport in Continuous Casting Mold. *ISIJ Int.* **2014**, *54*, 1314–1323. [CrossRef]
35. Takahashi, K.; Ando, M.; Ishii, T. Model of Gas Flow Through Porous Refractory Applied to an Upper Tundish Nozzle. *ISIJ Int.* **2014**, *54*, 304–405. [CrossRef]
36. He, Z.; Zhou, K.; Liu, S.; Xiong, W.; Li, B. Numerical Modeling of the Fluid Flow in Continuous Casting Tundish with Different Control Devices. *Abstract Appl. Anal.* **2013**, *2013*, 984894. [CrossRef]
37. Seshadri, V.; da Silva, C.; da Silva, I.; Junior, E. A Physical Modelling Study Of Inclusion Removal In Tundish Using Inert Gas Curtain. *Tecnol. Metal. Mater. Min.* **2012**, *9*, 22–29. [CrossRef]
38. Long, M.; Zuo, X.; Zhang, L.; Chen, D. Kinetic Modeling on Nozzle Clogging During Steel Billet Continuous Casting. *ISIJ Int.* **2010**, *50*, 712–720. [CrossRef]
39. Zhong, L.; Li, L.; Wang, B.; Zhang, L.; Zhu, L.; Zhang, Q. Fluid flow behaviour in slab continuous casting tundish with different configurations of gas bubbling curtain. *Ironmak. Steelmak.* **2008**, *35*, 436–440. [CrossRef]
40. Ramos-Banderas, A.; Morales, R.D.; de Barreto, J.; Solorio-Diaz, G. Modelling Study of Inclusions Removal by Bubble Flotation in the Tundish. *Steel Res. Int.* **2006**, *77*, 325–335. [CrossRef]
41. Yuan, Q.; Thomas, B.; Vanka, S. Study of transient flow and particle transport in continuous steel caster molds: Part II. Particle transport. *Metall. Mater. Trans. B* **2004**, *35B*, 703–714. [CrossRef]
42. Vermeulen, Y.; Coletti, B.; Blanpain, B.; Wollants, P.; Vleugels, J. Material Evaluation to Prevent Nozzle Clogging during Continuous Casting of Al Killed Steels. *ISIJ Int.* **2002**, *42*, 1234–1240. [CrossRef]
43. Zhang, L.; Taniguchi, S.; Cai, K. Fluid flow and inclusion removal in continuous casting tundish. *Metall. Mater. Trans. B* **2000**, *31*, 253–266. [CrossRef]
44. Sasai, K.; Mizukami, Y. Reaction Mechanism between Alumina Graphite Immersion Nozzle and Low Carbon Steel. *ISIJ Int.* **1994**, *34*, 802–809. [CrossRef]
45. Fukuda, Y.; Ueshima, Y.; Mizoguchi, S. Mechanism of Alumina Deposition on Alumina Graphite Immersion Nozzle in Continuous Caster. *ISIJ Int.* **1992**, *32*, 164–168. [CrossRef]
46. Shih, T.; Liou, W.; Shabbir, A.; Yang, Z.; Zhu, J.A. A new k-ϵ eddy viscosity model for high reynolds number turbulent flows. *Comput. Fluids* **1995**, *24*, 227–238. [CrossRef]
47. ANSYS Inc. *FLUENT 12.0, User's Guide*; Centerra Resource Park: Cavendish Court, Lebanon, 2009. Available online: https://www.afs.enea.it/project/neptunius/docs/fluent/html/ug/main_pre.htm (accessed on 4 January 2024).

MDPI AG
Grosspeteranlage 5
4052 Basel
Switzerland
Tel.: +41 61 683 77 34

Fluids Editorial Office
E-mail: fluids@mdpi.com
www.mdpi.com/journal/fluids

www.ingramcontent.com/pod-product-compliance
Lightning Source LLC
Chambersburg PA
CBHW041016270926
51850CB00024B/766
* 9 7 8 3 7 2 5 8 6 7 3 8 7 *